KT-385-484

Christic Church

T015025

ELECTRON OPTICS

P. GRIVET

Professor at the University of Paris

With the Collaboration of

M.-Y. BERNARD, F. BERTEIN, R. CASTAING,
M. GAUZIT AND A. SEPTIER

ELECTRON OPTICS

SECOND ENGLISH EDITION

IN TWO PARTS

Translated by P. W. HAWKES
Revised afresh for this Edition by A. SEPTIER

Part 2 INSTRUMENTS

PERGAMON PRESS

OXFORD · NEW YORK
TORONTO · SYDNEY · BRAUNSCHWEIG

Pergamon Press Ltd., Headington Hill Hall, Oxford
Pergamon Press Inc., Maxwell House, Fairview Park, Elmsford,
New York 10523
Pergamon of Canada Ltd., 207 Queen's Quay West, Toronto 1
Pergamon Press (Aust.) Pty. Ltd., 19a Boundary Street,
Rushcutters Bay, N.S.W. 2011, Australia
Vieweg & Sohn GmbH, Burgplatz 1, Braunschweig

Copyright © 1972 Pergamon Press Ltd.

*All Rights Reserved. No part of this publication may be reproduced,
stored in a retrieval system, or transmitted, in any form or by any
means, electronic, mechanical, photocopying, recording or otherwise, without the prior permission of Pergamon Press Ltd.*

First English edition 1965
Second (revised) English edition 1972
Library of Congress Catalog Card No. 72-132964

This book is a translation, with much revision and
addition, of *Optique Électronique*, originally
published in French by Bordas Éditeur, Paris

WITHDRAWN FROM
CHRIST CHURCH LIBRARY
CHRIST CHURCH
LIBRARY
OXFORD

Printed in Great Britain by A. Wheaton & Co., Exeter
08 016228 2

CONTENTS

PART 2. INSTRUMENTS

CONTENTS OF PART 1

PART 1. OPTICS

ix

PREFACE TO THE SECOND ENGLISH EDITION

WE HAVE tried to improve this new edition of *Electron Optics* for the benefit of two categories of reader. First of all, we have endeavoured to make it more useful to the graduating student. With this in mind we have, for example, added some introductory sections describing selected but nevertheless classical methods of calculating electric and magnetic fields; these indications are intended as a guide for the beginner who sometimes feels lost in the abundant specialized literature.

On the other hand, we thought it essential to keep the research worker (even the beginner) informed of the important recent progress in the field: this seems to us an essential prerequisite for a book which aims at helping scientists to form their scientific taste and choose their own path in research.

This proved to be a fascinating but difficult task because activity in the field of electron optics has been both important and successful during recent years. The familiar magnetic objective has been significantly perfected, mainly by its discoverer, Professor Ruska, while the practical introduction of superconducting lenses gives some realistic hopes of a breakthrough in the domains of high resolution and high energy: today, several careful measurements prove that a resolving power of 4 Å is already achieved and that one can distinguish lines spaced 1 or 2 Å apart: the traces of Bragg planes are clearly distinguishable and these results give good reason to hope that in a not too distant future the individual atoms of most solids will be made visible by high-voltage electron optics.

On the theoretical side, the phenomenon of interference between two electron beams has gained in interest, as it may be usefully compared to the superconducting interferometer, a more mysterious but also more practical instrument making use of the stability of Cooper electron pairs and producing beats in the time domain instead of fringes in space. For this reason, greater care has been taken to bridge the gap between refractive index theory and the theory of the phase shift for electron waves.

The new appearance of this second edition is again mainly due to the efforts of Dr. Septier and the new text was also very ably translated into English by the English specialist of electron optics, Dr. Hawkes. I am glad to thank them for their renewed efforts on behalf of this book, in the name of all our team. The success of the first edition induces me to express our gratitude to the founder of this team during the hard times of the forties, Academician Dr. Ponte. We hope that the qualities of this book reflect in no negligible part the precious lessons of his two teachers during the twenties, Professor L. de Broglie and Sir John (J. J.) Thomson: their creative enthusiasm still had a beneficial impact on us 40 years later.

PIERRE GRIVET

PREFACE TO THE FIRST ENGLISH EDITION

With the appearance of this new edition in English it is a pleasure to look back at the original French version of the book; the latter is not so very old, and yet how important have been the advances which have been made since its publication, in 1955 and 1958. The most fundamental sphere of application of electron optics, electron lenses, seemed at that time to have been fully explored, but nevertheless, the subject has a fresh appeal to young research workers. The methods of calculation have developed beyond recognition, for nowadays no-one hesitates to invoke the aid of an electronic computer, so that the atmosphere of slightly timorous respect in which the differential equations to be solved used to be enveloped has practically disappeared. The importance of these electronic computers has been shown by the major discoveries which have resulted from their use, the behaviour of the quadrupole lens, for example. The need to consider this new type of lens was imposed by the requirements of nuclear physicists; none of the ordinary methods was suitable for studying these lenses, but even so, their properties and their surprising qualities are now well known after less than five years' study—ordinary lenses, with their much simpler structure, had to wait almost two decades before being thoroughly understood.

Other pleasant surprises await us in the instrumental field. To begin with, the electron microscope is the most perfected instrument, having bettered the performance which the theoreticians were reasonably hoping for around 1949. In their caution, these latter were doubtless too severe in their estimates of the limiting resolving power which could be attained, for all over the world today, industrial instruments of various nationalities—English, German, Japanese, French and Russian—are in current use and reach and even slightly exceed the most optimistic of the estimates of the theoreticians of ten years ago.

Still more astonishing are the results which have been obtained with this microscope. Ten years ago, it was still thought that the electron microscope was capable of giving only a rather frozen view of reality, unnaturally stiff like the view which we are given of the life of prehistoric organisms by paleontology. But at Toulouse, Professor Dupouy has already begun to study living biological cells with the aid of very high voltages; and even in more ordinary laboratories, we have no longer to be satisfied with an examination of a replica or desiccated skeleton of reality. Proper sections can now be cut which show the real structure of the dead cell. The success is still more striking where solids are concerned; with an electron microscope

xiii

we can fasten on to the very life of a metal, by filming, for example, the evolution of the dislocations when the metal is heated; this can be performed directly on a very thin layer cut from a block of the metal being studied.

Metallurgy provides us with yet another example of a great success. The metallurgists have now whole-heartedly adopted the electron probe; with its aid, they are able to perform extremely detailed and accurate metallurgical analyses.

This analyser was invented in France in 1949, and since 1954 Castaing has made a highly perfected form available to the metallurgists. It has taken six years for metallurgists all over the world to become convinced of the excellence of this method, and 1960 marks the beginnings of world-wide employment of this micro-analyser. This example demonstrates how useful it always is to struggle for the propagation of scientific ideas and methods, even when they have received the most brilliant experimental confirmation in the laboratory.

In these present times it is always necessary to make a very great effort towards the diffusion of scientific thought and the renewal of experimental methods and to disseminate them among a wide public, as, in the last resort, it is this public which reaps all the benefit. It is for this reason that the senior author of the French edition feels justified in his delight at seeing that this book is going to present the ideas of the French school of electron optics to an English-speaking public, to a public whose language is that of the country in which the "electron" was born.

It is a pleasure to thank the many colleagues, both at home and abroad, who have graciously contributed to this work by helping us to clarify some difficult points, or by providing us with illustrations taken from their own personal work. From among them, we should like to single out Professors Castaing, Dupouy, Fert, Lallemand, Möllenstedt, Müller, Nozières and Pernoux, and Doctors Duchesne, Haine, Mulvey, Stohr and Wlerick.

We are also grateful to various industrial firms who furnished us with certain documents, and in particular to C.S.F. and O.P.L. (Paris) and Zeiss (Oberkochen).

This book, in its present form, owes much to the painstaking work of Dr. Septier, who brought it up to date. Dr. Hawkes has been more than a translator, and his profound knowledge of Electron Optics is present, if not apparent, in many parts of this book. Last, but not least, I should like to express my personal gratitude, and that of most of my colleagues in this enterprise, to Professor Marton; without his continual encouragement during the past fifteen years, much of the research and teaching included within these pages would perhaps never have been performed.

PIERRE GRIVET

PART 2

INSTRUMENTS

CHAPTER 11

ELECTRON OPTICAL INSTRUMENTS

11.1 LIGHT RADIATION
AND CORPUSCULAR RADIATION;
THE SIMPLE CHARACTER OF PHOTONS

In the first part of this work, we have taken generous advantage of the ideas and methods of light optics to develop the optics of electron lenses. Despite the deep-rooted analogy between these two disciplines, however, there are features which are peculiar to electron optics, the original nature of which we have already indicated (Chapter 6); it is only rarely that we can transfer directly into electron optics results which have been obtained for glass lenses. This same characteristic reappears, in a still more pronounced form, when we begin to consider applications, and is apparent as soon as we try to classify the various instruments of electron optics. An illustration of this is the "mass spectrograph" and the "velocity spectrograph"; in light optics, we speak only of the "spectrograph" without qualification, since the optical spectroscope is unique like the property of the radiation which it analyses.

A light source emits photons which are characterized by the frequency v of the electromagnetic wave which is associated with them. The energy of each photon is hv ($h = 6.6253 \times 10^{-34}$ joule sec^{-1}) and the velocity *in vacuo* is c ($c = 2.997 \times 10^8$ m sec^{-1}) whatever the frequency. In a medium of refractive index $n(v)$, the phase velocity falls to c/n, and is thus determined as soon as the frequency is known. A single type of spectrograph is adequate to analyse the radiation at the various frequencies. Further, a primary luminous source—an incandescent metal or a luminescent gas, for example—does not reveal its structure, does not display fine geometrical variations which might have provided an incentive to build a microscope to study emission phenomena.

11.2 MASS AND VELOCITY SPECTROGRAPHS;
DIFFRACTION DEVICES

Corpuscular radiation, which we shall now consider, is not wholly defined when the nature of the particles of which it is composed is known; for the nature of the radiation to be completely defined, the energy of the particles must also be known. In particle optics, therefore, two new types of instrument appear. The first of these, the *mass spectrograph*, sorts the particles according to their mass, or more precisely, according to the ratio of their charge to their mass, e/m, irrespective of their velocity. Instruments of this kind contain a magnetic prism which is sensitive to the quantity "e/m", in conjunction with an electrostatic unit which is insensitive to variations in e/m; the chromatic aberrations of the latter are useful, however, to compensate those of the magnetic prism and thus to provide a combination which is barely sensitive to the velocity dispersion of the particles which are being analysed. The mass spectrograph has no real analogue in glass optics.

There are "velocity filters", on the other hand, which are insensitive to the value of e/m, and which are used to ensure that the velocity of a beam of particles of various masses and charges is homogeneous. The most important members of this second class of instruments are those which are used to analyse the velocities of the beams of positive or negative electrons which are emitted by radioactive bodies; since these are the particles which constitute the β-radiation (β^+ or β^-), the instruments are known as *β-ray spectrographs*, and can be constructed from any prism or lens which has a large chromatic aberration. β-ray spectrographs may have either a magnetic or an electrostatic lens, therefore, (cf. § 7.4.3) and velocity spectrographs either a magnetic or an electrostatic prism. These instruments, which are the electron counterparts of optical spectrometers, will be compared and described in detail in the final chapter.

Grating spectrometers have been employed in optics since the work of Fraunhofer. An analogue of the grating is to be found in electron optics also, but here the grating is never artificial; instead, it is the natural three-dimensional grating which is formed by the regular lattice of atoms within a crystal. This, we recognize, is an inevitable consequence of the shortness of the de Broglie wavelengths (§ 6.1.5), which are of the same order of magnitude as those of X-rays. For this reason, electron diffraction cameras perform the same function as X-ray spectrometers, analysing the structure of a crystal with the aid of a beam of monoenergetic electrons, of known velocity. The diffraction mechanism of de Broglie waves in a three-dimensional crystal lattice is the same as that which governs the diffraction of light at the lines of a plane Rowland grating, but the field of applications of the former—crystal structure analysis— is as original as it is extensive. This technique is still undergoing rapid development.

11.3 CATHODE RAY TUBES; OSCILLOGRAPHS

A property peculiar to electron lenses and prisms is the basis for another category of instruments, oscillographs and the devices which are derived from them. Once a glass lens has been made, it is inflexible, and to focus the image, either the lens, the object or the screen must be shifted. In electron optics, on the contrary, the optical properties of the lenses can be modified simply by varying the potentials of the electrodes (or the currents in the coils). Focusing, therefore, is extremely straightforward, and it is also possible to alter the direction of a beam very rapidly by varying the deflection of a prism. As a result, instruments which are most inconvenient and of limited use in glass optics become wholly practicable and are widely used when we consider their electron optical counterparts.

The most striking example of such an instrument, one of the commonest of electron optical devices, is the *cathode ray tube*, very often no more than the electron equivalent of the inconvenient "mirror oscillograph", namely, the cathode ray oscillograph. The extremely small mass of the electrons is of unequalled utility as regards the lateral deflection; whereas mechanical devices are restricted to acoustical frequencies (of kc/s, or less) the cathode ray oscillograph is capable of tracing out the sinusoidal curve $x = x_0 \sin 2\pi \nu t$, for values of the frequency ν as high 10,000 Mc/s. Such extreme speeds are as yet rarely exploited, and the applications which are of the greatest everyday importance—radar and television—are still far from requiring this extreme rapidity from oscillographs; the qualities of which they do take the fullest advantage can be summarized as convenience, simplicity and sensitivity, the precise meaning of which will appear in the next chapter.

A number of very specialized applications of the cathode ray tube have been developed, in which the properties of the tube which we have just discussed are linked to the phenomenon of secondary emission. The electron beam is no longer incident upon either a fluorescent screen or a photographic plate, but upon a "target" which has a large secondary emission coefficient. The target is said to be "explored" with an "electron probe", and the intensity of the stream of secondary electrons provides the "signal" from which we can extract information about the target. Similarly, in television, we reconstruct the image of the "target" at which the camera is aimed, on the screen of the receiver with the aid of a series of amplifiers, a radio transmitter, and finally a "television oscillograph". Here, however, the target is artificial—the mosaic image formed by an iconoscope, for example—since it is the optical image of the scene which is transmitted. But in these other devices which are used in physics, we obtain the image of the structure of a natural target on an analogous principle. These *probe microscopes*, or *microanalysers*, which are of relatively recent development, are extremely valuable in metallurgy and crystallography.

11.4 MICROSCOPES

This brings us to a class of instruments which are much nearer to their light optical originals, *electron* and *ion microscopes*. But here once again, the transmission electron microscope, which is the most typical member of the class, makes it possible to observe new types of interaction within the object which create the contrast which is at the very foundation of the image. Further, it provides us with an original solution to the standard problem of the resolving power. No longer can we work with stigmatic lenses: the familiar Weierstrass conjugate points have no useful analogue in electron optics, and as a result, the role of the aberrations is most important. It is a triumph of electronics to have been able to show both in theory and in practice that despite this it is possible to form well-separated images of two points less than a millimicron (or 10 Å) apart, by a skilful balancing of the harmful effects of aberrations and diffraction, and by correcting defects due to mechanical shortcomings in the construction.

Electron (or ion) microscopy produces one further new field of applications—the microscopy of sources of electrons or ions, with which we can see all the physically interesting features of a natural structure. On the screen of an *emission microscope*, we are shown this new aspect of matter at a high magnification (which may reach 5000). This is microscopy in the proper sense of the word, as we can study such phenomena as allotropic transformations at a high temperature, or the rapid evolution of grain boundaries, in conditions which are poorly adapted to any other means of examination. Further, by employing the electron emission from a very fine point subjected to an electric field of several million volts per centimeter, we can examine the actual arrangement of the molecules or atoms which compose the surface of the crystal from which the point is formed at a magnification of more than a million times.

11.5 TELESCOPES AND IMAGE-CONVERTERS

Yet another instance of the diversity of electron optical methods is the *electron telescope* (or "image-intensifier"), not a particularly fortunate choice of name. In fact, it is principally of interest for the transformation of one luminous image into another, by means of a long chain of operations, the first of which employs the photoelectric effect. A faithful reproduction of the initial image eventually appears on the fluorescent screen of a cathode ray tube; the reproduction however, is an "improvement" upon its original, not only in that it may be magnified, but above all in that it may be 100 or 1000 times brighter. The function of the telescope, in short, is that of a "brightness amplifier". Further, the final image is visible whereas the initial image—which might well be produced by ultraviolet or infra-red light—need not necessarily have been; in this case, the telescope functions as an "image-converter".

In this descriptive summary of the instruments of electron optics, it is their distinguishing characteristics that we have stressed, mentioning no more than their fundamental properties and their essential parts. Naturally, "electronics"—a word which nowadays embraces a whole section of physics—finds its place in all these instruments, which owe their efficiency and convenience—their quality, in brief—to the individual electronic apparatus, interesting but involved, which is associated with each of them. Within the limits of such a study as this, it does not seem possible to describe all this electronics, and we shall in fact only consider the basic structure of the cathode ray tube, ignoring completely the special electronic circuits which are associated with it.

Only the truly optical features of electron optical instruments will be considered in the following chapters of this book. The lavishness of possible subject matter has already led to the banishment of accelerators which produce beams of particles of very high energy from the present work. The reader is referred to the relevant volume of Flügge's *Encyclopædia of Physics* and to *Linear Accelerators*, edited by P. Lapostolle and A. Septier (North-Holland, 1970).

The order in which we have set out the various instruments in this introduction will not be followed in the rest of the book; we shall begin with those instruments which most resemble their optical counterparts, and continue in order of increasing complexity as far as the mass spectrograph, the study of which is an excellent preparation for the instruments of nuclear physics.

CHAPTER 12

THE CATHODE RAY TUBE

12.1 HISTORICAL INTRODUCTION;

THE IMPORTANCE OF THE CATHODE RAY TUBE

The cathode ray tube is the oldest of the instruments of electron optics. As early as 1896, Braun built the tube which is named after him, in which the electrons were obtained by secondary emission, and gave a green spot at the end of the receiver. There were neither stops nor a fluorescent screen, as the "spot" was produced by the natural fluorescence of the glass. It was, however, possible to deflect the spot by applying a magnetic or electric field perpendicular to the beam, and on the screen the spot traced out curves from which the temporal variation of the applied field could be deduced; the first cathode ray oscillograph had been created.

The most fruitful advance was made by Wehnelt, who replaced the capricious secondary source (the behaviour of which is so dependent upon the pressure which prevails in the tube) by a hot, oxide-coated cathode. In addition, he introduced the additional electrode which bears his name in the present-day Continental literature (the grid, in English). After this, the cathode ray tube could be said to involve genuine "electron optics"— fine, bright spots could be obtained, the brightness of which could be adjusted by varying the potential of the grid. It became possible to trace out well-contrasted pictures on the screen, by breaking them down into the curves which the spot followed under the action of two time-varying potentials applied to the deviating plates. The intensity of the spot was controlled independently by a third potential, which was applied between the cathode and the grid. The principle of the television receiver had been discovered. This is the principle, too, of the radar receiving unit, so rapidly developed as a result of the exigencies of the second world war. With three potentials, which can be varied at will in time, either by amplifiers (in the case of the potential which originates at an aerial and that which acts on the grid) or by scanning oscillators (which displace the spot), a whole map of the region which is being explored by the radio beam can be produced on the fluorescent screen.

The importance of radar and television to-day—in wartime as in peace— has made this one of the most highly perfected techniques of the age. Cathode ray tubes are mass-produced by machines just like radio valves.

A whole mass of research, often enough empirical, has led to a considerable improvement of the properties of Braun's and Wehnelt's tubes—the fineness of the spot, its luminosity or the sensitivity—without introducing any fundamental modification of the basic design. A second lens has been introduced, which can be either electrostatic or magnetic, in such a way as to obtain a fine and brilliant spot, and the shape of the plates or coils which produce the deviation has been modified so that the spot remains as small as possible wherever it may be on the fluorescent screen. Again, great progress has been made in the chemistry and crystallography of fluorescent powders. Instead of 0·1 per cent, the efficiency with which the kinetic energy at the point at which the electrons hit the fluorescent screen is transformed into light can be of the order of 25 per cent, and a handsome range of colours is available, for colour television. In high tension oscillographs (functioning between 30 and 100 kV), the televised image is brilliant enough to be projected onto a cinema screen. Finally, technological advances (developments in vacuum techniques, and in sealing into glass, for example) have made possible a considerable reduction in the manufacturing cost of these instruments, which are virtually of everyday use in the sociological sense of the phrase.

So well stocked have the manufacturers' catalogues become that almost every conceivable requirement of the engineer, the physicist, or the television amateur is catered for. Even so, new types of tube appear on the market every year, for the inventive efforts of the physicists concerned keep the models available in a state of continual evolution. For this reason, a method of scaling, particularly well illustrated in the work of H. Moss (1950), has appeared which makes an ingenious use of dimensional relations to solve a number of the problems with which designers are faced. For example, suppose we have a direct vision cathode ray tube with an exceptional performance, which has a screen 38 cm in diameter; in what way must we modify the design to obtain the characteristics of a tube which is to be used in a projection television set, which has a screen with a diameter only one-third the size, but which is, of course, to provide no less detail in the image? The diameter of the spot will be reduced to one-third of its previous value, but the fact that the image is to be projected means that the luminous intensity must be greatly increased (perhaps some five times). The solution, however, has to be found subject to certain restraints—there are quantities, like the specific emission of the cathode, which cannot, unfortunately, be varied between very wide limits. The dimensional relations which have been established by Moss provide an acceptable solution which is more than a simple geometrical similitude: the geometrical dimensions of the envelope will be one-third of those of the original tube, but the size of the electrodes in the gun is reduced only in the ratio 1 : 0·63; the direct vision tube is operated at 8 kV, while the projection tube requires 28 kV. The current at the spot of the latter reaches 60 μA, whereas it had previously been 150 μA. The laws by which we pass from one situation to another are thus rather complex, and for details of this ingenious method of "generalized similitude" the reader is referred to the original article by Moss.

This study of cathode ray tubes will be restricted to the basic faculties of focusing and deflecting the beam. Nor shall we consider construction techniques, again referring the reader to the article by Moss (1950) mentioned earlier, and also to the book by Say (1954). A much quicker and more effective way of learning, however, is to "perform an autopsy" on a discarded tube, attempting afterwards to reassemble it with the aid of a vacuum pump.† Similarly, we shall not attempt the description of the various auxiliary devices which have to be added to a cathode ray tube to transform it into a cathode ray oscillograph, a television receiver, or a radar receiver. For further information on such points the reader is referred:

(a) to the books and articles by Rider and Uslan (1950), Bonvalet (1952), Wilson (1953) and Say (1954), on cathode ray oscillographs;

(b) to Soller, Star and Valley (1948) for radar receivers; and

(c) to the publications of Maloff and Epstein (1938), Zworykin and Morton (1954), Delaby (1951), Delbord (1951), Puckle (1951) and Grivet and Herreng (1953) for information about television receivers.

In these are to be found details of the construction of scanning oscillators, amplifiers, and stabilized sources, in short, of accessories, which are easier to manipulate and construct than the tubes themselves and are of simpler structure.

It is convenient to conclude this section by pointing out that the same form of cathode ray tube is to be found only slightly modified in the television camera. In the image iconoscope are to be found all the elements which have been mentioned earlier—gun, focusing lens, deflection system. The difference between the two instruments resides not in the electron optics, but in the surface on which the beam impinges. The spot no longer sweeps across a fluorescent screen, producing light, but instead, across a mosaic of photoelectric cells; its purpose is to measure the charge which has been acquired by the cathode of each of these photoelectric microcells as a result of the light which has been shone upon it. For this new function, an extremely fine spot is necessary, but on the other hand, only a low beam current— a few microampères, ten or a hundred times less than in an oscillograph— is required. It is due to this that the gun of an iconoscope is of an essentially different design; again, further information is to be found in the relevant references which have already been quoted.

---

† Opening a television tube, it must be realized, is a delicate operation. The gun cannot be recovered by breaking the tube, which would implode, as this would be a dangerous procedure. The tube has first of all to be filled with air, by producing a leak near the point at which the tube had been sealed off.

12.2 FORMATION OF THE SPOT

12.2.1 The General Appearance of the Optical System

A cathode ray tube is an instrument which is sealed off at a static vacuum. Only in a few oscillographs, which are designed to be used at very high tension (60 kV) to study rapid phenomena (of frequencies above a few Mc/s), is the vacuum a dynamic vacuum, maintained by a diffusion pump as in the electron microscope. It is worth mentioning that Dufour used a dynamic vacuum in one of the earliest cathode ray oscillographs, while his "scanning system" consisted in moving the photographic plate. The most recent developments in this field are those of von Ardenne (1956). We shall limit ourselves to studying sealed-off systems, protected by a glass envelope the shape of which is well-known. All the electrical connexions are led out through the base, with the exception of the "post-acceleration" lead the purpose of which is to maintain the fluorescent screen and the surrounding part of the tube at a high potential (up to 40,000 V). As Fig. 116a shows, this connexion is led out of the tube through a glass–metal seal, for the obvious reason of insulation.

It is no more difficult to achieve the vacuum of 10^{-6} mm Hg which prevails inside the tube than it is in the case of a radio valve, but it is less easy to maintain as the way in which the screen is made can lead to the production of volatile residues, which are not present in radio valves, and which have a most harmful effect upon the emissive properties of the cathode. To overcome this hazard, the tube is outgassed by a very careful baking, and an extremely active "getter" (made of one of the alkaline earths) is used, which absorbs the residual gas if it is heated while the tube is being evacuated. The envelope has to be able to withstand the stress due to atmospheric pressure over a considerable surface area, which creates difficult problems in glass or steel-and-glass technology (the base of a television tube 35 cm in diameter has to be capable of withstanding a force of the order of a ton weight).

The various parts of a typical tube (Fig. 116a) are, starting from the base:
—first an oxide-coated cathode K about 1 mm in diameter, in the form of a small disc of magnesium–nickel alloy (1 per cent Mg) coated with a mixture of barium oxide and strontium oxide. This is heated with a tungsten filament to the working temperature which is normally of the order of 1000°K, and is chosen in such a way that the mean emission at the cathode allows a long lifetime;
—a cylinder or ring, W, which is known as the "grid" or "Wehnelt"; in normal operation, this electrode is held at a potential of the order of a few tens of volts, negative with respect to the cathode;
—an anode, A_1, which accelerates the electrons to a few thousand volts and which in general consists of a metal cylinder together with a plane

plate in which a hole of a few millimetres diameter has been cut, through which the beam passes.

This combination represents the simplest type of gun, the triode gun, which makes the electrons converge to a very fine "crossover" in the neighbourhood of the hole in the anode. By varying the grid potential,

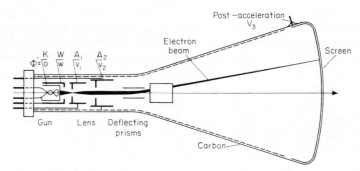

FIG. 116a. Diagram of a cathode ray tube with electrostatic deflexion and tetrode gun.

the crossover can be moved away from the anode, but the linear magnification of this immersion lens rapidly becomes very large (Chapter 12) and the crossover of a prohibitively large diameter.

Triode guns are rarely used directly, but rather, in conjunction with a supplementary electron lens. A second anode A_2 is placed beyond A_1 and the two form a two-tube accelerating lens; the potential V_2, measured with respect to the cathode, is higher than V_1. By varying the ratio V_2/V_1, the convergence of the lens is varied, and the position of the focus can be varied between wide limits; the whole system constitutes a "tetrode gun".

The electron beam then passes through a series of electron prisms which attend to the deflexion, and finally falls on a fluorescent screen; the glass envelope is rendered conductive by depositing a layer of carbon (aquadag or lamp-black), so that the beam travels through a region which is genuinely field-free from the static point of view. The end of the tube which carries the screen is sometimes insulated from this layer and raised to a very high potential which accelerates the electrons during the later part of their trajectory (post-acceleration) to energies which may reach 30–50 kV.

The electrical circuit is closed by virtue of a thin metallic layer which covers the fluorescent screen and which is connected to the source of high tension. This layer is transparent to electrons and increases the brightness of the screen, by returning any light which could have been emitted into the interior of the tube back towards the observer (Epstein and Pensak, 1946). For further information on this point, the reader is referred to § 12.2.5 on fluorescent screens; to § 8.4 on guns, and to § 8.1 on two-cylinder lenses. Complementary information about the guns which can be used in cathode ray tubes is to be found in the articles of Morton (1946), Field (1946) and Moss (1946, 1946a, 1950).

The whole tube assembly is enclosed within a mumetal (or permalloy) shield, which eliminates the perturbations which exterior parasitic magnetic fields could produce; such fields are in particular produced by transformers in the supply unit. An alternating field with peak amplitude 0·5 gauss, acting over a distance of 4 cm in the neighbourhood of a gun operating at 1000 volts is enough to displace the spot 4 mm on the screen 20 cm away, and thus to mangle the image completely.

Nowadays, guns of a rather more complicated design are employed, with which the different adjustments to the gun can be clearly separated. "Focusing" consists in bringing the image of the crossover of the beam onto a fluorescent screen; this crossover lies at the exit of the triode system formed by K, W and A_1, which here behaves as a strongly convergent lens. The focusing is carried out by varying the potential V_1 of the anode, leaving the final accelerating potential of the electrons (V_2) constant; V_1 is also the potential which extracts the electrons, however, and there is in consequence a slight variation of the emitted intensity. The brightness on the other hand, is adjusted by varying w, the potential of the grid; the action resembles that of a triode valve, in which the emissive area of the cathode is varied. This variation of w is accompanied by a slight change in the convergence of the triode system, and hence by an axial displacement of the crossover which entails a slight defocusing. The coupling between the two adjustments is small enough in ordinary tubes, however, not to be troublesome; this explains the widespread use of the tetrode gun.

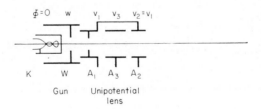

Fig. 116 b. Diagram of a pentode gun.

The interaction of brightness and focusing can be suppressed by replacing A_2 by a magnetic lens, situated outside the tube. Alternatively, the accelerating gap $A_1 A_2$ can be replaced by a genuine einzel lens by inserting a further electrode A_3 between A_1 and A_2. These two electrodes are then connected electrically and held at an accelerating potential V_2; the system can be focused by varying the potential V_3 of the central electrode A_3. The result is a pentode gun (see Fig. 116 b). The lateral extent being so great, the glass envelope has to be large in diameter (5–10 cm), and this in turn allows us to use lenses with large openings and hence to reduce the spherical aberration. There are also advantages in using large diameter guns so far as the modulation of the brightness of the spot is concerned, as we shall see later. Figures 116c and 116d show examples of such constructions. In Table 1, we quote some characteristic parameters of cathode ray tubes (tubes constructed by RCA).

TABLE 1

NUMBER	2 API-A	7 DP-4	16 AP-4	7 NP-4
PURPOSE	Oscillograph	Television set		Projection on a large screen
DIMENSIONS (in cm): Length Screen diameter	18 5·2	35 18	55 40	45 20
GUN: Cut-off potential Anode potential	− 60 1 000	− 27 410	− 40 300	− 125 625
FOCUSING (the first electrode is the anode of the gun)	3-electrode lens 1 000 V 250 V 1 000 V	4-electrode lens 410 V 6 000 V 1 200 V	Exterior Magnetic lens	2-electrode lens 625 V 16 000 V
SCREEN POTENTIAL: (P.A.: post-acceleration)	1 000	6 000	12 000 (P.A.)	80 000 (P.A.)
DEFLECTION	Electric; sensitivity: 0·11 mm/V (vert.) 0·13 mm/V (horiz.)	Magnetic; outside the tube (see the examples in § 12.3.4)		

The reader will find a comparison of the deficiencies and advantages of tetrode and pentode guns in an article by Darbyshire (1955).

Oscilloscopes with a very high resolution (a spot of the order of 0·1 mm diameter) are provided with an improved optical system. First of all, the tetrode gun must be selected in such a way that for a given distance between gun and screen, the linear magnification is as small as possible; experience shows that it is best to use a two-tube lens (the two tubes formed by A_1 and A_2) in which the diameter of A_2 is considerably less than that of A_1 ($d_2/d_1 = 0·33$, for example). The size of the final spot is still further reduced by using an einzel objective lens L in conjunction with $A_1 A_2$; L is placed

Fig. 116c. Cathode ray tube gun. (Photo: C. S. F., France – No. 504. 609. 63.)

FIG. 116d. The F8071 Thomson-CSF tube, with electrostatic focusing and deflexion, and with post-acceleration, for studying high frequency signals (up to
150 Mc/s). This tube is equipped with a deflexion amplifier using quadrupole lenses
(which can be seen in the detailed view of the gun). Observations are made on a
120 mm × 80 mm flat screen. (Photo: C.S.F.)

at a suitable distance between A_2 and the deflexion plates. The latter, as we shall see later, behave like weak cylindrical lenses, which results in over-focusing of the deflected spot and elongation of the spot in the direction perpendicular to that of the deflexion. This can be remedied to some extent by introducing in front of the plates a focusing element which is itself astigmatic, the horizontal and vertical convergences of which can be adjusted independently. At best, the spot is then constituted by the circle of least confusion which lies mid-way between the two real line foci of the beam. An elegant way of achieving this is to replace the central electrode of the objective L by two electrodes with elliptical openings, crossed, excited with two independently adjustable potentials (Cooper, 1961). In this way, we can obtain a spot, the diameter of which remains practically constant throughout all the useful part of the screen.

12.2.2 The Spot

(i) *The dimensions of the spot*

If we denote the ratio of the distances between the lens and the screen and between the lens and the crossover by K, and the focal length of the immersion lens which constitutes the gun by f, an application of the Lagrange–Helmholtz formula (§ 4.3.5) yields the diameter d_s of the spot,

$$d_s = 2Kf\sqrt{\frac{T}{11600\,\Phi_0}},\tag{12.1}$$

in which Φ_0 is the potential of the fluorescent screen and T is the temperature of the cathode. For the common kinds of television tube, the orders of magnitude are $T = 1160°\mathrm{K}$, $\Phi_0 = 10,000$ V, $K = 5$ and $f = 3$ mm, which leads to a value for the diameter in the region of $100\,\mu$. This has nothing to do with the spot into which the point is spread by the distribution of the thermal velocities of the electrons, and the effect of the aberrations of the lens has also to be superimposed. Nevertheless, the defects are not sufficiently serious to prevent the tube from being used in practice. In France, for example, the television image is broken up into 819 horizontal lines, which means that the spot on a screen 30 cm high must be $360\,\mu$ in diameter $\left(\text{that is }\dfrac{30 \times 10^4}{819}\right)$. The margin for the other defects is quite adequate. The plates, on the other hand, deform the spot much more markedly, above all when the spot is close to the edge of the field, a point to which we shall return later. We can now understand one of the reasons why the tube is operated with such a high potential at the screen, some 40 or more kV. As the spot diameter varies inversely as $\sqrt{\Phi_0}$, this is a way of obtaining a finer spot; we shall reconsider this in § 12.2.4.

(ii) *The intensity of the spot*

If i_c represents the current density emitted at the cathode, and if we assume that i_c is uniform inside the region of diameter d_0 over which electrons are emitted, then the current density in the spot, i_s, is given by the formula developed by Langmuir (1937) which we have already obtained in § 4.3.5:

$$i_s = i_c \left[\frac{11,600\ \Phi_0}{T} \right] \alpha_s^2, \qquad (12.2)$$

where α_s is the semi-aperture of the beam which produces the spot at the screen. In fact, of course, the current density is not uniform across the whole spot; Law (1937) suggested that the distribution law might have a Gaussian

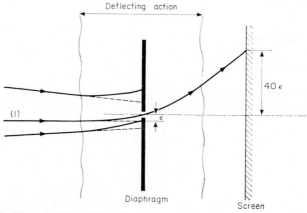

FIG. 117. The principle of Beam's arrangement: ray (1) would originally have struck the screen a distance ε from the spot—it is therefore deflected a distance ε. The deflecting action continues afterwards, however, and the ray falls on the screen at a point far from its centre.

form, which was subsequently verified by Dosse (1940) and by Jacob (1939). More recently, Beam (1955) has invented a simple way of producing an image of the spot on the screen magnified forty times. It is thus possible to measure the intensity very precisely (Fig. 117). The spot is moved with the aid of magnetic deflexion over a metal plate in which an extremely small hole has been punched (with a diameter of about 2×10^{-2} mm). This plate lies in the plane in which the spot would normally have been formed. For a given deflexion, part of the beam passes through the hole and impinges on the fluorescent screen which is placed a few centimetres beyond the plate. The deflexion device is still effective beyond the plate, however, so that the electrons move away from the axis as they travel between the plate and the screen, and the result, effectively, is a magnification. For example, consider the electrons which would have been 100 μ

from the middle of the spot; in order to pass through the hole, they have had to be deviated 100 μ across the plate, but as the electron beam continues to move away from the axis beyond the plate, the electrons strike the fluorescent screen some millimetres from the central point. We obtain, therefore, a magnified spot.

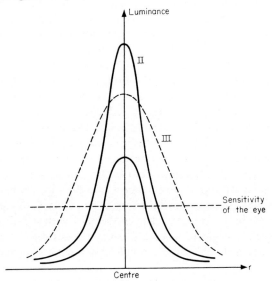

FIG. 118. The apparent broadening of the spot produced by increasing the intensity (II) and the genuine broadening due to space charge (III).

The current density clearly ought to be as large as possible. The current density i_c which is emitted cannot, however, exceed a certain value (0·25 A cm^{-2}) without the risk of exhausting the cathode and the danger of reducing its lifetime; (normally a cathode should last for at least 2000 hours). α_s can barely exceed 0·01, as otherwise the geometrical aberrations become too prominent, and naturally we cannot reduce T, as i_c is dependent upon T, being a rapidly increasing function of temperature. In practice, T has to remain in the neighbourhood of 1000°K, for the decrease of i_c outweighs the increase of $\dfrac{1}{T}$ if T is reduced. We can, therefore, only increase Φ_0, which is a second reason for using high anode potentials.

12.2.3 The Effect of Space Charge on the Beam

The mutual repulsion between the electrons also prevents an intense electron beam from being perfectly focused. This real effect is not, however, to be confused with a deceptive effect which is always observed and which we shall first of all describe. It is easy to verify that if the brightness of a television tube or an oscillograph is increased (by making the grid positive) the diameter of the spot appears larger. This impression can be explained in terms of the properties of the eye in the following way. The light distribution across the spot is represented by a bell-shaped curve, as shown in Fig. 118, and not by a uniform distribution; the eye, however, is sensitive to brightness only above a certain level, below which no impression is made. The apparent spot diameter, therefore, is obtained by drawing a line parallel to the abscissa axis at the height which corresponds to the threshold sensitivity of the eye; since the principal effect of increasing the brightness of the spot is to widen the distribution-curve for each value of the ordinate, it is plain that the diameter of the visible disc, obtained by the construction which has just been described, increases, quite irrespective of any broadening which may be produced by space charge effects.

Having disposed of this subjective illusion, we can pass on to an objective study of the broadening which space charge produces, which can be measured either with a Faraday cage, or by a number of other ingenious methods. We shall demonstrate that the space charge plays a role which is comparable with that of a weak divergent lens, supposed to be superimposed onto the lens of the cathode ray tube, but which cannot be adjusted. We shall examine the behaviour of the beam which leaves the lens with a radius r_0 and an angular aperture r_0', negative since the beam is convergent (see Fig. 119). Normally, the beam would form a point spot after a distance

$$z_0 = \left| \frac{r_0}{r_0'} \right|.$$

We assume that the potential in that part of the tube which follows the lens is constant, so that the electrons travel at a constant velocity. If this potential is represented by Φ_0, then the velocity, which is almost axial, is given by

$$v = \sqrt{\frac{2e\,\Phi_0}{m}} \simeq \frac{dz}{dt}.$$

Consider now an electron on the edge of the beam—it experiences both the electric field E which is created by all the other electrons in the beam, together with the magnetic field which is associated with the current to which a beam of electrons in motion is equivalent. The effect of this magnetic field we can ignore, for being proportional to $\left(\dfrac{v}{c}\right)^2$ it is very small, of the same order of magnitude as a relativity correction. The electric field, which is radial, can be calculated by applying Gauss' theorem to a small cylinder with axis Oz, length dz and radius r. The flux through the surface is given by

$$\text{flux} = 2\pi r \cdot dz \cdot E_r$$

$$= \frac{1}{\varepsilon_0} \varrho\, \pi r^2 \cdot dz.$$

The charge density, however, is related to the current I and to the axial velocity, and hence to the potential, in the following way:

$$\pi r^2 v \varrho = I,$$

$$\varrho = \frac{I}{\pi r^2} \sqrt{\frac{m}{2e\, \Phi_0}},$$

and consequently

$$E_r = \frac{1}{2\pi\, \varepsilon_0} \sqrt{\frac{m}{2e}}\, \frac{I}{\sqrt{\Phi_0}}\, \frac{1}{r}.$$

The lateral equation of motion takes the form

$$\frac{\mathrm{d}^2 r}{\mathrm{d}t^2} = \frac{e}{m} E_r = \frac{1}{2\pi\, \varepsilon_0} \sqrt{\frac{e}{2m}}\, \frac{I}{\sqrt{\Phi_0}}\, \frac{1}{r}.$$

On eliminating the time, remembering that the axial motion is uniform, we find

$$\frac{\mathrm{d}}{\mathrm{d}t} = \sqrt{\frac{2e\, \Phi_0}{m}}\, \frac{\mathrm{d}}{\mathrm{d}z},$$

and finally arrive at the equation for the trajectory

$$\frac{\mathrm{d}^2 r}{\mathrm{d}z^2} = \left(\frac{1}{2}\beta\right) \frac{1}{r},$$

with

$$\beta = \frac{1}{2\pi\, \varepsilon_0} \sqrt{\frac{m}{2e}}\, \frac{I}{\Phi_0^{3/2}} \simeq 3 \cdot 10^4\, \frac{I}{\Phi_0^{3/2}}. \tag{12.3}$$

Multiplying both sides by $\dfrac{\mathrm{d}r}{\mathrm{d}z}$ and integrating, we find, on substituting the initial conditions, that

$$r'^2 - r_0'^2 = \beta \log\left(\frac{r}{r_0}\right). \tag{12.4}$$

This equation has often been discussed in the literature—the reader is referred to the articles by Watson (1927), Zworykin (1933) and to a very detailed examination by Ivey (1954). We shall outline the method by which it may be integrated. We have

$$\frac{\mathrm{d}r}{\mathrm{d}z} = \beta^{1/2} \left(\log \frac{r}{r_0} + \sigma^2\right)^{1/2}, \tag{12.4(i)}$$

with

$$\sigma^2 = \frac{r_0'^2}{\beta}.$$

We write

$$\log \frac{r}{r_0} + \sigma^2 = \xi^2,$$

so that

$$r = r_0\, e^{-\sigma^2}\, e^{\xi^2},$$

$$\frac{dr}{dz} = 2r_0\, \xi\, e^{-\sigma^2}\, e^{\xi^2}\, \frac{d\xi}{dz}, \qquad (12.4(ii))$$

and from equation (12.4 (i)):

$$\sqrt{\beta}\, dz = 2r_0\, e^{-\sigma^2}\, e^{\xi^2}\, d\xi.$$

The origin is selected at $z = z_0$, so that

$$z - z_0 = \frac{2r_0}{\sqrt{\beta}}\, e^{-\sigma^2} \int\limits_{\sigma}^{\sqrt{\log\frac{r_0}{r} + \sigma^2}} e^{\xi^2}\, d\xi, \qquad (12.4(iii))$$

and tables of the integral

$$D(\eta) = \int\limits_0^{\eta} e^{\xi^2}\, d\xi$$

are available (Dawson, 1898; Terrill and Sweeny, 1944a, 1944b).

If the beam is initially parallel to the axis ($r_0' = 0$, $\sigma = 0$), the relation takes the simple form

$$z - z_0 = \frac{2r_0}{\sqrt{\beta}}\, D\!\left(\sqrt{\log\frac{r}{r_0}}\right), \qquad (12.4(iv))$$

since

$$D(0) = 0.$$

Given the radius r_0, the displacement and the factor β by which the beam is characterized, we can extract the value of $D\!\left(\sqrt{\log\frac{r}{r_0}}\right)$ from (12.4(iv)). If we consult the curve plotted in Fig. 119a, we can obtain the value of $\sqrt{\log\frac{r}{r_0}}$ on the abscissa axis, and hence $\frac{r}{r_0}$ or the widening of the beam.

If the beam is initially convergent (see Fig. 119b), there will be a point with abscissa z_m, where the radius of the cross-section is a minimum, r_m, and the slope zero. From (12.4 (i)),

$$r_m = r_0\, e^{-\sigma^2} = r_0 \exp\left(-r_0'^2/\beta\right). \qquad (12.5)$$

This would be the radius of the spot if only the space charge acted. In reality, the effect due to the aberrations of the lens and the gun must be added. The abscissa of the minimum is given by:

$$z_m - z_0 = \frac{2r_0}{\sqrt{\beta}}\, e^{-\sigma^2} \int\limits_{\sigma}^{0} e^{\xi^2}\, d\xi,$$

or since σ is negative,

$$z_m - z_0 = \frac{2r_m}{\sqrt{\beta}} \int_0^{|\sigma|} e^{\xi^2} \, d\xi = \frac{2r_m}{\sqrt{\beta}} D(|\sigma|). \qquad (12.5(i))$$

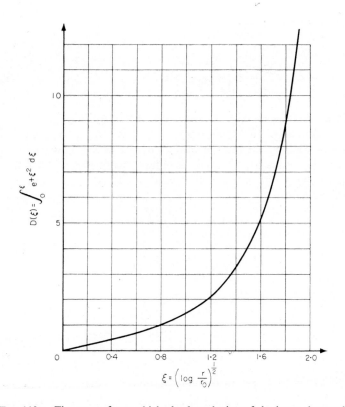

FIG. 119a. The curve from which the broadening of the beam due to the effect of space charge can be calculated.

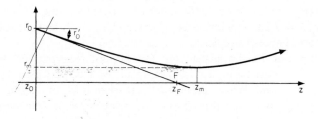

FIG. 119b. The appearance of the outermost ray of the beam: the zone of minimum cross-section (waist).

Equation (12.4 (iii)) takes the form

$$z - z_0 = \frac{2r_0}{\sqrt{\beta}} \, e^{-\sigma^2} \left[D \left(\sqrt{\log \frac{r}{r_0} + \sigma^2} \right) - D(\sigma) \right].$$

But

$$\log \frac{r}{r_0} + \sigma^2 = \log \frac{r}{r_m},$$

and hence

$$z - z_m = \frac{2r_m}{\sqrt{\beta}} \, D \left(\sqrt{\log \frac{r}{r_m}} \right). \qquad (12.6)$$

We therefore first calculate β, then σ, and finally r_m. Figure 119a gives $D(|\sigma|)$ and equation (12.5(i)) then provides z_m. For $z > z_m$, we are brought back to the preceding case, as the initial conditions are now $r = r_m$ and $r' = 0$ at $z = z_m$. Equation (12.6) is in effect identical to (12.4(iv)).

If the space charge did not act, the beam would converge onto $z = z_F$ (Fig. 119b). We have

$$z_F - z_0 = \frac{r_0}{r_0'},$$

and

$$r_m = r_0' z_0 \, e^{-\sigma^2} = \sqrt{\beta} \, \sigma z_0 \, e^{-\sigma^2}.$$

From (12.5 (i))

$$z_m - z_0 = 2\sigma z_0 \, e^{-\sigma^2} D(|\sigma|),$$

and finally

$$\frac{z_m - z_0}{z_F - z_0} = 2 \frac{r_0'}{\sqrt{\beta}} \exp \left(\frac{-r_0'^2}{\beta} \right) D \left(\left| \frac{r_0'}{\beta} \right| \right). \qquad (12.7)$$

In general, $\dfrac{z_m - z_0}{z_F - z_0} \neq 1$, except for one particular value of σ; we consider a beam for which r_0 and r_0' remain constant in $z = 0$, and we decrease σ, increasing the factor $\left(\dfrac{I}{\Phi_0^{3/2}} \right)$. The ratio $\dfrac{z_m - z_0}{z_F - z_0}$ is at first equal to unity, increases to about 1·3, and subsequently decreases and becomes much smaller than unity for very small values of σ (see Fig. 120). The maximum occurs for $\sigma^2 = 2·28$. The influence of space charge will be lessened when σ is greater and hence β smaller.

Since I cannot be diminished too far without reducing the brightness of the spot, Φ_0 must be increased. Yet again, we meet a reason for using a high screen potential. We shall see later that when it is not possible to work with a very high potential throughout the whole tube, we can nevertheless "post-accelerate" the electrons in the neighbourhood of the screen alone. The equations of motion above are no longer applicable in a region of accelerating electric field; Moss (1945), Sangster (1955) and Moak (1960) have studied quantitatively the case of beam-widening under the action of space charge for the case of a constant electric field.

In conclusion, we should point out that the influence of space charge is progressively less as the beam diameter is larger. The space charge therefore has little effect during the passage through the lens which follows the gun.

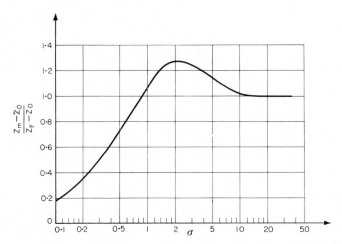

Fig. 120. The position of the minimum cross-section with respect to the geometrical focus. (The influence of space charge increases when σ^2 decreases.)

12.2.4 Modulation of the Intensity of the Spot

Now that we have seen how a fine spot can be produced by the electrons, and reviewed the various causes due to which it may be broadened, we must analyse the mechanism by which we can vary the intensity of the current which falls on the spot. Modulating the spot is not in itself a difficult problem, but difficulties appear when we consider how to modulate the intensity while at the same time keeping the spot as fine as possible. In practice, we adjust the potential of the grid. The potential of this electrode is usually negative, a few tens of volts; if its potential is varied, the intensity varies correspondingly, and a characteristic curve can be plotted which relates the beam current to the modulation of the potential. This curve is similar to the characteristic of a triode valve; an example is shown in Fig. 121. The grid acts upon the brightness by modifying simultaneously both the area of the cathode from which the electrons are emitted and the emissive power. The experimental investigations of Moss (1946a) show that the diameter d_c of the part of the cathode which is actively emitting is proportional to the "conduction potential", that is, to the difference between the

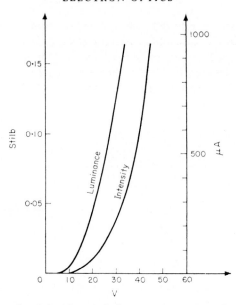

FIG. 121. An example of the characteristic curve of a cathode ray tube (10 BP 4);
this curve gives the intensity and brightness of the spot a s a function of the Wehnelt
(grid) potential, measured relative to the cut-off value.

actual grid voltage and the value at cut-off; we shall denote the potential
at cut-off by Φ_c. If Φ_w is the grid potential and d_w its diameter, d_c is given by

$$d_c = d_w \frac{\Phi_c - \Phi_w}{\Phi_c}.$$

The cut-off potential is related both to the geometry of the gun and to the
anode potential; experiment shows that it is proportional to the latter,

$$\Phi_c = K\Phi_0. \tag{12.8}$$

In modern tubes, the dimensions of the electrodes are chosen in such
a way that the negative cut-off potential is of the order of a few tens of
volts, while the maximum current is obtained when the grid potential is
zero. The spot can thus be controlled with the aid of a simple voltage
divider, so named because it takes no current and therefore provides
negligible power. Also, we avoid the danger of drawing "grid current"
from the beam. The current emitted at any point of the cathode depends
both on the temperature at the point and on the space charge in the vicinity
of the point, exactly as in a diode valve; the electron gun is of course no
more than a variant of the latter. The characteristic curve of a diode valve,
which relates the intensity I to the potential of the plate, falls into two parts

(Fig. 122). One section consists of the curve known as the Langmuir region which is described by

$$I_c = p \, \Phi_0^{3/2},$$

while the other is horizontal, and corresponds to saturation. The current at saturation is given by Richardson's law, which states that

$$I_s = I_0 T^2 \exp\left(-\frac{e \, \Phi_s}{kT}\right),$$

in which Φ_s is the work function of the cathode.

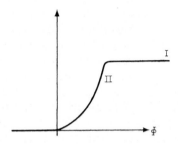

FIG. 122. The characteristic of a diode: I, saturation conditions; II, space charge limited conditions.

Returning to the electron gun, we find that here too the emission can become saturated if the temperature is low (900°K); this is apparent as the potential of the grid becomes ineffective as a regulator of the emission. This is, therefore, an exceptional situation which is only of interest if we are exploring the properties of the tube; the emission is constant, and is governed by Richardson's law. In normal conditions, on the contrary, the characteristics are chosen in such a way that the intensity at saturation is far higher than the values which space charge allows when the grid potential is varied over the whole range of useful potential, between cut-off and full aperture. The mean emission is then given by a law similar to Langmuir's; the coefficient p is known as the "perveance" of the gun. It is not a particularly useful measure of the quality of the guns which are used in cathode ray tubes, since in these it is the increase in the emitting area which plays the essential role, while p only varies over a rather narrow range. This parameter is less useful in this context than in connexion with the high-intensity guns which are used in klystrons.

Since the mechanism which governs I_s is so complex, it is difficult to obtain a theoretical estimate of the intensity of the spot. The intensity can, on the other hand, be measured, and the measurements which Moss (1946a) and Jacob (1952) have made verify satisfactorily that I_s obeys a law of the form

$$I_s = K \frac{(\Phi_c - \Phi_w)^{7/2}}{\Phi_c^2} , \qquad (12.9)$$

in which I_s is the total intensity at the spot. K varies with the dimensions of the gun. The law which relates the potential by which the intensity is controlled to the intensity at the spot is not, therefore, linear. The linearity of this curve has been improved in certain modern guns without any serious loss in sensitivity by constructing the gun in a special form which has been studied in detail. It must be admitted, however, that the eye is extremely tolerant towards the defects which are present in a television picture, for example, as a result of the curvature of the modulation characteristic. The acceleration and the focusing of electrons are clearly separated functions in a pentode gun; with such a gun, therefore, we can obtain a final spot of which the position is almost independent of the grid potential. Throughout the process of modulation, the size of the spot remains constant; in the present state of the technique, it seems that it is by the distribution of the emission velocities of the electrons as they leave the cathode that the fineness of the spot is limited (Dufour and Wendt, 1954).

If a pentode gun is used, another way of modulating the spot is feasible. In the central zone of the unipotential lens which follows the triode gun, the electrons are considerably slowed down; if a supplementary negative potential is applied to the central electrode, the lens can be gradually transformed into a mirror (see Chapter 8), the outer part of the beam being halted first. Electrons which do not manage to pass the potential barrier are reflected back towards the cathode. It has been found necessary to design lenses which satisfy the following two requirements:

(i) The modulation characteristics should be as linear as possible, and very steep, so that a potential of about ten volts suffices to cut off a beam which has been accelerated through several hundred volts.

(ii) The spot must not be defocused during the modulation.

This can be achieved by employing a highly convergent lens. The divergent beam which emerges from the gun converges again into an extremely fine crossover at the point in the lens where the potential has a minimum, and then diverges again as it leaves the lens. A supplementary accelerating lens converges the beam onto the final screen. If the central electrode includes a stop of very small diameter, the ability of the lens to cut off the beam is considerably enhanced, and in addition, the optical properties of the two parts of the lens (before and after the diaphragm) are virtually independent. We can then hold the exit electrode at a potential appreciably higher than that of the anode of the gun. A particularly elegant solution of the problem is due to Schlesinger (1961); the cut-off potential of the lens is about 12 volts for 500 volt incident electrons. The focusing does not vary during the modulation, and for the same current, the spot which is obtained is finer than that of an ordinary tube modulated by the grid.

12.2.5 The Fluorescent Screen

Finally, the electron spot has to be converted into a spot of light, the luminous intensity of which must vary as nearly as possible proportionally with the electron intensity. In § 1.4, we have described the essential properties of the fluorescent materials. Here, we shall give the complementary results which concern cathode ray tubes in particular, and with the aid of which, the efficiency with which the kinetic energy of the electrons is transformed into light can be considerably increased.

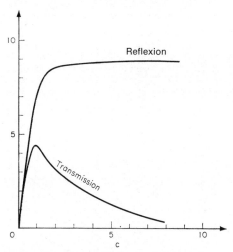

FIG. 123. The yield of a fluorescent screen (in candles per watt) as a function of the surface density c of the fluorescent substance (in mg cm^{-2}).

The fluorescent screens of cathode ray tubes are peculiar in that the image is observed by transmission through the layer and through the glass support, while in a microscope, for example, the observer usually looks at the face of the screen on which the electrons fall. Thin layers of fluorescent material have to be used, therefore, as a great deal of the light will otherwise be lost. Figure 123 shows the way in which the luminous yield at the screen depends upon the thickness of the fluorescent powder, and reveals that there is an optimum thickness, which varies with a whole set of parameters, but in particular with the size of the grains of powder.

For a given thickness of the fluorescent layer, the luminance of the screen is an increasing function of the accelerating potential and of the electron intensity in the beam. This latter cannot be increased very markedly without the risk of shortening the lifetime of the tube considerably. Further, Fig. 124 shows that the fluorescent layer rapidly becomes saturated, and the efficiency with which the energy is transformed into light falls as the intensity rises. The potential of the screen, therefore, must be increased—this is one of the reasons why high accelerating potentials are used (Fig. 125). The

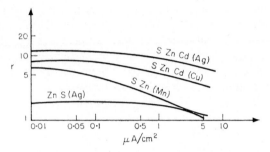

FIG. 124. The reduction of the yield of a fluorescent screen with increasing current density (for a beam accelerated to 6000 volts).

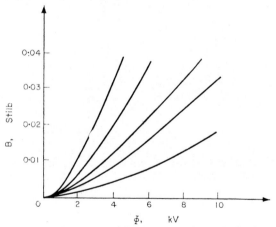

FIG. 125. The variation in the brilliance of a fluorescent screen (a television screen P_4 which fluoresces white) as a function of the accelerating potential of the beam.

The current densities corresponding to these curves are 1, 2, 3, 5 and 10 μA cm^{-2}.

yield at the screen diminishes though, when the potential is increased. The fluorescent products which are available commercially are mixtures; the composition of the ingredients is obtained empirically, and is denoted by various conventional notations (P_1, P_2 in America, CR_1, CR_2 in England). Table 2 lists the essential properties of some of these powders. The reader will find a host of details in the book by Leverenz (1945) and in the articles by Hopkinson (1946), and Nguyen Tien Chi (1952). A fluorescent material consists of a base of some crystallized mineral material (a semi-conducting oxide or sulphide) which contains a small quantity of (metallic) activating material. The crystallized substance which forms the base has a considerable effect upon the efficiency of the transformation of energy, as Fig. 126 illustrates. The spectrum of colours which is obtained is usually described by a bell-shaped curve (Fig. 127). To obtain white, which is vital for tele-

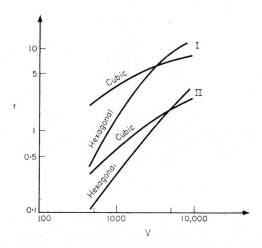

Fig. 126. The influence of the crystal symmetry of the base; I, zinc and cadmium sulphides activated with silver; II, zinc sulphide activated with silver.

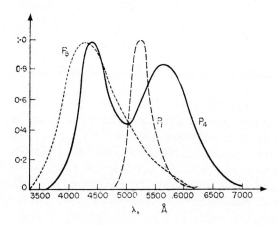

Fig. 127. The light spectra of fluorescent powders; P_1 gives a green spot, P_4 a white spot, and P_5 a blue spot.

vision tubes, two powders have to be mixed, and the spectrum has two maxima. The development of colour television has resulted in a great deal of research directed in particular towards the production of substances which fluoresce red.

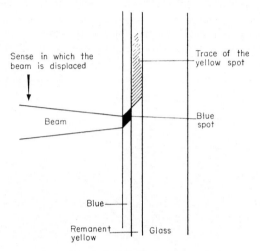

Sense in which the beam is displaced

Beam

Trace of the yellow spot

Blue spot

Blue

Remanent yellow

Glass

FIG. 128. A double fluorescent screen.

When the screen is struck by the electron beam, light is emitted, and a steady state is established as long as the bombardment continues; when the latter ceases, the luminous intensity decreases according to one or other of two laws:

$$I = I_0 e^{-t/\tau}$$

or

$$I = I_0 \frac{1}{1 + t/\tau}.$$

The value of the time constant, τ, which measures the "remanence" of the screen can vary between wide limits. In television, this remanence plays an important part, for if it is too short, one frame fades away before the next one appears, and the screen flickers; if on the contrary it is too long, a number of images are mixed together.

Many screens to-day consist of a double layer, a design which was introduced in connexion with radar using the P.P.I.† system where each region of space is explored periodically. In Fig. 128, an example of a double layer is illustrated; the electrons impinge on a layer with a short persistence

† "Plan position indicator": the instrument traces out on the screen of the cathode ray tube what is virtually a map of the region which the beam explores.

TABLE 2

	BASE	ACTIVATING SUBSTANCE Atomic concentration with respect to the base (%)	COLOUR Wavelength in Å at the intensity maximum	PERSISTENCE TIME(sec) during which the intensity is more than 1% of its original value	YIELD Candles per watt at	
					6000 V	1000 V
P_1	Zinc silicate	Mn (0.25%)	Green 5250	0.05	9	3.5
P_4	Zinc sulphide	Ag (0.01%)	White 4440 5250	5×10^{-3}	10	2
P_5	Zinc sulphide 48% and cadmium sulphide 52%	Ag (0.01%)	Blue 4300	10^{-5}		
P_{11}	Calcium tungstate	Ag (0.01%)	Blue 4580	0.05		
P_{13}	Zinc sulphide Magnesium silicate	Mn (0.15%)	Red 6740	0.1	2	0.4

Note: Extensive further information is to be found in Bril, A., and H. A. Klasens (1952) *Philips Tech. Rep.*, **7**, 401; Kroger, F. A. (1956) *Ergebn. Exakt. Naturwiss.* **29**, 61.

which emits a blue light (zinc sulphide activated with silver). The light which is emitted then passes through the second layer and to the observer, this primary brilliance indicates the instantaneous position of the spot. Another part of this flux of primary light—the ultra-violet—is absorbed in the second layer, and produces a highly remanent fluorescence, yellow in colour, in this film of zinc and cadmium sulphide activated with copper. This provides a reminder of the appearance of the map between the times t and $t + \tau$, which is useful in radar stations where the geography is fixed (at aerodromes) or slowly moving (at sea) while the positions of the aeroplanes or ships are changing rapidly. This double layer can also serve the purpose of in-

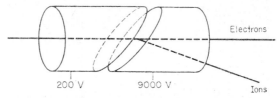

Fig. 129. An ion trap. A magnetic field perpendicular to the plane of the figure is maintained by a permanent magnet (see also Fig. 116c).

creasing the light yield at the screen; the electrons provoke ultra-violet radiation in the first layer, with a high efficiency both as regards the transformation of energy and the number of quanta. These ultra-violet rays then produce—again, quite efficiently—visible light in the second layer by photo-luminescence. In this way, as much as 10 per cent of the electron energy can be transformed into light, We should recall that the efficiency can be increased further by returning towards the front the fraction of the light which is emitted backwards, by the aid of a very thin metal layer which is spread across the bottom of the screen (Epstein and Pensak, 1946).

Finally, it is necessary to take precautions against the "ion spot" which is produced by the negative ions emitted by the cathode. These ions form a halo round the spot after the tube has been in use for a few hours, which diminishes the contrast considerably and is intolerable in television tubes (Schaeffer and Walcher, 1943; Broadway and Pierce, 1939). This halo is removed by building an "ion-trap" near the gun. This consists of two opposed deflexion systems, one electric, the other magnetic. The intensity is adjusted in such a way that the deflexions cancel one another for electrons; they will not, therefore, compensate one another for ions, which are far heavier and will be lost to the wall of the tube. Figure 129 is a diagrammatic sketch of a trap, in which an electrostatic two-cylinder lens is obtained by cutting a cylinder "obliquely". This lens will deflect the beam, and the compensation is provided by a small permanent magnet placed outside the tube.

This trap also eliminates the positive ions which are created along the path of the beam of electrons by collisions with residual gas molecules; otherwise, there would be a risk that the latter might damage the emissive layer, as they are accelerated back towards the cathode.

12.3 DEFLEXION OF THE BEAM

12.3.1 Deflexion Plates

After leaving the second focusing lens, the beam passes through two pairs of plates which produce deflexions in the directions parallel to Ox and Oy. In oscillographs which are to be used as measuring devices, the useful deflexion is only of the order of a few degrees, and can be obtained electrically or magnetically. In television tubes, on the contrary, which have

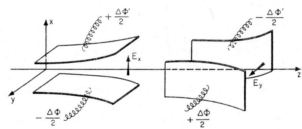

FIG. 130. The structure of electrostatic prisms which provide deflexion in the Ox and Oy directions.

to be as short as possible, the deviation is much larger ($\pm 45°$ is usual, and recently, $\pm 55°$ has been reached). We shall only consider the theory for more modest angles; to produce a tube capable of providing these high deflexions requires a laborious experimental step-by-step procedure.

In Fig. 130 an electrical system is shown. Two electrodes, symmetrical about the plane yOz, produce a field of which the principal component is directed along Ox; the equipotential plot is kept most nearly symmetrical if the plate potentials are disposed *symmetrically*, that is, if one is at a potential $+\dfrac{\Delta\Phi}{2}$ with respect to the potential of the last electrode from which the beam emerged, and the other at $-\dfrac{\Delta\Phi}{2}$. In this situation, the axis Oz remains at a constant potential, namely the potential of the layer of aquadag which is linked electrically to the final electrode, Φ_0. This arrangement introduces complications into the supply circuits which we should often prefer to avoid, however, and generally speaking this would be done by holding one plate at the potential $\Delta\Phi$, and joining the other to the layer of aquadag. This is the *unsymmetrical* arrangement which is frequently used, but which produces important defects, for the beam which is travelling in the neighbourhood of Oz has first of all to change its axial velocity, as within the plates the potential is approximately $\Phi_0 + \frac{1}{2}\triangle\Phi$.

The appearance of the lines of force in such a system in these two cases is shown in Fig. 131. It is evident that even though the lines of force are similar inside the prism, their appearance is quite different outside and beyond the prism.

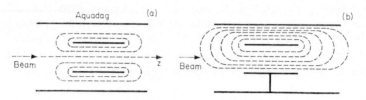

FIG. 131. Two possible ways of supplying electrostatic prisms: (a) One plate is held at $+\frac{1}{2}\triangle\Phi$, the other at $-\frac{1}{2}\triangle\Phi$: this is the symmetrical arrangement. (b) One plate is tied electrically to the aquadag layer, the other is held at a potential $\triangle\Phi$; this is the unsymmetrical arrangement.

The potential function can be expanded in powers of x and y, the coefficients being functions of z. Taking into account the symmetry of the system and the fact that the potential satisfies Laplace's equation, the potential of the first prism in the symmetrically excited arrangement takes the form

$$\Phi(x, y, z) = \Phi_0 + \triangle\Phi[\varphi_1(z)\, x + \varphi_3(z)\, x^3$$
$$- \{3\varphi_3(z) + \tfrac{1}{2}\varphi_1''(z)\}\, xy^2 + \cdots], \quad (12.10)$$

while in the unsymmetrically excited arrangement, it takes the form

$$\Phi(x, y, z) = \Phi_0 + \triangle\Phi[\varphi_0(z) + \varphi_1(z)\, x + \varphi_2(z)\, x^2$$
$$- \{\varphi_2(z) + \tfrac{1}{2}\varphi_0''(z)\}\, y^2 + \cdots]. \quad (12.10')$$

The functions $\varphi_1(z)$ and $\varphi_2(z)$ play a role here equivalent to the role of the axial potential in electron lenses. They are characteristic of the plate system, and with their aid the trajectories can be calculated to first order, and the corrections to this first-order approximation—the aberrations—can be estimated. Magnetic deflectors consist of windings situated outside the tube—to modify and regulate them is therefore a straightforward affair. In tubes used for television, where the optimum conditions with a wide deflexion can be obtained only empirically, magnetic deflexion is preferable to electric deflexion as it is less easy to experiment with the latter. Further, magnetic coils have the advantage of occupying comparatively little space along the axis, which means that they are "superposable"; a beam which is deflected in the Ox and the Oy directions over the same axial region will have less aberration. Figure 132 represents an example of a magnetic deflector. The system is characterized by the components of the magnetic induction, expanded as a power series in x and y. For the case shown in Fig. 132, for example,

$$B_x = \mu_0 n I \left[k_1(z) + k_2(z)(x^2 - y^2) - \frac{k_1''(z)}{2} y^2 + \cdots \right],$$

$$B_y = \mu_0 n I \left[- 2 \left\{ k_0(z) - \frac{k_1''(z)}{2} \right\} x y + \cdots \right], \tag{12.11}$$

$$B_z = \mu_0 n I \left[k_1'(z) x + \frac{k_2'(z)}{3} x^3 - \left\{ \frac{k_2'(z)}{3} - \frac{k_1''(z)}{2} \right\} x y^2 + \cdots \right].$$

The winding consists of n turns, carrying a current I; the functions $k_1(z)$ and $k_2(z)$ are determined by the geometry of the system.

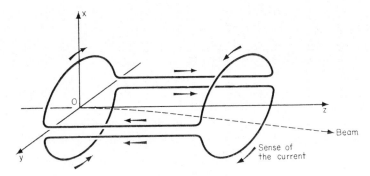

FIG. 132. Diagram of a magnetic prism. The electron beam is deflected towards the positive y direction.

12.3.2 The Trajectories in Electron Deflectors

The deflexion which is produced by a deflector system depends upon which ray is considered. The deflexion of a ray parallel to the axis will differ, therefore, from the deflexion of an inclined ray; and the deviation of a ray which enters the system at the axis will differ from the deviation of a ray at the edge of the beam. The electron beam will no longer be conical when it leaves the deflector, so that the spot which is obtained on the screen will be somewhat complicated in shape, instead of the point which is desirable. Deflexion systems introduce considerable aberration; we shall not examine the calculation of the trajectories in the general case, but simply give an example of a special case in the next paragraph. For further information, the reader is referred to the detailed publications by Wallraff (1935), Picht and Himpan (1941), Glaser (1949), Hutter (1947, 1948), Deserno

(1953) and Wendt (1954). Here, we shall mention only a few details of the form of the traces which are found at the fluorescent screen, that is, of the aberration figures.

If we restrict the series expansions to terms of the first order, we can show that the prism produces a deflexion θ_x which is the same for all the rays. The incident and the deflected rays, produced, intersect in a plane perpendi-

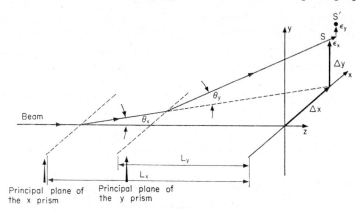

Fig. 133. Deflexion of a trajectory in a cathode ray tube.

cular to the axis which is the *principal plane of the deflector* (see Fig. 133). For a symmetrically fed electrostatic system, for example, it is not difficult to show that

$$\theta_x = \frac{\Delta\Phi}{\Phi_0} \int_{-\infty}^{\infty} \varphi_1(z)\, \mathrm{d}z \tag{12.12}$$

(cf. § 12.3.3 ii) in which $\varphi_1(z)$ is the function defined in formula (12.10). If we denote the distance between the principal plane and the screen by L_x, the spot is displaced through a distance

$$\Delta X = \theta_x L_x, \tag{12.13}$$

while the plates which produce a deflexion in the perpendicular direction displace the spot through

$$\Delta Y = \theta_y L_y. \tag{12.13'}$$

If we now take into account terms of higher orders in x and y, the ray will be described more accurately—the point S' at which it intersects the fluorescent screen will have coordinates $\Delta X + \varepsilon_x$, $\Delta Y + \varepsilon_y$. The corrections ε, which are clearly small, will depend upon which trajectory we are considering, while the trajectory is determined by its slope (x_0', y_0') where it enters the deflexion system. The slope (x_0', y_0') is adequate, for the coordinates of the ray at the point at which it enters the prism are necessarily $x_0' L$

and $y_0' L$, as a result of the requirement that the spot shall be formed at the screen (L is the distance to the screen). The magnitudes of the correction terms will also depend on the deflector, that is, upon θ_x and θ_y. We can write, therefore,

$$\varepsilon_x = f(\theta_x, \theta_y, x_0', y_0'),$$
$$\varepsilon_y = g(\theta_x, \theta_y, x_0', y_0').$$

To proceed further, we must make a calculation similar to the one in § 6.2, where the aberrations of electron lenses were analysed. ε_x and ε_y are expanded as a power series in the four variables, but fortunately there are a number of simplifying conditions which limit the number of terms of this expansion. First, if the deflectors are switched off, $\theta_x = \theta_y = 0$, the spot lies at the centre of the screen, and $\varepsilon_x = \varepsilon_y = 0$. The series, therefore, contains no terms independent of θ. Further, terms of the first order in θ_x and θ_y cannot be present, as it is they which produce the original deviations $\triangle X$ and $\triangle Y$. The expansion must therefore begin with second order terms,

$$\varepsilon_x = A\,\theta_x^2 + B\,\theta_y^2 + C\,\theta_x\,\theta_y + D x_0'\,\theta_x + E x_0'\,\theta_y + F y_0'\,\theta_x + G y_0'\,\theta_y,$$
$$\varepsilon_y = A'\,\theta_x^2 + B'\,\theta_y^2 + C'\,\theta_x\,\theta_y + D'\,y_0'\,\theta_y + E'\,y_0'\,\theta_x + F'x_0'\,\theta_y + G'\,x_0'\,\theta_x, \quad (12.14)$$

and contains fourteen coefficients. The formulae from which these coefficients can be calculated are to be found in the articles quoted earlier. Provided the deflectors are perfectly symmetrical, it is possible to cancel all the coefficients; since this perfection of the symmetry includes both the electrical and the geometrical properties, the potential must be fed symmetrically to the system.

The terms which depend upon x_0' and y_0' introduce *astigmatism* into the beam, but not field curvature. The circle of least confusion is at the image plane. There is no defect of the same nature as coma (which would be produced by terms containing $x_0'^2$) nor is there spherical aberration (which is a specifically third order aberration). On the other hand, there are terms in θ alone which displace each point of the spot and hence represent *distortion*. If θ_x were varied smoothly keeping θ_y constant, the spot would trace out a straight line so long as this distortion was negligible. By varying θ_y with θ_x constant, and conversely, it would be possible to trace out a pattern of squares on the screen. In reality, the effect of distortion is to deform this pattern of squares into a pattern of the form shown in Fig. 134. This defect would be particularly obvious in television tubes, for its effect upon the image is one to which the eye is especially sensitive, used as it is to judging proportions accurately (when we think of proportions in terms of the canons of aesthetics). In addition, it is not possible to base precise electrical measurements on the displacement of the spot when this aberration is present. Distortion, therefore, is a defect which we endeavour at all costs to eliminate.

These second order aberrations can, fortunately, be corrected by using two perfectly symmetrical electrical supply units with the same mid-point; this is the solution which is adopted in television sets. In this case, ε_x and ε_y contain third order terms only, and we find

$$\varepsilon_x = a\,\theta_x^3 + b\,\theta_x\,\theta_y^2 + c\,x_0'\,\theta_x^2 + d\,x_0'\,\theta_y^2 + e\,\theta_x\,\theta_y\,y_0'$$
$$+ f\,x_0'^2\,\theta_x + g\,y_0'^2\,\theta_x + h\,x_0'\,y_0'\,\theta_y, \qquad (12.15)$$
$$\varepsilon_y = a'\,\theta_y^3 + b'\,\theta_y\,\theta_x^2 + c'\,y_0'\,\theta_y^2 + d'\,y_0'\,\theta_x^2 + e'\,\theta_x\,\theta_y\,x_0'$$
$$+ f'\,y_0'^2\,\theta_y + g'\,x_0'^2\,\theta_y + h'\,x_0'\,y_0'\,\theta_x.$$

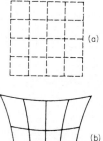

Fig. 134. Deformation of the scanning network by second order (trapezoidal) distortion.

Fig. 135. Deformation of the scanning network by third order distortion in a symmetrical system.

The terms which contain θ alone again correspond to a distortion, but are less serious than their second order counterparts, as the effect is not trapezoidal; Fig. 135 gives a general idea of the deformation which third order distortion produces upon the basic square pattern.

The terms which contain x_0' and y_0' to first order give rise to *astigmatism*, which is accompanied by *field curvature*. It can be reduced either by using a curved screen, or by varying the convergence of the lens at the same time as the spot is deflected (Hoadley, 1951).

The terms which contain quadratic combinations of x_0' and y_0' have an effect upon the spot which is similar to that of *coma* in lenses. Figure 136 shows the appearance of the spot which is to be seen on the screen. There is, however, no aberration analogous to spherical aberration, as this would require a term in $x_0'^3$.

FIG. 136. Some examples of the spots obtained in a cathode ray tube in which the spot (a) given by the gun has been deformed by astigmatism and coma.

Formulae from which the coefficients of third order aberration can be calculated are to be found in the articles which have been cited earlier. There are thirteen of these coefficients although there were sixteen second order terms, as certain natural equalities appear: $h = g'$, $h' = g$ and $e = e'$. This theory has been verified in a remarkable way by Wendt (1954), who used an extremely fine pencil, which can be thought of effectively as a single ray, rather than a complete beam. A rotating magnetic field turned this beam about the Oz axis in such a way as to describe a conical surface, with its point at the screen. It is clear that if we introduce a deflecting system, it will act on this pencil just as it would act on the various rays which lay in each of the positions occupied by the pencil. The point of the pencil, therefore, traces out the outline of the spot on the screen which can be photographed easily.

12.3.3 An Example of an Electrical Prism

We shall give some details of the simplest of the electrical deflectors which consists of two plane parallel plates, a distance d apart. They are excited symmetrically ($\Phi_0 + \frac{1}{2} \triangle \Phi$ on one plate, $\Phi_0 - \frac{1}{2} \triangle \Phi$ on the other).

(i) *The end correction*

We assume that a uniform electric field exists within the prism for a certain distance, beyond which it is zero. Recknagel (1938) has shown that this is a valid model of the real situation provided the length of the region of uniform field is regarded as being slightly longer than the length l_0 of the plates. (This is similar to the "equivalent length" which we have defined for quadrupole lenses.)

The behaviour of the lines of force in this model is shown in Fig. 137, together with the meaning of the correction $\triangle l$. Recknagel showed that

$$\triangle l = 0 \cdot 318 d \left(1 - \log 0 \cdot 636 \frac{d}{l_0} \right), \qquad (12.16)$$

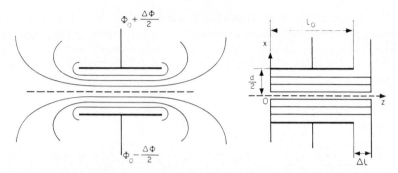

FIG. 137. Recknagel's model for the electrostatic prism: (a) the actual appearance of the equipotential surfaces; (b) their appearance in the model.

which gives a correction of the order of $0{\cdot}7d$ for $d = 0{\cdot}2\,l_0$. We shall write

$$l_0 + \Delta l = l,$$

and in the useful region, the potential is given by

$$\Phi = \Phi_0 + \frac{\Delta\Phi}{d}\,x.$$

(ii) *The equations of motion*

The lateral motion is described by the simple equation

$$m\frac{\mathrm{d}^2x}{\mathrm{d}t^2} = e\,E = e\,\frac{\Delta\Phi}{d},$$

which gives

$$\frac{\mathrm{d}x}{\mathrm{d}t} = \left(\frac{\mathrm{d}x}{\mathrm{d}t}\right)_0 + \frac{e}{m}\,\frac{\Delta\Phi}{d}\,t. \tag{12.17}$$

Combining this with the conservation of energy relation, we find

$$\frac{1}{2}\,m\,v^2 = \frac{1}{2}\,m\left[\left(\frac{\mathrm{d}x}{\mathrm{d}t}\right)^2 + \left(\frac{\mathrm{d}y}{\mathrm{d}t}\right)^2\right] = e\left(\Phi_0 + \frac{\Delta\Phi}{d}\,x\right).$$

As the electron trajectory is always inclined only slightly with respect to the plates, we can assume that $\mathrm{d}x/\mathrm{d}z$ is small compared with unity, and write

$$\left(\frac{\mathrm{d}z}{\mathrm{d}t}\right)^2 \simeq \frac{2e\,\Phi_0}{m}\left[1 + \frac{\Delta\Phi}{\Phi_0}\,\frac{x}{d}\right]\left[1 - \left(\frac{\mathrm{d}x}{\mathrm{d}z}\right)^2\right], \tag{12.18}$$

$$\left(\frac{\mathrm{d}z}{\mathrm{d}t}\right) \simeq \sqrt{\frac{2e\,\Phi_0}{m}}\left\{1 + \frac{1}{2}\,\frac{\Delta\Phi}{\Phi_0}\,\frac{x}{d} - \frac{1}{2}\left(\frac{\mathrm{d}x}{\mathrm{d}z}\right)^2 - \frac{1}{8}\left(\frac{\Delta\Phi}{\Phi_0}\,\frac{x}{d}\right)^2 + \cdots\right\}.$$

This equation displays the importance of the correction terms which even in this simple model are appearing little by little. If we neglect all the correction terms in the formula, we shall produce a "zero order" optics, which will give the deflexion and the principal plane. It can be shown quite straightforwardly that the trajectory is a parabola, which implies that the principal plane lies at the centre of the useful part of the field. The magnitude of the deflexion is

$$\theta = \left(\frac{dx}{dz}\right)_f - \left(\frac{dx}{dz}\right)_i = \frac{1}{2}\frac{l}{d}\frac{\Delta\Phi}{\Phi_0}. \tag{12.19}$$

(iii) *Sensitivity*

By the sensitivity of a cathode ray tube is meant the potential difference which displaces the spot through its own width. Starting from formula (12.1) which gives us the width of the spot, we eliminate θ, and find

$$\sigma_E = \frac{4Kf}{L}\frac{d}{l}\sqrt{\frac{T}{11,600}}\sqrt{\Phi_0}, \tag{12.20}$$

in which L is the distance between the screen and the principal plane. The sensitivity of a cathode ray tube is in the region of one volt per diameter, and there would be an advantage in reducing Φ_0, since σ_E has to be as small as possible. This would reduce the brightness, however, and the influence of space charge would be accentuated.

(iv) *The importance of the maximum deflexion*

From the point of view of the electrical supply to the deflectors, it would be advantageous to use long plates, close together. This cannot be pursued to too extreme a limit, however, as the beam has to be able to reach the whole of the screen, and the angle θ has to be able to reach some maximum, θ_m, which is fixed by the dimensions of the tube. Considering Fig. 138a, we see that $\theta_m \simeq d/l$, so that if we increase l/d—and with it, the sensitivity— θ_m is diminished. For this reason, a great deal of research has gone into the study of more complicated types of system, with inclined or even curved plates. A full discussion of this point is to be found in the book by Klemperer (1953). With plane plates inclined at an angle α (see Fig. 138b), the lines of force are approximately circular, and we find

$$\theta \simeq \frac{1}{4\tan\alpha}\log\left[1 + \frac{2l}{d}\tan\alpha\right]\frac{\Delta\Phi}{\Phi_0}, \tag{12.21}$$

and

$$\theta_m \simeq \frac{d}{l} + 2\tan\alpha. \tag{12.21'}$$

The full analysis shows that for a given value of θ_m, the sensitivity is an increasing function of α. In practice, the plates consist of a horizontal part and an inclined part, as illustrated in Fig. 138c. It would be still more advantageous to use curved plates (Maloff and Epstein, 1938), parallel to the beam at the point at which its deflexion reaches a maximum, θ_m (see Fig. 138 d). The equation which the cross-section of the plates must satisfy is given by

$$z - z_0 = 2y_0 \left(\frac{V_0}{V_d} \right)^{1/2} \int_0^{\sqrt{\log\left(\frac{y}{y_0}\right)}} e^{u^2} \, du \qquad (12.21'')$$

in which $2y_0$ represents the initial distance between the plates, Φ_0 is the accelerating potential and Φ_d the peak deflexion potential measured with respect to Φ_0.

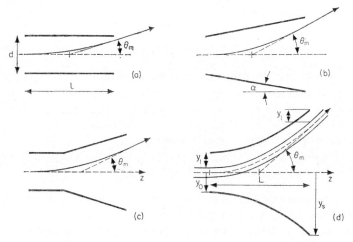

FIG. 138. Different kinds of electrostatic prism. (a) A prism with parallel electrodes. (b) A prism with oblique electrodes. (c) The prism which is employed in the majority of tubes. (d) A prism with curved electrodes.
For the same deflexion θ_m, the sensitivity increases in the order (a) to (d). With the systems (c) and (d), very high values of θ_m can be attained.

For $z - z_0 = L$, we have $y = y_s$.

Since the integral which appears in equation (12.21'') is the same as the integral which expresses the broadening of the beam which space charge produces (see § 12.2.3), we can calculate the form of the plates numerically with the aid of Fig. 119 a.

(v) *The aberrations*

The results become more complicated when we take into account the aberrations. For example, suppose that the beam is of width x_0 when it enters the prism, the deviation θ of the electrons varies from one ray to another *because the potential is not the same for all the rays.* Calculation shows that the emergent beam is more convergent than the incident beam, but only in the plane *xOz*. The converging power of the equivalent lens would be given by

$$\frac{1}{f} = \frac{1}{2} \frac{l}{d^2} \left(\frac{\triangle \Phi}{\Phi_0} \right)^2. \tag{12.22}$$

This is superimposed onto the convergence of the principal lens in the tube, but as it is produced by a cylindrical lens, the result is an astigmatism which can easily be seen in certain oscillographs when the spot is near the edge of the screen. This is the second order astigmatism which we mentioned in the general study.

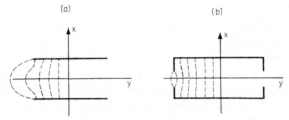

FIG. 139. The production of a field which is as uniform as possible over the whole depth of the prism. Prism *b* has a larger useful zone than prism *a*.

If the second deflector is also perfectly symmetrical, we can see how it might be capable of producing a certain convergence in the other plane, which would partially compensate the astigmatism of the beam; the spot would still be broadened, however, as the crossover would be formed in front of the screen. As a result, if the trace of the beam is circular and as small as possible at the centre of the screen, elsewhere it will be elliptical in the directions *Ox* and *Oy* and practically circular along the directions at 45° to the axes.

We shall not pursue the study of the aberrations of this model any further; were we to do so, we should find all the general terms of § 12.3.2. In conclusion, we should nevertheless mention two other possible causes of aberration.

If the beam is wide and the plates too narrow, the edges of the beam travel in a region in which the lateral overlap field disturbs the distribution of field in the deflexion system.

If the horizontal and vertical deflexion plates are too close together in the *Oz* direction, the space between the two pairs of plates behaves like an astigmatic lens of very inferior quality (Cooper, 1961). This defect, which is apparent when the deflexion is strong, can be removed by separating the two pairs of plates by an earthed double shield.

12.3.4 Magnetic Deflectors

The simplest version, although it is rarely used, is the electromagnet, which produces a uniform magnetic induction B over a distance l, and has no effect outside this region. In §5.1.1 the reader will find a description of the principle behind the calculation of θ

$$\theta = \sqrt{\frac{e}{2\,m}}\, l\, \frac{B}{\sqrt{\Phi_0}}. \qquad (12.23)$$

The sensitivity here is characterized by the value of B which displaces the spot through its own width on the fluorescent screen, and we find

$$\sigma_M = \frac{Kf}{l}\sqrt{\frac{2m}{e}}\sqrt{\frac{T}{11{,}600}}. \qquad (12.24)$$

This expression does not contain Φ_0. We should remember, however, that the usual type of magnetic system is of the form illustrated in Fig. 132. The field inside is far from uniform, therefore, and calculation of the trajectories is complicated. Very fortunately, it is easy to make measurements, and excellent magnetic deflexion devices have been achieved by a series of successive adjustments (Woroncow, 1946; Wendt, 1954). As an example, we shall quote the characteristics of the system which is used in a radar tube. The number of turns is 2000, wound in the form shown in Fig. 132, the diameter is 7·5 cm and the length 5·6 cm; a current of 1·34 mA is necessary to deviate the beam through 1° when the electron beam is accelerated through 5 kV (Soller, Starr and Valley, 1948).

12.3.5 The Scanning Mechanism and the Writing Speed

Save in exceptional cases, the potentials which are applied to electrostatic plates do not remain constant, any more than do the currents which govern magnetic systems. The spot therefore will move about the screen, tracing out a curve which can be photographed. The speed with which the spot moves along this curve is known as the "writing speed"; and as soon as the sensitivity of the tube and the expression which describes the applied potential as a function of time are known, its calculation presents no problem. This speed is characterized by an "index" τ, the time which the spot takes to move through its own width. In the case of an electrical system, fed with a potential $\triangle\Phi \sin \omega t$, we find

$$\tau = \frac{\sigma_E}{\omega \cdot \triangle\Phi}. \qquad (12.25)$$

The majority of tubes, provided they are supplied with a suitable frequency, are capable of providing a satisfactory record in an interval of the order of 10^{-8} sec. If the frequency of the supply is increased any further, it is found that the sensitivity begins to diminish, which is a result of the transit time as we shall show in the next paragraph; it is this phenomenon which provides a natural limit to the writing speed.

It is to be noticed, too, that if the spot takes 10^{-8} sec to move through its own width, 300 μ say, its linear velocity is some 30 km s^{-1}. Few phenomena demand such extreme writing speeds, but they are useful nevertheless to display the form of the steep wavefronts on transmission lines, or of those provided by sources of decimetre waves, for direct observation.

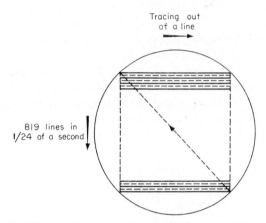

FIG. 140. The scanning system in a television tube. The return track, which is shown as a dotted line, is in fact rendered invisible by applying an appropriate polarization to the Wehnelt during the return.

At the present time, the electron intensity at the spot, the efficiency of the fluorescent screen and the sensitivity of the plate can achieve so high a quality that the writing speeds which can in consequence be attained are sufficiently rapid for the decrease in sensitivity resulting from the effect of transit time to become perceptible. We shall return to this point in § 12.3.6.

In a cathode ray tube which is to be used in a television receiver, a potential (or a current) which has a saw-tooth variation in time is applied to the system which provides the horizontal deflexion; the spot moves from one side of the screen to the other in a time T, then flies back to its starting point in a negligible time. Another saw-tooth potential with a period 819 times as long is applied to the other prism, so that, as Fig. 140 shows, the spot explores the whole screen by tracing out 819 practically horizontal lines in a time 819T. This time is chosen in such a way that 25 images are formed every second† (the frequencies of the two scanning oscillators,

---

† We neglect the small difference between two successive images which results when the return is interlinear.

therefore, are 25 c/s and 20,475 c/s). At each moment, only the potential of the grid need be varied to obtain a suitable brightness at the corresponding point on the screen, where we can watch a black and white image being traced out.†

In a cathode ray tube which is to be used as the panoramic receiver of a radar set, the potentials (or currents) which are applied to the two deflectors are of the form

$$\Delta\Phi_x = \varrho(t) \sin \frac{2\pi t}{T},$$

$$\Delta\Phi_y = \varrho(t) \cos \frac{2\pi t}{T},$$

in which $\varrho(t)$ is a saw-tooth function of period T', small in comparison with T. The spot describes a star-shaped curve, as Fig. 141 shows. The signal which is received by the radar set is applied to the grid and hence governs the brightness of the spot at each point on the screen. For further details, the reader is referred to the book by Soller, Starr and Valley (1948).

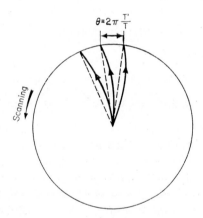

FIG. 141. The scanning in a panoramic receiver.

If, finally, the cathode ray tube is to be used in an oscillograph, the signals which are being studied are applied to the deflectors after suitable amplification. The spot traces out a curve which describes the way in which one of the phenomena depends on the other. If the signals are periodic, and if the periods are simply related, the curve is closed and can be examined with ease thanks to the persistence of the screen. Otherwise, we have to resort to photography, and we should have to ensure that the trace was sufficiently brilliant for photography to be possible. From the article by

† The numbers in this paragraph relate to French television standards (Ed.).

Hopkinson mentioned earlier, we have extracted the following examples—for a blue fluorescent screen (zinc sulphide), the writing speed of a spot 1 mm in diameter and 20 μA in intensity cannot exceed 10 km sec^{-1} for electrons at 3000 eV; this limiting velocity is practically independent of the emulsion. With a green screen, the velocity cannot exceed 5 km sec^{-1} with an orthochromatic emulsion, or even 2 km sec^{-1} with a panchromatic emulsion.

If the phenomenon is periodic, the problem is far less difficult. An exposure of one second is sufficient with a blue screen, although it has to be extended to 5 seconds with green screens. Further details are to be found in the book by Puckle (1951). In modern cathode ray tubes with post-acceleration, such as those with which high quality commercial oscilloscopes are equipped, the exposure time with blue screens is very short, 1/50 or 1/25 of a second.

12.3.6 Transit Time as a Factor Limiting High Writing Speed

If we increase the frequency of the tension or current which is providing the deflexion, the spot will clearly move more rapidly across the screen, but a limit soon appears. To pass between the deflecting plates of length l, an electron moving with velocity v will take a time l/v which has to be negligible in comparison with the period T of the applied potential. Only in this case is it legitimate to regard the latter as constant during the time of transit, and to apply the electrostatic arguments of the preceding paragraph. Moreover, the "transit angle", φ, defined by

$$\varphi = 2\pi \frac{l}{v\,t} = 2\pi \frac{l}{T} \sqrt{\frac{m}{2e\Phi_0}} = 1{\cdot}05 \times 10^{-5} \frac{\varrho}{T\sqrt{\Phi_0}}, \qquad (12.26)$$

becomes important only when the frequency is very high. For a potential of 3000 V, and plates 3 cm long, the angle is 6×10^{-3} for an applied frequency of 100 kc/s. For 1 Mc/s, it reaches $0{\cdot}060$ and for 100 Mc/s, $0{\cdot}60$. At this last frequency, it is obvious that the actual deflexion in the oscillograph will bear little relation to the deflection which is calculated by electrostatic methods, and it is in fact considerably less.

We shall examine the simple case of an electrostatic system in a uniform field, but the results can be extended to any deflexion whatsoever without major modification. Once again, we consider the equation of motion in the lateral direction, this time supplying the system with an alternating potential of period T. We find

$$m \frac{\mathrm{d}^2 x}{\mathrm{d}t^2} = e\,E_0 \sin \frac{2\pi}{T}(t + t_0).$$

The time origin is selected to be the moment at which the electron enters the condenser; at this instant, the field has the value $E_0 \sin \frac{2\pi}{T} t_0$. The longitudinal motion remains uniform,

$$\frac{dz}{dt} = \sqrt{\frac{2e\Phi_0}{m}}.$$

From these equations, we deduce the trajectory equation and the deflexion angle as the beam leaves the plates; for the latter, we find

$$\theta = \frac{l\,E_0}{2\Phi_0} \sin\left(2\pi\frac{t_0}{T} + \frac{\varphi}{2}\right)\frac{\varphi/2}{\sin(\varphi/2)}, \qquad (12.27)$$

whereas the value which we should have found if we had used a static field of magnitude equal to the mean value in traversing the plates, $E_0 \sin\left(2\pi\frac{t_0}{T} + \frac{\varphi}{2}\right)$, would have been

$$\theta_0 = \frac{l\,E_0}{2\Phi_0} \sin\left(2\pi\frac{t_0}{T} + \frac{\varphi}{2}\right).$$

The actual sensitivity, taking the transit angle into account, is therefore

$$\sigma = \sigma_0 \frac{\varphi/2}{\sin(\varphi/2)}, \qquad (12.28)$$

in which σ_0 is the static sensitivity. σ, we recall, has to be kept as small as possible, since it represents the potential which is necessary to displace the spot through its own width. But σ increases with φ; so long as the transit time is not too long, we have effectively

$$\sigma = \sigma_0(1 + \varphi^2/24), \qquad (12.28')$$

a decrease of 1 per cent in the sensitivity for $\varphi = 0.48$. If φ becomes appreciable, the result is catastrophic. If, in particular, φ reaches 2π, the instrument is useless, as its sensitivity is zero; an infinite potential would be necessary to move the spot through its own width. This extreme case can be observed with ordinary instruments by feeding the plates from a transmission line and a source with a frequency corresponding to decimetre waves.

Only one simple method of overcoming this problem exists, which consists of reducing the transit-time by increasing the potential at which the tube operates. As we have already seen, however, this has repercussions on the static sensitivity, which is reduced. By operating a cathode ray tube at 70,000 V, and moving the screen sufficiently far away not to lose too much sensitivity, potentials at frequencies up to 30 Mc/s can easily be examined. To go beyond this figure is, however, extremely difficult, and it is better to search for solutions of a new kind, by modifying the structure of the tube itself. Ingenious solutions have been suggested by Lee (1946), Owaki et al. (1950), Smith et al. (1952) and Bloom and von Forster (1954), which enable tubes to be built with a sensitivity which is still appreciable at 30,000 Mc/s— of the order of 0.1 mm V^{-1} for a beam accelerated to a few keV. The writing speed reaches some 15,000 km/s. Most of these tubes employ the properties of travelling waves to obtain deviation, so that the principle behind their mode of operation resembles that of the linear accelerator. The H.F. signal which is to be studied is led into the instrument along a Lecher line which passes into the interior of the envelope. The electric field associated with this

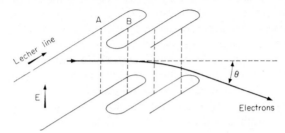

FIG. 142. The principle of "resonance deflexion" in cathode ray oscillographs for very high frequencies.

wave provides the deviation. As Fig. 142 shows, however, the wave which has to follow the transmission line travels along a far longer path than the electrons which move in a straight line. At *A*, the beam will be very slightly deflected, after which it moves in a straight line until it encounters the wave again at *B*. The transit time of the wave between *A* and *B* is chosen in such a way that the electrons see an electric field at *B* which is in phase with the field they had seen at *A*. The beam is deflected again, therefore, in phase with the deflexion produced at *A*; by using a long enough line, we can obtain the required overall deflexion. Further details are to be found in a recent exposition of this topic by Hardy, Jackson and Feinberg (1956). At the present time, instruments are available with a writing speed of the order of 20,000 km/s.

12.4 THE POST-ACCELERATION

During this examination, we have come across numerous reasons for adopting as high a value of the anode potential as possible. The diameter of the spot will be less since the thermal velocity dispersion will have less effect (§ 12.2.2), the effect of space charge will be imperceptible (§ 12.2.3), and the brightness at the fluorescent screen will be intensified for a given value of the current (§ 12.2.5). Unfortunately, the accelerating potential also has an effect upon the sensitivity (§ 12.3.4) and as we have seen, an increase in Φ_0 reduces the sensitivity of an electrostatic tube, but has no effect upon the sensitivity of a tube with magnetic deflexion. It is, therefore, of the greatest interest to operate a magnetic tube at a very high potential but, for an electrostatic tube—which would become less sensitive—this is not so.

Another inconvenience is that in a commercial instrument, it is dangerous to use high voltages (perhaps 50 kV) when the current is not negligible (reaching many milliampères). To hold the cathode at such a potential is complicated—it is to be remembered that in the normal type of tube, the screen, the layer of aquadag and the anode are all earthed and it is the cathode alone which is connected to the high tension—since the cathode has also to be heated, and the beam modulation has to be introduced; both of these require either a highly insulated transformer, or H.T. condensers, and these are expensive pieces of apparatus.

To overcome these problems, Schwartz (1938) and later Rogowski and Thielen (1939) suggested that the original arrangement should be used, with an accelerating potential of a few thousand volts; only the screen and the region immediately adjacent to it should be held at a very high voltage with respect to earth, as shown in Fig. 116. This is known as "post-acceleration", and entails making a cut in the layer of aquadag near the screen, and supplying the latter with a potential of 25 or 30 kV. After deflexion, therefore, the electron beam is subjected to the influence of another electron lens, effectively a two-cylinder lens, across which the potential changes from the accelerating potential Φ_0 to the post-acceleration potential Φ_a, both measured relative to the cathode potential. The diameter of the spot at the screen, as a scrutiny of the analysis given in § 4.3.5 shows, is now related to the potential Φ_a; the formula for the spot diameter is still given by equation (12.1), but with the screen potential inserted, thus

$$d_s = \frac{A}{\sqrt{\Phi_a}}$$

(in which A is a constant independent of the potentials).

The brightness of the screen is excellent, since it is governed by the potential of the screen itself. The space charge aberration too is diminished, as the electrons move more rapidly over the last part of their trajectory where the space charge is most intense. Finally, the technical problems which the construction of this system creates are lessened, since the steady voltage which has to be applied to the screen only requires a "post-acceleration terminal" (Fig. 125). What we have now to do is to confirm that this modification to the optical system has not reduced the deflexion sensitivity.

We shall demonstrate that the deflexion which the prism has to produce, in order to displace the spot through its own width when it is acting on a beam in a region in which the potential is Φ_d, remains unaffected by the behaviour of the potential beyond the deflector. Whether the potential of the screen Φ_a is equal to Φ_d, as in ordinary tubes, or whether it is considerably larger than Φ_d, as in post-acceleration tubes, is irrelevant. In the latter case, the way in which the potential changes from Φ_d to Φ_a is also of no importance. The result is independent of the way in which the post-acceleration potential is applied, as it is of the type of focusing and deflexion which can be either magnetic or electrostatic. It is not easy to demonstrate this property by extending the argument which was used to establish the value of the sensitivity if the structure of the post-acceleration field is unknown. An indirect line of reasoning due to Pierce (1941), on the other hand, is of general validity. We consider two diametrically opposed points on the spot, A and B, shown on a vastly exaggerated scale in the transverse

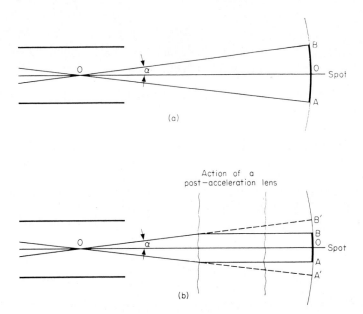

FIG. 143. The effect of post-acceleration on the beam: (a) without post-acceleration. (b) the effect of the post-acceleration lens.

direction in Fig. 143. The rays OA and OB which have passed through the centre O of the deflecting prism define the angular aperture 2α of the pencil of rays passing through O which would be produced by a diaphragm with a small hole placed at O. This is an obvious result if there is no post-acceleration, as in this case the rays OA and OB are rectilinear, but it is still true even if there is post-acceleration, despite the fact that the rays OA and OB are curved, for in this case A and B are the images of the points A' and B', which lie on the "virtual spot" which is formed by the part of the oscillograph which precedes the post-acceleration system so that OA' and OB' are straight lines. But if the deflector bends the rays through an angle 2α, it is clear that the rays which would have passed through O and lain between A and B will now lie outside AB; in particular, the ray which would have passed through A now intersects the screen at B. We can repeat the argument for every other position O' of the point O within the deflecting prism, so that 2α, which is the natural aperture of pencils at the centre of the deflector, is equally the value of the deflexion to which the whole beam has to be subjected to displace it through its own diameter. The value of α, however, depends only upon Φ_d, the potential in the region of the deflector, and the cathode temperature, as we have already shown (§ 4.3.5); if C is a constant determined by the gun, therefore, we have

$$\alpha = \frac{C}{\sqrt{\Phi_d}}.$$

If the deflector is electrostatic, the voltage σ_E which is required to produce a deflexion α is [cf. § 12.3.3, formula (12.19)]:

$$\sigma_E = \alpha \frac{\Phi_d}{C_E} = \frac{C}{C_E} \sqrt{\Phi_d} \tag{12.29}$$

(where C_E is a geometric constant, and depends only upon the deflector) so that σ_E depends not upon the screen potential Φ_a but upon the mean potential at the deflector, Φ_d. We conclude, therefore, that the sensitivity of electrostatic tubes can be improved by introducing post-acceleration, as the mean potential at the deflector, Φ_d, can be kept small for a given value of the brightness of the spot on the screen.

For a magnetic system, on the other hand, the current σ_M which is required to produce a deflexion α is given by the formula

$$\sigma_M = \frac{\alpha \sqrt{\Phi_d}}{C_M} = \frac{C}{C_M} \tag{12.30}$$

(cf. § 12.3.4), so that the sensitivity finally depends only on the cathode temperature (through the constant C) and the geometry of the system (through the constant C_M). It is completely unaffected by the presence or absence of post-acceleration; the technique is still useful, however, as a means of avoiding the risk of breakdown between the electrodes of the gun.

Finally, we should point out that we have not examined the defects which the post-acceleration lens produces in the marginal region; such defects do appear, and in order to reduce them, we try to use as uniform a field as possible, which means that the lens is extremely weak. For this reason, Lambert and Field (1946) apply their post-acceleration potential progressively along ten rings of aquadag in such a way that the equipotential surfaces are as nearly as possible regularly spaced concentric spheres. Sponsler (1955) has made a further improvement by replacing the rings by a helix of aquadag round the tube. The perturbation of the Gaussian rays which this produces is extremely small.

CHAPTER 13

THE EMISSION MICROSCOPE

13.1 INTRODUCTION

The image which is formed by an emission microscope is created by electrons emitted by the surface of the object itself. No illuminating system is required, therefore, and microscopes of this kind are simpler than the usual types. The electron optics possesses original features, and the sphere of application is specialized, for with an emission microscope, massive specimens can be examined directly, once they have been rendered emissive. This type of microscope is valuable, also, in that the structure of massive solids can be studied—instead of examining simply a rigid replica or an ultra-thin layer of the substance (see Chapter 20), the surface itself and its evolution can be examined.

Historically speaking, the first electron microscope was in fact an emission microscope; with the discovery of electron optics, the designers of cathode ray tubes destined for oscillographs were able to form an image of the emitting cathode on the fluorescent screen. The first photographs of this phenomenon were published in 1931–2. Subsequently, with the research that was directed towards improving the image, the oscillograph (which was ill adapted to play this new role) became clearly separated from the emission microscope, in which it became possible to obtain a resolving power at first equal to, and more recently much better than that which can be obtained with an optical microscope.

13.2 OPTICAL PROPERTIES

13.2.1 The Initial Velocities

In order to have a standard of comparison, we shall first mention the properties of a transmission microscope (see Chapter 15). Here, the electrons arrive at the objective with a high velocity and at a very small inclination, which can be adjusted at will with the condenser. The dispersion of the velocities about their mean value is very small, and the consequent chromatic aberration is negligible in the great majority of cases.

In an emission microscope, on the contrary, the electrons *emitted* at some point of the object have a very small velocity, and the most probable velocity may vary between a few tenths of a volt and a few volts according to the particular case in question. The angles at which the electrons are emitted can lie anywhere between 0 and 90°, and no simple means of reducing this angular aperture exists. These features are unfavourable as far as the clarity of the image is concerned, and the resolving power of emission microscopes is undeniably inferior to that to be obtained with a transmission instrument.

13.2.2 The Resolving Power

By calculating the trajectories of the electrons which emerge from an object point, the minimum *diameter* of the beam in the neighbourhood of the Gaussian plane which corresponds to the object (§ 4.2.4) can be calculated; the resulting theoretical resolving power is (see § 8.3.2):

$$\delta_{th} = k \frac{\Phi_0}{E_0}$$

in which Φ_0 is the most probable energy of the emitted electrons, measured in electron-volts, and E_0 is the electrostatic field at the point on the cathode which is being considered; k is a numerical coefficient, close to unity. In the case of thermally emitted electrons with $\Phi_0 = 0.1$ V, δ_{th} is of the order of 400 Å if a modern objective for which $E_0 = 30$ kV cm^{-1} is used. If we calculate the distribution of electron intensity across the beam in the vicinity of the image plane, either by taking the diffraction into account (Recknagel, 1943) or by using geometrical optics and investigating the shape of the caustic of the pencils of electrons (Septier, 1955) which is valid for small values of Φ_0, we obtain a more favourable value of the theoretical limit. k is found to be of the order of 0.1 or 0.2, so that in the example discussed earlier, δ_{th} is some 40 or 80 Å. In reality, this theoretical limit is unattainable without an objective stop because the spherical aberration of the objective itself intervenes.

We might consider improving this resolving power by stopping down the elementary pencils at the crossover of the beam with a hole a few microns in diameter (Möllenstedt and Düker, 1953; Fert and Simon, 1956, 1957; Simon, 1959). We shall summarize briefly the main results of the study of the role of this aperture which Fert and Simon have made. The objective is regarded as the combination of a purely accelerating section, length $\triangle$, followed by a convergent lens L (see Fig. 144a). The results will still be approximately valid for normal immersion objectives.

We let α_0 be the angle at which a ray leaves the axial object point P; at the entry to the lens L, this angle has become α_1. If we place a diaphragm of radius r at the exit of the lens, and if we suppose that the real image of P is formed at infinity, the angle α_1 cannot exceed an upper limit which is given by $\alpha_{1m} = r/a$ where a is the distance between L and the virtual image P_1 of P which is formed by the accelerating system.

If we make allowances for the defocusing action of the hole in the accelerating anode, we find $P_1 P = \dfrac{\Delta}{3}$ and the magnification of the virtual image at P_1 is $G_1 = 2/3$. The slope α_1 is given by the relation

$$\sin \alpha_0 \sqrt{\overline{\Phi_0}} = \frac{2 \cdot}{3} \alpha_1 \sqrt{\overline{\Phi}},$$

in which Φ_0 represents the energy with which the electrons leave, and Φ the accelerating potential; $\Phi \gg \Phi_0$.

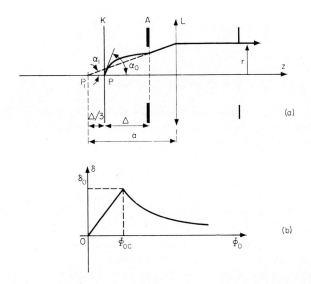

FIG. 144. The resolution of an objective with a stop. (a) The path of the outermost ray which leaves P through the simple objective formed by an accelerating region KA followed by a thin lens L and a diaphragm of radius r. The limiting angle is $\alpha_0 < 90°$. (b) Variation of the resolution δ as a function of the energy at emergence, Φ_0.

To the limiting angle α_{1m} corresponds a critical energy Φ_{0c} such that $\Phi_{0c} = \dfrac{4}{9} \alpha_{1m}^2 \Phi$. If the electrons which emerge from P have an energy less than Φ_{0c}, none will be halted by the diaphragm; if $\Phi_0 > \Phi_{0c}$ however, the diaphragm will stop all electrons which emerge at angles such that

$$\sin \alpha_0 > \frac{2}{3} \alpha_{1m} \sqrt{\frac{\Phi}{\Phi_0}}.$$

The diameter of the aberration disc in the plane of the circle of least confusion can be calculated from the formula:

$$\delta = 1 \cdot 2 \frac{\Phi_0}{\Phi} \Delta \sin \alpha_0 (1 - \cos \alpha_0)$$

(see, for example, Septier, 1954a). If $\Phi_0 < \Phi_{0c}$, therefore, α_0 is not limited by the diaphragm, and we have

$$\delta = 1 \cdot 2 \frac{\Phi_0}{\Phi} \Delta = 1 \cdot 2 \frac{\Phi_0}{E_c} .$$

If $\Phi_0 \geqq \Phi_{0c}$, α_0 is restricted by the diaphragm to the value given above:

$$\delta = 0 \cdot 8 \sqrt{\frac{\Phi_0}{\Phi}} \Delta \alpha_{1m} \left(1 - \sqrt{1 - \frac{4}{9} \alpha_{1m}^2 \frac{\Phi}{\Phi_0}} \right) .$$

For $\Phi_0 \gg \Phi_{0c}$, we should have

$$\delta \simeq 0 \cdot 178 \Delta \sqrt{\frac{\Phi}{\Phi_0}} \alpha_{1m}^3 .$$

Figure 144b shows the variation of δ as a function of Φ_0, with the diaphragm present. The resolving power is clearly given by the maximum value, δ_m of δ, which can be calculated with $\Phi_0 = \Phi_{0c}$:

$$\delta_m = 1 \cdot 2 \frac{\Phi_{0c}}{\Phi} \Delta \simeq 0 \cdot 5 \Delta \alpha_{1m}^2 ,$$

or

$$\delta_m = 0 \cdot 5 \Delta \frac{r^2}{a^2} .$$

When a diaphragm is present, the resolution is independent of the velocity dispersion of the electrons emitted by the cathode.

With an objective for which $\Delta = 2 \cdot 6$ mm, $a = 6$ mm, $\Phi = 20$ kV, and $2r = 30 \mu$, we find $\alpha_{1m} = 2 \cdot 5 \times 10^{-3}$, $\Phi_{0c} = 0 \cdot 06$ V and $\delta_m = 90$ Å.

The influence of the spherical aberration of L is of the order of 40 Å (for a focal length of 7 mm with $C_s = 140$ mm)—the stop gives rise to a diffraction aberration which is of the same order of magnitude. With such a system, therefore, we can hope in theory to reach a resolution of the order of 100 Å. The experimental resolution obtained by Fert and Simon (1956) is of the order of 250 Å. This value is eight times better than that to be obtained with the best light microscopes, but forty times worse than that of the best transmission electron microscopes.

13.2.3 The Objective

The electrons leave the object with a very low energy, Φ_0, and arrive at the screen with an energy Φ_i which corresponds to the accelerating potential. We return to the concept of refractive index $n = \sqrt{\Phi}$; in object space,

n_0 is small—$0 \leqq n_0 \leqq 2$ according to the particular case in question—while in image space, n_i is large—of the order of 100 or 200 in modern instruments. Here, therefore, there is a large disparity between the indices at the entry and the exit, unlike the situation in transmission optics where $n_0 = n_i$. By analogy with the light optical situation in which this disparity between the indices is to be found (always, a far smaller disparity) the objective of emission instruments is known as an "immersion objective".

The objective performs two distinct functions, as it serves both to accelerate the electrons which are emitted, and to provide a real magnified image of the object. From the very beginning, both of the two types of objective which are possible have been used—either the lens can perform both operations simultaneously, in which case it can only be electrostatic, or alternatively it can be "mixed", when it consists of a part whose function is solely to accelerate, followed by an ordinary lens, which is more often magnetic, with which the beam is focused. It was with the aid of the former of these two types of objective that various research workers were able, between 1940 and 1941, to obtain a higher resolving power than the limit of the light microscope, of the order of tens of millimicrons (100 Å) (Mahl, 1936; Johannson 1933; Mecklenburg, 1942; Boersch, 1942). Kinder (1944), however, proved that a mixed objective could give equivalent results. At present, the best purely electrostatic solution at a given operating potential seems to be a mixed objective with an electrostatic lens (Septier, 1953b).

(i) The electrostatic immersion objective (Fig. 144)

This is the name by which the earliest, and now the most common type of electrostatic objective is known. Originally, its construction was very crude—the electrodes simply thin sheets of molybdenum—but gradually, improvements inspired by the study of transmission microscope lenses have been made, with the result that the electrodes of modern objectives of this type, operating at high tension, are machined and centred to an extremely exacting degree of precision. There are only two electrodes, which lie in front of the plane cathode K which is in fact the surface of the object. The first electrode W (which is commonly known as the grid or Wehnelt, by analogy with the gun in a cathode ray tube) is mechanically rigidly linked to the anode A, with the aid of insulating supports, but is held at a potential very close to that of the cathode. In general, the anode is earthed, and is joined to the positive terminal of the high tension; the insulated cathode is connected to the negative terminal. The object-cathode can be moved in any direction; by moving it perpendicular to the axis, the specimen can be explored, while a displacement along the axis provides a coarse focusing of the image at the screen E. The fine focusing is achieved by varying the potential difference v between the grid and the cathode slightly; v reaches at most a few hundred volts when the accelerating

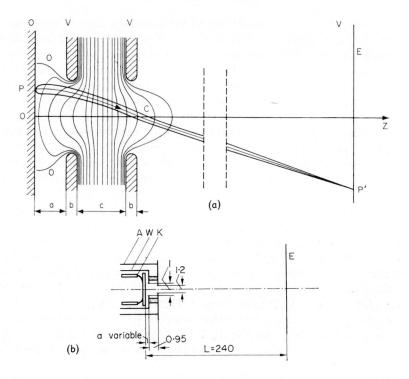

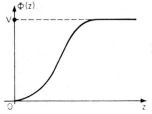

FIG. 145. (a) Diagram of the objective of Johansson's microscope (1933). (b) Diagram of the electrostatic objective. The appearance of the equipotential surfaces and of the electron trajectories which originate in a point P; P' is the image of P and C is the crossover.

FIG. 146. The behaviour of the axial potential distribution in an immersion objective, in focus.

potential of the anode is of the order of tens of kilovolts. The potential of the grid is usually negative with respect to that of the cathode.

The two functions of the lens, acceleration and focusing, are intimately bound together in this case. Figure 145 shows a map of the potential in such a system, through which the path of an elementary pencil is indicated. If the rays which emerge from a point object are to be reunited into a point image, the equipotentials which the rays cross must be curved convexly towards the cathode, and further, the axial potential distribution must be of the form illustrated in Fig. 146, in that the second derivative in the vicinity of the object must be positive. Experiment shows that when the system is well focused, the electrostatic field E_0 at the cathode is weak. If we denote the accelerating potential by V, and express E_0 in volt cm^{-1}, the latter can never, it is found, exceed V volt cm^{-1}; this imposes a limit on the theoretical resolving power (Septier, 1954). The field E_0 increases when the dimensions of the objective are scaled down, but this cannot be continued indefinitely because of the danger of continuous discharge, and because of the additional aberrations which appear when the radius of the holes is reduced too far. In good conditions, the minimum dimensions of an objective with a high quality factor operating at about $V = 30$ kV are

$a = 0.2$ to 0.4 mm   for the object–grid distance,

$\left. \begin{array}{l} b = 0.2 \text{ to } 0.3 \text{ mm} \\ R_1 = 0.65 \text{ to } 0.7 \text{ mm} \end{array} \right\}$ for the grid,

and

$\left. \begin{array}{l} c = 2 \text{ mm} \\ R_2 = 0.25 \text{ to } 0.5 \text{ mm} \end{array} \right\}$ for the anode.

In the optimum situation, the grid potential is close to zero. The focal length, in this case, is of the order of 2·5 to 3 mm, which allows direct magnification of 200 or 300 times in a long enough instrument; this is often adequate.

There are certain applications, however, in which we have to make allowance for the bulk of the parts which are merely accessories in light optics, but which are indispensable to the metallurgist; the interelectrode spacing has to be increased, E_0 decreases in consequence, and so does the resolving power. Hence the high tension V must be increased as far as the limits defined by the possibility of breakdown between the electrodes permit.

(ii) *The objective with a magnetic lens, in which the two functions are separated*

The plane cathode lies in front of an accelerating anode with a very small opening, behind which is placed a weakly convergent magnetic lens. We consider a typical point P (Fig. 147) of the object which is emitting electrons; as far as the lens L which follows the accelerating unit is concerned, the electrons originate in a virtual object P' which lies, to a first approximation, at a distance $2a$ from L, as the trajectories are parabolae. For a system operating at 30 kV, we find $a \geqq 2$ mm. As P' has to be effectively at the focus of L, we shall have at the minimum $f \sim 5$ or 6 mm, making allowance for the thickness of the lens. To obtain a good image when the diameter of the opening in the anode is of the order of a millimetre, the plane cathode must be surrounded by a guard-ring held at the same potential, but placed slightly in front of it (this latter is to avoid field curvature). This guard-ring has a slightly unfavourable effect upon the axial potential, in that the field E_0 is slightly diminished. If, on the contrary, the opening in the anode is a few tenths of a millimetre across, and if the latter is thin, a plane cathode can be used; the electric field E_0 is then virtually equal to the ratio Φ_i/a.

Nowadays, a shielded objective of the standard type is used, feebly excited. The direct magnification provided by the objective alone is slight owing to its long focal length, and if a magnification of the order of 1000 is

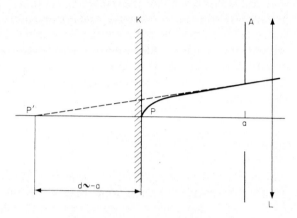

FIG. 147. A trajectory in an accelerating system.

required, a second lens has inevitably to be introduced as projective. An excellent microscope can be constructed in this way, which possesses the further advantage of being capable of functioning at much higher accelerating potentials than the objective described earlier (Fert and Simon, 1956).

(iii) *The wholly electrostatic objective, in which the functions are separated* (§ 8.3.2)

This system works on the same principle as the lenses of the preceding paragraph. Once again, the accelerating system is separated from the lens itself which follows it. This lens is a weak three-electrode unipotential lens. For a given limit V of the accelerating potential, this arrangement is an improvement on the standard type of objective, as the field E_0 is far greater (Septier, 1954; Düker and Illenberger, 1962).

13.3 THE VARIOUS TYPES OF MICROSCOPE

Modern microscopes are optically equal in quality, whether they be magnetic or electrostatic. A clearer classification is achieved by considering the way in which the electrons are emitted by the object, and we shall distinguish between:

(a) microscopes in which the emission is thermionic,

(b) microscopes which use secondary emission, and

(c) microscopes in which the electrons are emitted photoelectrically.

The difference between these instruments lies not in their optics but in the arrangement of their cathodes. Finally, we shall also examine some very different kinds of microscope, some of which have a specialized optical system which can cope with the emanations from radioactive materials, while others have no lenses at all, like the line or point microscope.

13.3.1 The Thermal Emission Microscope

This is the oldest type, but only a small number exists in the whole world at the present time.

(i) *The range of emitters*

A wide variety of emissive objects can be examined:

Pure metals and refractory metals which emit spontaneously at high temperatures: molybdenum, tantalum, niobium and tungsten. The minimum emission for which an image directly magnified a few hundred times can be obtained is reached when the metal is at a temperature of the order of 1200 to 1500°C. A powerful heating system is necessary therefore, and it is useful to be able to reach a temperature as high as 2500°C if the recrystallization of these refractory materials is to be studied.

Metals which contain impurities which diffuse to the surface and which have in addition an intense emissive power—thoriated tungsten and thoriated molybdenum, for example, from which the cathodes of certain vacuum tubes are constructed. The thorium is always present in only very small quantities, but distributed over the surface of the material, it creates a pronounced reduction of the potential barrier at the surface of the metal on which it lies, so that an intense emission can be obtained at far lower temperatures than with pure metal.

Metals which are virtually non-emissive (because they melt before they begin to emit freely) *with activation of the surface.* The main agents which are used for activation are the alkaline earth metals (barium and strontium) and caesium; they form a monomolecular layer over the surface of the object, uniformly distributed so that all the details of the underlying surface are still present. The surface can be examined at about 800°C when it is covered with a layer of barium or strontium, which is laid down by evaporation *in vacuo* within the body of the microscope itself (the carbonate, or any other pure salt which on heating gives off the metallic constituent without any harmful impurities, is used). The metal of the object must be able to support a temperature of about 1200°C, however, which permits "activation". To obtain an acceptable emission, and to render the underlying metal apparent, many cycles 800° ⇄ 1200° are sometimes necessary;

if the activation is to be accurately reproducible, and above all if the observations are to be of long duration, the microscope must be very thoroughly evacuated (to a pressure of less than 10^{-5} mm of mercury). In a new activation technique using barium, described and discussed in detail in an article by Heidenreich (1955), a solution of barium formate in a mixture of formic acid and isopropyl alcohol is used. Perfect polishing and cleaning of the surface which is to be activated are vital conditions for success, and the final result depends on the nature of the metal. Iron, nickel, copper, cobalt and chromium give intense and easily visible emission; the technique described above cannot be used with caesium, which is too easily oxidized, unless the whole operation is conducted inside a highly evacuated sealed-off glass envelope. Alternatively, the object can be plunged into caesium vapour while it is actually being examined; the vapour is either produced in an adjacent oven which is incorporated into the apparatus, or by heating a tungsten filament covered with caesium to about 700°C. The observations themselves are made at temperatures of the order of 500°C.

Complex emitters, such as cathodes formed from the oxides of the alkaline earth metals, together with a thick or thin oxidized layer; L-type cathodes, which consist of a piece of porous tungsten, through the pores of which the barium which is spread in the form of the oxide on the other face of the specimen can diffuse; thorium oxide cathodes, in which the metal support is covered with thorium oxide; cathodes consisting of the borides of the rare earths, obtained by sintering; metal cathodes directly alloyed to barium (Ni–Ba alloy, for example); these examples by no means exhaust the possibilities.

Composite objects, which contain numerous substances juxtaposed together, of which at least one is emissive; *at the emission temperature, the ensemble must be a conductor*, as the surface of the cathode has to be an equipotential if focusing is to be possible. In this way, it has been possible to examine such materials as the complex alumino-chrome agglomerates, and also many sorts of alumina which are conducting above 1000°C.

Finally, by simply changing the polarity, it is possible to transform a microscope which had been operating with thermionic emission into a *microscope using thermally emitted ions*. It is then possible to examine the surfaces of poor conductors which emit positive ions spontaneously (Couchet, Gauzit and Septier, 1951, 1952), for example the complex silicates which comprise the "steatites".

(ii) *The origin of the contrast*

The contrast in the image is attributed to a number of different causes. (a) The irregularities of the surface can be seen, and it is often difficult to interpret the appearance of the image when the relief is very pronounced (of the order of a few hundredths of a millimetre, that is). The depth of field of an immersion objective is, in fact, extremely small—theoretically it is zero—and in addition, even the smallest surface irregularity behaves like an elementary diverging or converging lens. The contrast can become inverted according to which plane in the neighbourhood of the surface corresponds to the focus—for example the edges of the irregularity shown in Fig. 148 appear dark or bright according as the focus corresponds to the plane P or the plane Q. A sharp ridge appears as a dark line with two bright streaks parallel to it, and a similar phenomenon is seen if the irregularity is pointed.

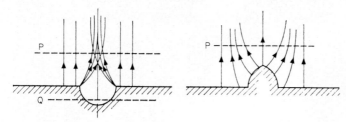

FIG. 148. The distortion of the image which is produced by surface relief.

(b) In the case of highly emissive cathodes formed from juxtaposed grains, an effect is produced by space charge—certain points of the image are dark not because they correspond to non-emissive zones, but because there is a slight depression on the emitter at which the emission is sufficiently intense to create enough space charge to repel the emitted electrons.

(c) On a specimen which has been polished and, after etching off the superficial amorphous layer produced by the polishing, annealed, the different crystals are visible in the form of grey regions, ranging between light grey and completely black. This can be explained quite simply, as the work function Φ_s of the metal varies slightly according to the orientation of the crystal face. If the emission can be described by Richardson's law, $i = A T^2 \exp(-e \Phi_s/kT)$, it can be seen immediately that a change $\triangle \Phi_s$ of a few tenths of a volt is sufficient to alter the intensity of emission by a factor 10. Exact measurements on single crystals have shown that Φ_s can vary up to 20 per cent about its mean value, so that $\triangle \Phi_s$ can exceed 1 volt for tungsten (Müller, 1955). Moreover, the emission microscope can be used to measure these variations, $\triangle \Phi_s$, by keeping the temperature fixed, and bringing the image of successive crystals onto a Faraday cage. The problem is to identify the faces being studied. The contrast is better at lower temperatures near the point at which the emission begins to appear; when T is increased, however, all the crystals emit, and the completely black regions disappear. The different shades become uniform for reasons which are imperfectly understood at present—among these, the properties of the fluorescent screen play a part at high intensities.

Fig. 149. A niobium surface, displaying marked contrast between the different crystal faces, and gentle surface relief; $G = 600$ (Photo: Septier).

Fig. 150. Platinum crystals. The enlarged inset represents a part of the region enclosed by a black rectangle, and shows the slip lines which can be seen in the dark crystal; in the main figure, $G = 500$. in the inset $G = 6,000$ (Photo: Septier).

In Figs. 149 and 150, the appearance of various recrystallized metals is illustrated. The appearance of the surface of the activated metals is the same, and it seems likely that the activating material is selectively absorbed over the different facets of the crystals, and that the potential barrier Φ_s, which is generally speaking reduced by the presence of this activating agent, varies appreciably according to the crystalline orientation of the face which is being examined.

In conclusion, we should mention that in the microscope using thermally emitted electrons, just as in the metallographical light microscope, the surface requires a special preparation—in order to avoid relief which would produce "ghosts", the surface has to be given an extended polishing (first mechanical, and then electrolytic), and after this, it must be cleaned chemically, by cathode sputtering, or more simply by raising it to a high temperature *in vacuo* (Gauzit and Septier, 1951).

(iii) *Application to the study of electron emission processes*

When it was first developed, this type of microscope was used to try to throw some light on the mechanism of electron emission itself—the structure of the emitter, the evolution of the latter with time, and the effect of aging. The work of Brüche and Johannson (1932, 1934) on oxide-coated cathodes should be mentioned particularly in this context. In these early experiments the image was well contrasted but only magnified about 65 times. In Fig. 151 (obtained by Septier), the same type of image is shown, but far more highly magnified. It was in the course of their research that Brüche and Johannson discovered (in 1933) that the crystals of the metal could be made visible by covering the surface with a very small quantity of barium, which was obtained by heating barium oxide. Schenk (1935) measured the relative emissivity of different crystals of nickel activated with barium; he found that the most extreme variation of potential was in the ratio 6 : 1, which corresponds to a variation in the height of the potential barrier of 0·2 V. Mahl (1936) examined the influence of absorbed oxygen on the thermal emission at thoriated tungsten and molybdenum cathodes.

The diffusion mechanism by which thorium travels from the interior of a specimen of tungsten or molybdenum to the surface has been directly observed by Brüche and Mahl (1935, 1936). The thorium travels rapidly from the interior to the surface, through pores which lie not only at the junctions of the grains but actually traverse them. The thorium distributes itself around the emitting "crater", and then evaporates. At the optimum temperature the surface is completely covered, the thorium which arrives more than compensating the thorium lost by evaporation. At low temperatures the emission is poor as the diffusion does not proceed smoothly; very high temperatures also have a deactivating effect, as the evaporation outweighs the diffusion. The beneficial effect on the emission of a preli-

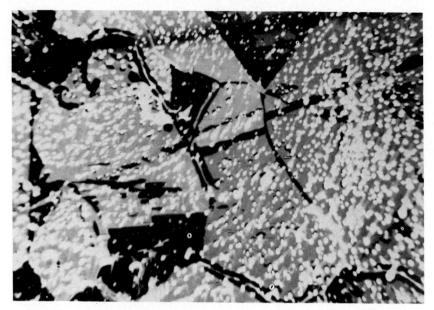

Fig. 151. Armco iron daubed with barium carbonate. The emissive points are grains of carbonate. The crystal boundaries are just beginning to appear (Photo: Septier).

minary carburation of the support has been demonstrated by the same method. The migration of the thorium to the surface follows privileged directions before it disappears as a result of evaporation. All these emission mechanisms had been thought of many years before by Langmuir and his students, but their proofs were most indirect and were based upon a complicated theory, whereas in the microscope, the proof is direct and striking, and is perfectly credible without reservations.

More recently, Stranski and Suhrmann (1947) have studied directly the electron emission of the crystallized surface of tungsten and its relation to the structure and also the mechanism by which foreign atoms are adsorbed. The adsorption proves to be selective, and depends upon the atomic diameter of the impurity, the orientation, and the potential barrier at the crystalline face.

For crystallography and metallurgy, Brüche and Knecht (1934, 1935) have developed a more supple technique for barium activation, in which the barium is vaporized in the vicinity of the object so that the vapour condenses on the surface being observed. In this way, we can, for example, watch the allotropic $\alpha \rightleftharpoons \gamma$ transformation which occurs at about 900°C in iron. The transformation of an individual crystal of the specimen can be followed as the temperature is slowly varied, as Burgers and Ploos van Amstel (1937, a, b; 1938 a, b) for example, have shown; the same authors have also followed the transformation of zirconium from the α-phase to the β-phase at 600°C, and have recorded photographically the recrystallization of ferro-nickel. Recently, Rathenau and Baas (1951) have taken up the same research but with a high resolution microscope capable of high magnification, and studied in particular the growth of individual grains. They have attempted to elucidate the mechanism by which the junctions of the grains move during the recrystallization in a large number of alloys, amongst which the ferro-nickels are the simplest to observe.

Another possibility is to examine sections of wires, in order to determine the variations in the structure between the outer surface and the centre (Septier, Gauzit and Baruch, 1952)—in this way, it is possible to measure the diffusion of nitrogen in a niobium wire (Fig. 152). Using this same technique, it is likewise possible to study the diffusion of one metal through another, and precipitation and segregation in alloys.

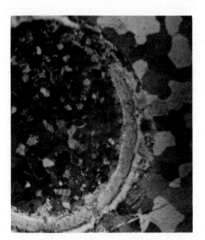

Fig. 152. Cross-section of a niobium wire, examined with a thermionic emission microscope. The wire is included in a small niobium plate. Right: pure niobium wire. Left: niobium wire nitridized in ammonia vapour. The layer of nitride (which is invisible optically) is clearly visible.

13.3.2 Secondary Emission

(i) *The basic principle*

Every surface which is bombarded by particles with a certain energy emits electrons which are known as "secondary electrons". This emission is related to the energy of the incident primary particles, but for the majority of substances, the ratio d between the secondary current and the primary current varies in the same way whether the primary particles be electrons or ions. d reaches one for incident energies between 0·5 and 1 kV, then exceeds this value, but for high energies (in the region of tens of kilovolts) it generally falls to a value below unity. The multiplication factor d varies with the nature of the body which is being bombarded (varying between 1 and 15 for the whole range of polycrystalline materials) and with the nature of the crystalline face which is encountered.

The possibility of using secondary electron emission to form electron images was demonstrated by Zworykin in 1933. Meschter (1938) made improvements to the earlier instrument after Behne (1936) had studied thin films by transmission; these attempts met with little success, however. The fact is that in all these attempts, some of the primary electrons reach the fluorescent screen, and since they cannot be focused at the same time as the secondary electrons, they confuse the image which is produced by the latter, by producing either a continuous parasitic illumination or intense parasitic spots. The resolving power which can be attained is very poor (a few tenths of a millimetre) as a result of the extreme velocity dispersion, and the project was therefore abandoned.

Recently, the method has been taken up once again by Bayh (1958) and with a highly developed experimental arrangement, he has been able to obtain extremely well-resolved images with 15 keV primary electrons (see § 13.4.2).

Mahl (1938) obtained an image of the cathode of a gas discharge tube with a pierced anode; the image which he obtained was of a poor quality for a number of reasons but the experiment is of interest as it points to the possibility of obtaining emission of secondary electrons when the primary particles are ions. A similar, though improved process has recently been used successfully by a number of workers (Möllenstedt and Düker, 1953; Septier, 1953a), who bombard the surface with *positive ions*. The ions which are reflected by the surface cannot get to the fluorescent screen, and the images prove to be excellent. The velocity dispersion of the electrons which are emitted is amazingly small (Möllenstedt and Düker, 1953) in contradiction to the opinion which had been current before these experiments. The half-width of the spectrum is only 2 V, which means that we can hope for a resolving power comparable with that of the light microscope without having recourse to a small stop, which is so very difficult to place accurately at the crossover. Figure 153 gives some examples of the values of the coefficient of secondary emission for a few common metals.

The ions which impinge on a surface also give rise to positive and negative secondary ions. The latter are predominantly hydrogen ions, H^-, and if the secondary electrons are eliminated by placing a weak magnetic field at the exit of the objective, it is possible to obtain an image of the surface which is being bombarded with the aid of these H^- ions (Bernard and Goutte, 1958). These ions originate in a layer adsorbed on the surface of the specimen; their emission is independent of the nature of the incident ions, and occurs whatever the specimen may be when the vacuum is 10^{-4} or 10^{-5} mm Hg.

The positive ions which are emitted by a surface which has been perfectly cleaned and placed in a very good vacuum (10^{-6} to 10^{-7} mm Hg) are mostly ions which are characteristic of the specimen, which have been torn away from the substance itself (Castaing and Slodzian, 1960). They too can be used to form an image of the surface; if we eliminate all the unwanted ions with a magnetic deflector, we can even form the image with one category of ions alone, which thus provides a map of the distribution of the corresponding element in the specimen (Castaing, Jouffrey and Slodzian, 1960; Castaing and Slodzian, 1962). The instrument with which images of this type can be obtained will be described later (§ 13.4.3).

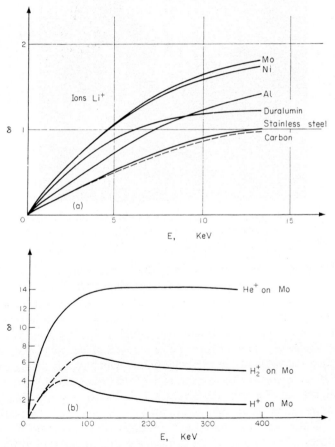

FIG. 153. The secondary emission coefficient of different metals under the impact of ions of energy E. (a) Couchet (1954). (b) Hill, Buchner, Clark and Fisk (1939).

(ii) *Contrast (secondary electrons)*

Just as in thermal emission, the origin of the contrast can be attributed to a host of possible causes.

The relief. For normal bombardment, the only effect produced by the relief is that of a weak lens, and the appearance resembles that of the similar phenomenon in thermal emission. When the bombarding particles are incident obliquely, however, we benefit from a *shadowing effect* (the obliquity

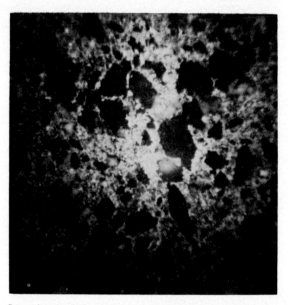

Fig. 154. Secondary emission by ion bombardment. The appearance of a complex chrome–aluminium; $G = 300$ (Photo: Septier).

is usually of the order of 30° or 45°); the bombardment is less intense or non-existent on the slopes which are "invisible" from the source of the primary particles, and the more important relief throws a shadow across the neighbouring part of the surface. Nevertheless, we should repeat once again that this relief must remain very small if the resolving power is to be acceptable—as a general rule, we may say that the relief should be no higher than the value of the resolving power which we are trying to attain.

The different nature of the constituents of the object in the case of a composite object. In Fig. 154, the appearance of a chrome–alumina mixture is shown, in which the regions of alumina are lighter, but observations are still few in number; they have been obtained with specimens at ordinary temperatures. How to display the influence of the orientation of the crystal faces is not known except during the few seconds immediately after the ion bombardment begins. The object has to be heated to a temperature of a few hundred degrees in order to eliminate the superficial parasitic layer; this layer, which is a consequence of the ion bombardment itself, consists principally of a deposit of carbon which is produced by the breaking down of organic molecules which are to be found inside the apparatus (in such substances as oils, greases, or solvents). A partial elimination can be achieved by heating the specimen to about 200° both before and during the experimental observation. Yet another difficulty appears, especially with those metals which are particularly susceptible to oxidization—all the contrast between the different grains is obliterated by a thin layer of

FIG. 155a. The surface of perlitic steel. The image obtained by secondary emission with ionic bombardment (Photo: Keller, see Möllenstedt and Hubig, 1954). Direct $G = 580$; final $G = 2550$.

oxide. Figure 155a is a photograph of the surface of a perlitic steel at 150°C, in which the different grains and the relief of the surface are perfectly visible (Möllenstedt and Hubig, 1954; Fert and Simon, 1956, 1957).

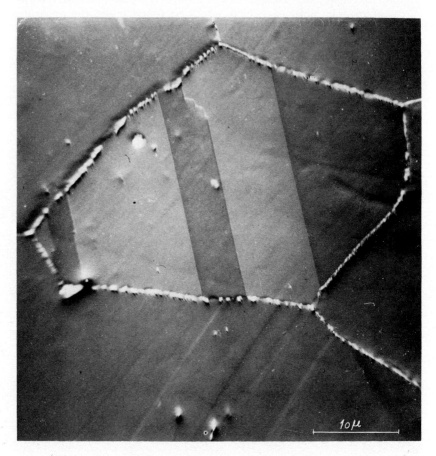

Fig. 155b. The image of the surface of chrome steel, obtained by secondary emission; $G = 2400$ (Photo: Simon).

(iii) *Contrast (secondary ions)*

The experimental work of Bernard and Goutte (1958) has shown that the contrast which is obtained with H^- ions is of twofold origin: the relief, and the nature of the metal—we could easily distinguish islands of gold evaporated onto aluminium, as the "emissive power" of the adsorbed layer (where the emission is of negative ions) would depend upon the nature of the metal.

The positive ion emission which is obtained with a purely electrostatic optical system (in which the convergence is independent of the mass) ought to be capable of forming very highly contrasted images of a heterogeneous specimen, as the emissive power for each of the characteristic ions is very different according to the various metals: the emission is particularly intense for silicon, aluminium and nickel, and weaker for beryllium and copper. Alkali compounds also give very intense emission. If we filter the beam behind the objective, and allow only one category of ions to pass, we obtain a very well-contrasted image.

(iv) *Applications*

The examination of surfaces with the aid of secondary emission widens the field of application of the emission microscope—it is possible nowadays to examine conducting surfaces at temperatures above a limit as low as 150 or 200°C, which is of great interest metallurgically where specimens of the light alloys are concerned. The resolving power, which is of the order of 2000 Å, is inferior to that which can be obtained with a reflexion instrument, but the data which are to be obtained are numerous and adequate for the study of recrystallization or of inclusions of foreign bodies.

If we form the image with each category of emitted positive ions successively, we can perform a kind of chemical analysis of the specimen.

13.3.3 Photoelectric Emission

(i) *The basic principle*

Photoelectric emission was first introduced in 1933 to obtain an image of the object without heating it. It was thought that by irradiating a clean and polished surface with intense ultra-violet light, it would be possible to produce a flux of electrons sufficiently dense to produce an appreciable magnification, and it was hoped that the contrast in the image of the crystals would appear just as in the case of thermal emission. Numerous experimental problems obstructed the development of this technique, which has in consequence been slow, and successive attempts have run up against very weak emission, and inhomogeneity of the velocities too marked for the resolving power to be other than poor, but above all, against the problem of rendering the crystals visible.

(ii) *Applications*

Zinc is a very good emitter of photoelectrons, and it was this element which was used in the earliest experiments, by Brüche (1933). The surface, which was freshly polished, was vividly illuminated by ultra-violet light, and the electrons which were emitted were accelerated through 30 kV and then focused with a magnetic lens. On the image, which is only very slightly magnified, only the shadow relief produced by oblique illumination can be discerned. Later, Pohl (1934) and afterwards Mahl and Pohl (1935) showed how unfavourable is the influence of monomolecular surface layers of impurity and of the absorbed gas (or oxides) which are invisible in an ordinary microscope. Only the relief is visible when the specimen is cold; even after cleaning the specimen by prolonged red-heat, and subsequently cooling it, only the outlines of a few crystals can be made out on specimens of platinum and nickel, although the magnification is slightly higher (30 instead of 6). With the aid of a cathode made of many metals, the same workers were able to show that a high tension discharge in air at a very low pressure reduces the emissivity (except for the noble metals, gold, silver and platinum) whereas the same discharge in hydrogen increases it by a large factor. Finally, Gross and Seitz (1937) succeeded in making the crystals of a nickel specimen visible, in two different ways. In the first method, they varied the properties of the gas at the surface; at the beginning of the cooling which is carried out *in vacuo*, the gas is adsorbed to a different extent at the different crystalline faces which have been heated to red-heat. The contrast which is produced by this procedure rapidly vanishes, however, as the layer of adsorbed gas becomes too important at low temperatures. In the other method, a polished and outgassed specimen of nickel is covered with a thick layer of barium, held for a considerable time at a temperature of about 1000°C to obtain a well-contrasted image of the nickel with the thermal electrons, and only after this is it cooled again. On irradiating it with ultra-violet light, quite as good an image is obtained, with the same contrast between the different crystals. After the specimen has been cold for some time, the crystals fade away, and only the relief remains visible, but with fresh heating, which gets rid of the gas adsorbed by the barium, the crystals reappear. This is a difficult technique and is only suitable for the relatively refractory metals, as a temperature in the neighbourhood of 1000°C is necessary for outgassing and activation. It was abandoned after 1937.

Recently, however, Huguenin (1954, 1955, 1956) has developed a general procedure for examining cold specimens with a high magnification. He is able to destroy the amorphous surface layer without recourse either to thermal outgassing or to activation with barium, by scrutinizing carefully the ways of "cleaning" the surface for which the crystalline contrast is visible. The standard procedure of cathode sputtering in an inert gas like argon or a reducing agent such as hydrogen is inadequate; the grains are visible, but almost without contrast, as in fact the object is enveloped in a very fine layer of metal dust—the metal is that from which the other electrode of the discharge system is composed. By using a powerful ion gun, the contrast becomes visible as this drawback is avoided. Two examples of photographs obtained in this way are shown in Figs. 156 and 157. The resolving power is of the order of a micron on these pictures; this is a valuable method at low temperatures, as metals which are particularly susceptible to oxidization—uranium, for example—for which all the other techniques break down, can now be examined.

FIG. 156. The appearance of a copper surface after ionic scouring; photoelectric emission. $G = 100$ (Photo: Huguenin).

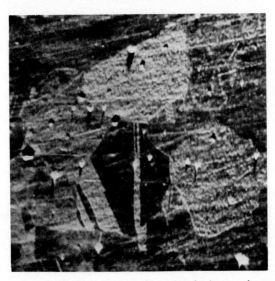

FIG. 157. The appearance of a silver surface after ionic scouring; photoelectric emission. $G = 100$ (Photo: Huguenin).

13.3.4 The β-ray Microscope and the Recoil Nucleus Microscope

A few attempts have been made to form an image of the surface of certain radioactive substances by using the natural β-emission. They have, however, run up against a number of difficult problems.

Firstly, the intensities are very small; exposures of many hours or even several days are necessary, although the magnifications are always small (between 1 and 10).

In addition, the particles emitted have a high velocity—β-rays have energies between 100 keV and 1 MeV; the magnetic lenses, therefore, must be able to accept rays emitted into a large solid angle but still be strong enough to focus electrons with an energy near 1 MeV within a reasonable distance.

Finally, the velocity spectrum is often very wide, so that the chromatic aberration is enormous.

In this context, we mention the articles by Marton and Abelson (1947) and by Barker, Richardson and Feather (1950); the latter obtained the image of a brass disc activated with radiothorium B, magnified seven times, after an exposure of four and half hours. The electrons which are used have an initial velocity of 148 kV and a particularly narrow velocity spectrum, which allows the resolving power to be 6 μ.

In this rudimentary form, such microscopes as these present no practical interest, as modern direct auto-radiographs (in which a special photographic emulsion is placed in direct contact with the specimen) have a resolving power of a fraction of a micron, far better than that of a β-ray microscope.

Recently, attempts have been made to form the image of a radioactive layer emitting α or β-rays by making use of the recoil atoms which leave the surface in the form of positive or negative ions with small energies as the elementary particles are emitted (Stühmer, 1960; Willax, 1960). The layer to be examined does not produce an intensity high enough to give an image which is visible to the eye on a fluorescent screen, and the optical adjustments are therefore made with a massive ion source situated in the plane of the layer (the first electrode of the immersion objective). As "objects", the authors used metallic plates activated with the emanations from thorium; they obtained ion images of the surfaces activated in this way with magnifications between 10 and 40 times, with exposures of the order of two hours, and accelerating potentials between 10 and 30 kV.

13.3.5 The Importance of the Vacuum in an Emission Microscope

We have already seen how fundamental is the role which is played by the surface phenomena so far as the appearance of contrast in secondary emission microscopy or photoelectric emission microscopy is concerned, as the basic mechanism of these types of emission is controlled by these phenomena. Within instruments of this kind, therefore, it is most important that the vacuum should be as good as possible and also as "clean" as possible, in the sense that it should be free of vapours which might condense or dissociate; this prevents the surfaces which are being examined from becoming contaminated, either by adsorption of gas or, above all, by the formation of a layer of carbon which will have originated in the molecules of grease or oil which always diffuse into the vacuum chamber, even when low temperature traps are employed.

At the present time, it seems highly desirable to employ the very latest developments of high vacuum technology when an emission microscope is being constructed.

The vacuum chamber and the optical system should be capable of being entirely outgassed by heating to some hundreds of degrees centigrade, which means that the whole assembly must be made of stainless steel, the joints must be metallic—made of aluminium or indium rather than of neoprene—and the insulators either quartz or a ceramic substance.

"Clean" pumps must be used, which will not produce troublesome vapours in the vacuum chamber. This requirement is best satisfied by the titanium pumps which appeared on the market a few years ago; the action of these pumps consists essentially in the adsorption of the residual gas by titanium vapour (which is similar to the effect produced by a "getter" in pumping down a radio tube). In certain types, ionic pumping is used as well—an intense electron beam falls onto a titanium plate (Huber, Schroff and Warnecke, 1959) which slowly evaporates and at the same time, ionizes the residual gas. The ions which are formed, together with the vapour which is loaded with adsorbed gas, are collected by a well-cooled outer electrode. If the pumping rate of the pump is to be increased, we can equally well use the evaporation of titanium droplets exposed to intense electron bombardment (Holland and Laurenson, 1959). Rates of the order of $2000 \, l \, sec^{-1}$ are to be expected, and a vacuum limited to about 10^{-8} mm Hg.

In another type which is simpler to build and operate, the titanium is vaporized by cathode sputtering in a cold discharge; the discharge takes place within a tube which resembles that of a Penning gauge (a gauge which is used to measure low pressures and is among the more familiar vacuum measuring apparatus) provided with a titanium cathode. The gas is ionized by the discharge, and the ions which are collected at the cathode set off an intense cathode sputtering. By connecting a number of discharge units in parallel, it is possible to obtain high pumping rates (up to $4000 \, l \, sec^{-1}$) which remain steady between 10^{-5} and 10^{-8} mm Hg (Jepsen, 1959). These pumps start at 10^{-2} mm Hg; here once again, we can obviate vapours from the oil by replacing the normal mechanical pump by an absorption pump in which the recipient is filled with activated carbon dipped in liquid nitrogen. When the pressure has fallen to 10^{-2} mm Hg, the connexion between this pump and the space to be evacuated is cut, and the titanium pump is set in operation. With such a pumping system as this, a vacuum of 10^{-7} or 10^{-8} mm Hg can be obtained without any vibration, and this latter property is another valuable asset in electron microscopy.

In conclusion, we should mention certain "molecular" rotary pumps which have recently been developed, with which "clean" vacua of at best about 10^{-6} mm Hg can be obtained, and which pump at several hundreds of litres per second; unfortunately, the speed of the rotation entails a certain danger of vibration, and if a very high resolution is desired, this is inconvenient.

13.4 SOME ACTUAL INSTRUMENTS

13.4.1 The Thermal Emission Microscope

The model which we shall describe is a microscope which was designed by J. Vastel and P. Grivet, and perfected by Septier and Gauzit (1950). It was intended that it should be convenient to use, and that *changing the specimen should be rapid*; further, that it should be easy to subject the latter to various thermal treatments, as metallurgical research requires a variety of temperatures. Figure 158 shows the appearance of the instrument. The head, which contains the object support and the devices for heating and shifting the object, can be separated from the body of the instrument by means of a lever. The object, which has to be prepared in the form of a thin plane disc 5 mm in diameter, is simply placed in a groove in the object support. It is heated by electron bombardment of its upper face, by means of an auxiliary electron gun which provides up to 50 watts, and with which a maximum specimen temperature of 2500°C can be attained. The object can be moved, *in vacuo*, as far as the point at which it would be withdrawn from the case (Fig. 159), and in this position, it can be activated conveniently in barium vapour which flows across its surface from a small lateral oven. It is also possible to bake the metal for a protracted period without polluting the electrodes of the objective. As an arrangement of this kind is sensitive to vibrations, which would limit the resolving power to 2000 Å, it is necessary to provide the upper part of the grid with an insulating refractory ring, on which the base of the object-support rests during observations.

Figure 160 represents the second type of objective studied by Septier (1953b). The projector lens, which is one of the symmetrical unipotential type with focal length $f = 16$ mm, has apertures of large diameter (17 mm), which leave the electrons a wide channel when only the objective is being used; the magnification of the latter is already between 100 and 250 (according to the distance between the object and the grid) and its field is very high. When the second lens is brought into action, the total magnification is between 1000 and 2500. The potential is normally between 20 and 30 kV, and the resolving power is in practice of the order of 1000 Å.

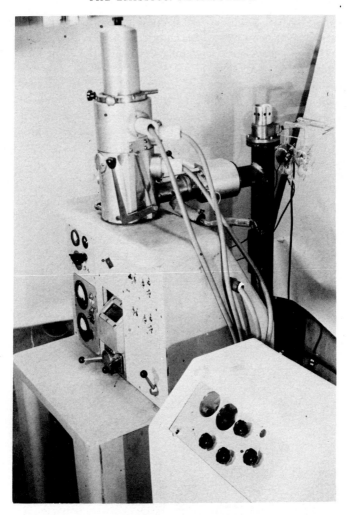

FIG. 158. The thermal emission microscope in use at the Laboratoire d'Électronique et de Radioélectricité (Paris). General appearance.

An instrument of the same kind but with various mechanical improvements has been described by Couchet (1954).

Heidenreich (1955) described a microscope similar to that of Mecklenburg (1942), using a double projector lens and an indirect cathode heating system, which was to be used for examining different metals at temperatures between 600 and 1000°C, after activation with barium. The resolving power was estimated to be 1000 Å.

An original type of thermal emission microscope has been described by Popp and Walcher (1957); the cathode consists of a thin biological section which is placed on a flat plate, heated by the Joule effect. In the immersion objective, we collect the *alkali ions* Na^+ and K^+ which are emitted by the heated tissue, and an image magnified 100 times is formed on a fluorescent screen. The resolution obtained is about $0.5\,\mu$, and the images obtained are very rich in detail. To reduce the exposure time, these authors lay down a very thin layer of Apiezon oil on the surface of the photographic emulsion (using Ilford Q_1 plates, for example); with a current density of the order of 10^{-10} A cm^{-2}, the exposure time is only between a half and three minutes.

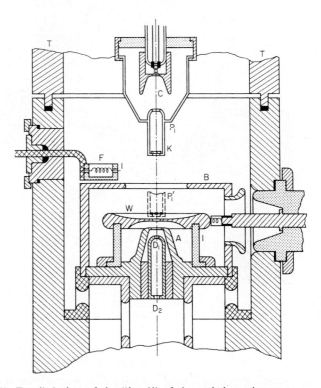

Fig. 159. Detailed view of the "head" of the emission microscope.

P_1' : the specimen-holder in the observation position.
P_1 : the specimen-holder in the activation position.
F : oven
C : heating gun.
W, A : objective.
D_1, D_2: diaphragms.

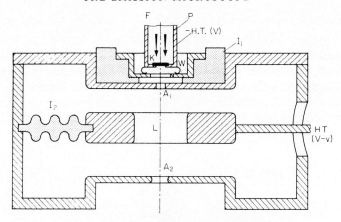

FIG. 160. Purely electrostatic mixed objective (Septier, 1954).

F : the electron beam which heats the cathode K.

K : the cathode (a metal disc) which is placed in the object-support P.

W : guard ring, held at the same potential as K.

A_1, A_2: anodes (30 kV).

L : electrostatic lens (at the potential $V - v$).

I_1, I_2 : insulators.

13.4.2 The Secondary Emission Microscope

The microscope of Möllenstedt and Düker (1953) is an electrostatic instrument, with two stages, giving a direct magnification of 200. Figure 161a shows the arrangement of the immersion objective and of the gun with which the object is bombarded with ions. The whole "head"—the object-support, cathode and grid—is held at a potential of 40 kV, together with the accelerating electrode of the ion gun. The grid has a conical form so that the ions which bombard the object at an angle of 30° with the object plane can pass. The anode which is connected electrically to earth is mechanically an integral part of the head.

Beneath the anode is a diaphragm in which is cut a 25μ hole, together with its centring mechanism, the function of which is to reduce the aperture of the elementary pencils and to make the resolution better; despite 2 volts velocity dispersion, Möllenstedt has succeeded in obtaining extremely fine images by the introduction of this innovation and a resolving power better than 1000 Å. The ions are produced in a gas discharge gun, and are subjected to the accelerating potential of the full high tension; slower ions would be preferable, as the resulting electrons would have less velocity dispersion, but supplying the ion gun would be less convenient.

An improved version of this instrument has been described by Düker (1960). A microscope with which it is possible to obtain images by thermal emission, secondary emission (ions → electrons) and photoelectric emission has been built by Bas (1960). Finally, Fert *et al.* (1958) have used a device with a magnetic objective to obtain resolutions of the order of 300 Å.

In the secondary emission microscope in which the primary bombardment consists of 15 kV electrons, which has recently been described by Bayh (1958), the arrangement resembles that used by Möllenstedt (Fig. 161a): the ion gun is simply replaced by an electron gun. By setting an aperture diaphragm of extremely small diameter (15 μ) just below the anode, a resolution of about 1000 Å can be obtained.

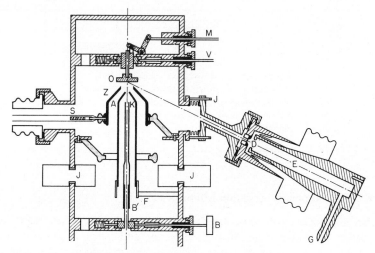

FIG. 161a. The immersion objective and the ion gun, for secondary emission microscopy (Möllenstedt and Düker).

A   : anode (electrode earthed).
B, B': apparatus for adjusting the contrast diaphragm.
D   : exit opening for the ions.
E   : ionization chamber.
G   : input for the gases.
J   : apparatus for controlling the ion beam.
K   : the contrast diaphragm.
M, V : apparatus for varying the position of the object O.
S   : high tension lead.

13.4.3 The Positive Secondary Ion Microscope

A diagram of this instrument is given in Fig. 161b (Castaing and Slodzian, 1962). The primary ion beam (protons) bombards the plane surface of the specimen, and the positive ions which are emitted are then accelerated by a four-electrode immersion objective followed by a stop. They next cross the most delicate part of the whole apparatus, which is the homogeneous field magnetic prism analyser which filters out the various different masses. The inclination of the entry and exit faces defines a stigmatic system for two particular points, F and F', which lie in symmetrical positions with respect to the prism. The point F coincides with the crossover of the

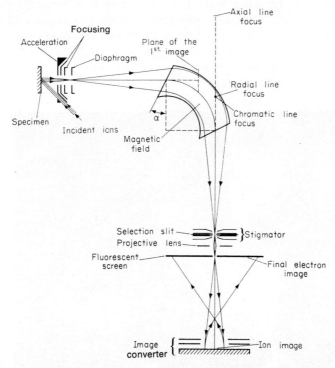

Fig. 161 b. Analytical microscope, using secondary ion emission (positive ions, characteristic of the metal), with an image-transformer (Castaing, Slodzian).

objective and a device to correct astigmatism is placed at F'. A projective lens magnifies the image which the objective has produced, and focuses it onto the perfectly polished cathode of a second immersion objective; this latter slows the ions down and behaves as an image converter (see Chapter 3 and § 14.4.2). The secondary electrons which are emitted are accelerated in this same objective, and form a magnified image on a fluorescent screen which is placed beneath the projective lens. In this way, the brightness of the image can be enhanced, a necessity for the feeble emissions of characteristic ions from the types of specimens which are used. Figs. 161c, d and e show the differences in the appearance of the surface of an Al–Mg–Si alloy when the silicon, the magnesium and the aluminium ions are successively selected.

13.4.4 A Photoelectric Emission Microscope

Huguenin (1955) constructed, during 1954 and 1955, an electrostatic two-stage instrument, which was wholly enclosed in a demountable metal casing. The direct magnification can vary between 100 and 200 (the objective magnification is 10, while that of the projective is between 10 and 20). The objective with plane electrodes (Fig. 162) has to operate with a long distance between the object and the grid (about 1 cm) to allow the illumination to enter through a lateral window at an angle of about 20°. The other dimensions have to be increased in the same ratio if the quality

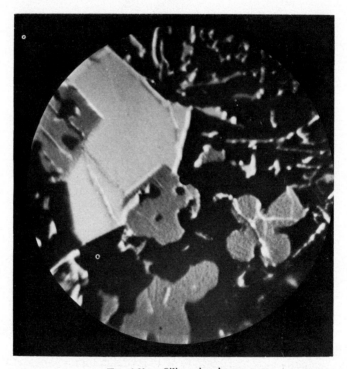

FIG. 161c. Silicon ion image.

of the objective is not to suffer, so that the holes in the electrodes are one centimetre in diameter; under such conditions, the focal length is of the order of 15 mm. The ion gun can be moved about *in vacuo*, and can be brought under the anode; the object is lowered until it is in contact with the grid, and the surface cleaned for rather a long time, 5 or 10 min, according to the nature of the metal being examined, with an ion intensity between

100 and 200 μA (or alternatively, a current density between 5 and 10 mA cm^{-2}). By using hydrogen, a much finer etching can be achieved than with argon. When the specimen is ready for observation, the ion gun is retracted, the object raised 1 cm above the grid, and the ultra-violet light is switched on. The insulation of the grid must be carefully protected from the metal particles which are released at the object while it is being etched by the ions. The body of the microscope is blackened inside, bent, and provided with a series of blackened diaphragms, all designed to prevent the ultra-violet light which is reflected by the object and the metal parts of the objective from reaching the fluorescent screen and masking the electron image.

FIG. 161 d. Magnesium ion image.

Perfected versions of the electrostatic photoemission microscope with which a high resolution (1000 Å) can be obtained, in which an aperture diaphragm is placed beneath the anode of the immersion objective, have recently been described by Koch (1958) and Bethge *et al.* (1958). The source is once again a high pressure mercury vapour lamp, and the operating potential is in the range 20–40 kV. The objective and the general arrangement recall those of Möllenstedt and Düker (Fig. 161a).

All these emission microscopes are described in a recent article by Möllenstedt (1963).

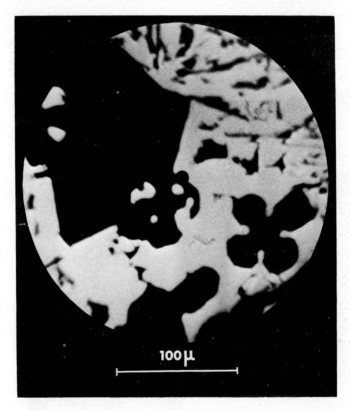

FIG. 161e. Aluminium ion image.

13.5 THE FILAMENT MICROSCOPE AND THE SPHERICAL (OR POINT-PROJECTION) MICROSCOPE

Besides the emission microscopes in the proper sense, much simpler instruments have been conceived in which a highly magnified image is produced without using any electron lenses at all. The accelerating potential of several kilovolts is simply applied between the cathode (which is either a point or a wire) and a curved conducting fluorescent screen spherical or cylindrical in shape; the cathode constitutes the centre or the axis of the screen.

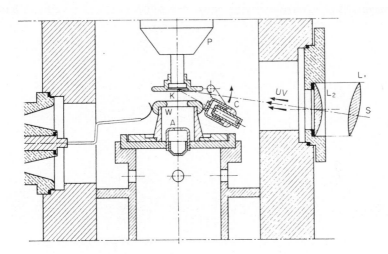

FIG. 162. Diagram of a photoemission microscope (Huguenin, 1955).
S   : ultra-violet source.
L_1, L_2: quartz optical system.
P   : object support.
K   : object; W: Wehnelt; A: anode.
C   : (retractable) ion gun for scouring the surface.

This type of instrument owes its inception to Johnson and Shockley (1936) and to Johnson (1938) who were studying the electron emission of poly-crystalline filaments and the adsorption of activating atoms at crystals of known orientation. The filament which is to be studied, of radius a, is stretched along the axis of a cylindrical glass tube of radius b. The inside wall of the tube is covered with a layer of some fluorescent substance, in which a wire bent into the form a helix of large pitch is embedded (Fig. 163); it serves the purpose of an anode. A potential difference of a few thousand volts is applied between the wire, which is maintained at a high temperature, and the surrounding helix. Electrons which are emitted by the wire travel radially in straight lines towards the wall of the tube, where they form a pseudo-image of the wire. In a cross-section, the magnification G is given by the ratio of the two radii b/a, which can be considerable, but in a meridian plane the magnification always remains unity. This produces a specific deformation of the image, as all the details appear to be stretched out in the transverse direction. The extraction field E_0 at the surface of the wire increases rapidly as the radius a is reduced ($E_0 \simeq K/a$); this explains why the electrons which are emitted in all directions at any given point are very quickly forced back towards the mean trajectory of the pencil, and hence, despite their initial divergence, produce a very small spot at the image. The transverse resolving power can be better than a micron, although

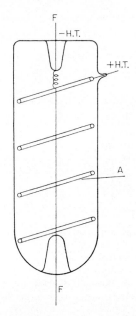

FIG. 163. "Wire" microscope.

the longitudinal resolution is only one or two tenths of a millimetre. Nevertheless, so exciting and so valuable was the information which this first instrument provided that the technique has since been considerably developed. The lack of symmetry in the magnification was first eliminated by Martin (1939) who replaced the wire and cylinder by two concentric spheres, as shown in Fig. 164. The object may be a single crystal of tungsten, for example, polished into a sphere and then etched away to display the various faces. This hollow sphere is heated from within by electron bombardment, and is placed at the centre of an evacuated spherical glass envelope of which the inner surface is coated with fluorescent material. A ring placed between the sphere and the wall, and held at a potential of a few kilovolts, accelerates the electrons which are emitted. As before, the magnification is given by the ratio of the radii of the envelope and the object. The instrument has been used (Benjamin and Jenkins, 1940) to measure the variations in the emissivity of the different crystalline faces, and to follow the modification to the emission which is produced by selective adsorption of various substances, some activating agents, others contamination.

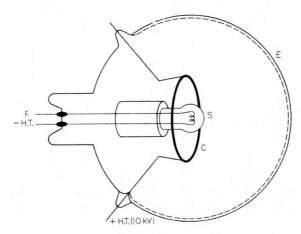

FIG. 164. "Sphere" microscope.

Finally, Müller has raised the magnification to an extraordinarily high value, after protracted study and improvement of the instrument (since 1937). The radius of the sphere he has reduced in the extreme, so that the microscope becomes the point-projection instrument which is widely used at the present time, and which we shall describe in detail in the next section.

13.6 THE POINT-PROJECTION MICROSCOPE

13.6.1 Description

A fine tungsten point, which is obtained by chemical action on a wire $^5/_{100}$ mm in diameter (the metal is etched away by liquid sodium nitrite) is mounted in a glass tube the other end of which is almost spherical and is coated with the fluorescent screen as in a cathode ray tube (Fig. 165). A second wire is welded to the first close to the point P, and the two wires are led out of the tube through seals; the point, therefore, can be held at a high temperature by passing a current of a few ampères. A large potential difference can be applied between the point and the fluorescent screen,

and for this, it is profitable to cover the fluorescent material *E*—after it has
been spread over the surface of the tube—with a thin layer of collodion (a few

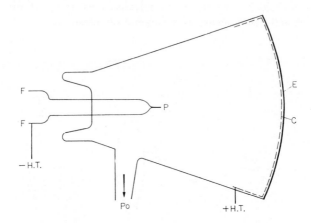

FIG. 165. The principle of the point-projection microscope (shown schematically).

hundredths of a micron thick) over which a metallic layer *C* about a
hundred Ångströms thick is spread by evaporation *in vacuo*. The sealed-off
high tension lead is also coated during the evaporation process, so that
the potential at the screen is well defined. In addition, the metallic layer
behaves like a mirror, and returns in the forward direction part of the
light which is emitted by the screen towards the interior of the tube; this
produces a marked increase in the brightness (§ 1.4.2).

The tube is evacuated with the greatest possible care, using a mercury
vapour diffusion pump with a liquid air trap. The whole operation must be
carried out extremely carefully, and in particular, the whole assembly of
tube and trap must be dried for at least 24 hours, heating the metal wires
to red heat at the same time, if possible. After this, the tube must be baked
again for one or two days with the trap immersed in liquid air. The tube
is now ready for use, as the vacuum will be of the order of 10^{-10} mm Hg.
Alternatively, we can work with a sealed-off tube by vaporizing a "getter"
in the tube before isolating it from the pump. We shall see below how the
quality of the vacuum in the tube can be estimated in a simple way. By
applying a potential difference of a few kilovolts between the point (which
is the negative electrode) and the fluorescent screen, we obtain an extremely
large electric field at the surface of the point of the order of V/r, *V* being
the total potential drop, and *r* the radius of curvature of the point. The
radius of curvature of the tip of the point can easily be reduced to less
than a micron, so that if *V* is 10 kV, the field at the point exceeds $10^8 \, \mathrm{V\,cm^{-1}}$.

This is a strong enough field to tear electrons out of the tungsten point; these electrons follow effectively radial trajectories, and form a kind of image of the surface of the point on the fluorescent screen. The magnification of this image is huge, as it is clearly given by the ratio of the radii of curvature of the fluorescent screen and the point which is emitting the electrons; for $r = 1\,\mu$ and $R = 10$ cm, the magnification is 10^5.

13.6.2 Emission from a Clean Metal Point

If we heat the point to a high temperature (about 2500°, say) for a few seconds, and then apply a high tension of a few kilovolts, we see a regular pattern of bright patches and dark regions on the fluorescent screen

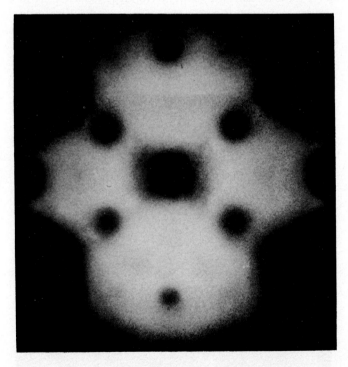

FIG. 166a. Bare tungsten point, radius 3500 Å; the equilibrium array obtained after annealing at 1300°K (Photo: E. W. Müller).

(Fig. 166a). The emissions of electrons, therefore, is not isotropic—rather, its spatial distribution is dictated by the orientation and the symmetry of the crystal lattice at each point which is emitting. In fact, the point consists of a single tungsten crystal, as a result of the smallness of its diameter. Certain crystallographic directions correspond to maximum emission, others to no emission at all. An attempt to explain the presence of these directions in which there is no emission has been made by Müller, by relating them to the direction in which propagation of the conduction electrons within the body of the lattice is forbidden; these are the directions in which there is Bragg reflection of the electron waves at the lattice planes. Unfortunately, such an interpretation would require the conduction electrons to have most unreasonable energies—to explain the lack of emission in the [310] direction, for example, the electrons would have to have an energy of 37·4 eV, whereas calculation of the Fermi energy gives 5·8 eV if there is one conduction electron per atom, 9·2 eV if there are two.

Subsequently, Müller (1939, 1943), Benjamin and Jenkins (1940), Smoluchowski (1941) and Ashworth (1951) have suggested a variety of explanations. As we are at present primarily concerned with the use of field emission in microscopy, we shall not consider the emission from a bare metal surface further.

FIG. 166b. A thin layer of barium, absorbed onto a hemispherical tungsten crystal, 2000 Å in radius. The mean coating is 1/20 of a monatomic layer thick (Photo: E. W. Müller).

13.6.3 The Adsorption of Foreign Substances at the Surface of the Point

If, after outgassing the point by heating it to a high temperature, we lay down a very thin layer of some foreign substance on the surface of the point by evacuation under a vacuum—a metallic film of caesium or copper for example—we can study the distribution of the foreign atoms over the surface of the point by applying the high tension and examining the modifications to the cold emission which ensue. This distribution can be modified by heating the point to a moderate temperature, but even at the normal temperature numerous scintillations are to be seen which are produced by migration of the foreign atoms across the surface in an attempt to group themselves into a more stable configuration (Fig. 166b). It is possible, therefore, to study the adsorption of the substance which is added to the tungsten surface; this adsorption is particularly favoured at those points of the surface at which the local orientation of the lattice is such that the mutual arrangement of the tungsten atoms allows the foreign layer to be in epitaxy. Increasing the temperature of the point accelerates the processes of migration and rearrangement, but at too high a temperature the layer begins to evaporate and the image produced by bare tungsten reappears. In a similar way, more complex processes, such as the oxidization of a fine metallic deposit, can be studied (Ashworth, 1948).

13.6.4 Estimation of the Quality of the Vacuum in the Point-projection Microscope

It would be extremely difficult to determine the pressure of the residual gas within the microscope by any of the usual methods, without the aid of a Bayard–Alpert gauge, as this pressure has to be far below 10^{-8} mm Hg if a serious study of some phenomenon is to be undertaken. Fortunately, however, the behaviour of the microscope itself enables us to make a rapid estimate of the residual pressure; we might say that the instrument plays the role of an almost unlimitedly sensitive gauge for its own benefit.

If we examine the pattern which is produced by the electrons which are emitted when the temperature of the point is high enough to denude its surface, the diagram changes progressively; it becomes hazy, bright points appear in the field of view, the overall brightness falls, and the image settles down into an equilibrium configuration perturbed by rapid local intensity fluctuations.

This phenomenon must be interpreted as the result of the adsorption of a continuous film of impurities—oxygen for example—at the tungsten surface. If a known pressure of oxygen has been admitted beforehand into the tube (although never more than 10^{-6} mm Hg) it is found that the duration of the period during which the pattern is settling down into its equilibrium position is inversely proportional to the pressure. Conversely, by measuring this "contamination time", we obtain a rapid absolute— if approximate—measure of the residual pressure in the chamber, even though it would have been too small to be measured by the usual methods. If it is only an order of magnitude that is required, we can consider that a monomolecular layer is sufficient to bring the pattern into equilibrium, and that this layer is formed atom by atom by chance molecular impacts at the point; each molecule which strikes the point is assumed to adhere to it without rebounding. Reasoning in this way, Ashworth (1951) found that a contamination time of 1 minute corresponds to a residual pressure of 10^{-8} mm Hg, 100 min to 10^{-10} mm Hg, and so on.

13.6.5 High Resolution Microscopy with Field Emission

(i) *Examination of the individual adsorbed particles*

We examine the dark parts of the pattern which correspond to the non-emissive regions of the point, immediately after the point has been out-gassed at high temperature. We can then watch the brilliant points appear and after a very short interval (less than a second), separate out into two adjacent spots; these turn around one another at first, and then either vanish or separate completely into two brilliant spots which stray independently about the field of observation. This behaviour can be explained by ascribing it to the adsorption of a singe oxygen molecule, which then breaks up into two atoms attached to adjacent sites in the tungsten lattice; subsequently, these two atoms can separate completely and either disappear as a result of evaporation, or migrate towards an emissive region where their image is swamped by the general brightness. A confirmation of this interpretation is provided by the experimental result that the spot does not divide into two if the residual gas is a monatomic gas such as argon.

This remarkable result suggests that the resolving power of the image which is obtained by field emission is sufficiently high for two emissive points to be distinguished when they are only separated by the distance between two adjacent tungsten atoms, about 2 or 3 Å. As we shall see, the structure of the field emission microscope is such that a resolving power of this order is indeed attainable, at least in favourable cases.

(ii) *The resolving power of the point-projection microscope*

In order to establish the limit of resolution, either we can consider electron trajectories and the geometrical aberrations only, or we can take into account the diffraction of the electron waves. We shall perform the geometrical analysis first.

We assume that the electrons are emitted from some given point of the surface—which we suppose to be completely spherical—of the tungsten crystal, with an initial velocity u_t corresponding to an energy $e\,V_t$, and that their initial direction is arbitrary. If u_t were zero, the electron would follow a radial path and arrive at some point P of the spherical screen (which is held at potential V). The non-zero initial velocity smears the impacts out over a small circle, centre P, diameter D; we can calculate the latter by taking into account the fact that the electrons which arrive at the rim of the circle must have left the sphere of radius r tangentially, and subsequently travelled along a hyperbolic path (as the repulsive force is of the form of a Coulomb force); the angle, 2θ, between the asymptotes of this hyperbola is very small, and one of its foci lies at the centre F of the tungsten crystal. When the electron hits the screen, it is virtually travelling along the asymptote $O\,H$, and is moving with velocity u_R (Fig. 167a).

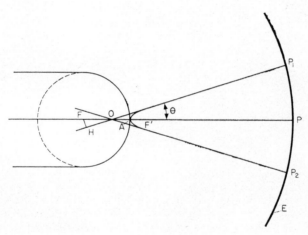

FIG. 167a. The appearance of the trajectories in the vicinity of the point.
(Note: $OH = OA$.)

Conservation of momentum gives us the relation $FH \cdot u_R = FA \cdot u_t$, and the geometrical properties of the hyperbola give $OF \cos\theta = OA$ or $FA = OF(1 + \cos\theta)$ and $FH = OF \sin\theta$. From these relations, we obtain

$$\frac{FH}{FA} = \frac{1 + \cos\theta}{\sin\theta} = \tan\frac{\theta}{2} = \frac{u_t}{u_R} = \sqrt{\frac{V_t}{V}},$$

$$D = 2R\theta \simeq 4R\tan\frac{\theta}{2} = 4R\sqrt{\frac{V_t}{V}}.$$

We have still to estimate V_t, and here the calculation becomes much less precise. The velocity of the thermal motion of the electrons (which corre-

sponds to an energy of about 0·02 eV at normal temperatures) is not the
only velocity to contribute to the tangential motion of the electrons as they
leave the cathode. The major part of this tangential velocity is produced
by distortion of the electrostatic field in the vicinity of the emissive point
(a single adsorbed atom, for example). The energy $e\,V_t$ produced in this way
has been estimated to be 10 eV by Ashworth (1951). For $V = 20$ kV and

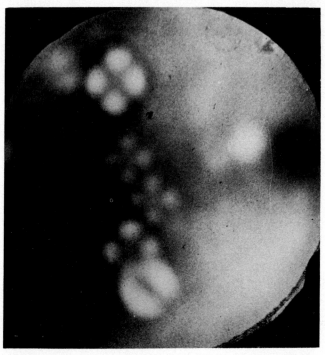

Fig. 167b. Phthalocyanine molecules absorbed on a clean tungsten surface
between the (011) plane (at the bottom of the photograph) and the (112) plane
(at the top). All the molecules lie parallel to the [011] direction (Photo: E. W. Müller).

$R = 5$ cm, we obtain $D = 4 \times 5 \sqrt{\dfrac{1}{2000}} = 0\cdot44$ cm. If we regard the spots
on the screen as being just separated when the distance between their
centres is just greater than their radius, and if we denote the magnification
by $G = R/r$, the resolving power ε is given by $\varepsilon = \dfrac{0\cdot22}{G}$ cm; for $r = 50$ Å
(which is an extreme case), with $G = 10^7$, the resolving power is $\varepsilon = 2\cdot2$ Å.

Benjamin and Franklin (1940), starting from a completely different point
of view, have obtained an estimate of the resolving power taking the dif-

fraction of the electron waves into consideration. They regarded the electrons which arrive at the screen as having originated in a virtual aperture traced on the spherical surface of the cathode. If we denote the diameter of this aperture by d, but otherwise use the same notation as before, the diameter of the spot at the screen will be dR/r in the absence of diffraction. We can, however, regard the angular aperture of the beam as seen from the screen, as being d/R, so that the effect of diffraction will be to enlarge the diameter of the spot by an amount $\lambda R/d$, in which λ is the wavelength associated with the electrons as they arrive at the screen. The total diameter of the spot, therefore, is $D = R\left(\dfrac{d}{r} + \dfrac{\lambda}{d}\right)$, which is a minimum for $d = \sqrt{\lambda r}$. We can consider this value, then, as defining the best resolving power. With $V = 20$ kV, $\lambda = \sqrt{\dfrac{150}{V}} = 0\cdot1$ Å and $r = 50$ Å, the resolving power is $2\cdot2$ Å and the minimum diameter of the spot is $0\cdot44$. This fits in well, besides, with the fact that the electrons which form the image have an energy of about 10 eV when they leave the object or equivalently, a wavelength of 4 Å, and are therefore incapable of distinguishing between details less than half a wavelength apart, that is, about 2 Å.

We might try to combine these two ways of examining the resolving power, by regarding the emitting atom as a small aperture of diameter d, emitting electrons in all directions. This localization into a small region about the point on the cathode which is emitting electrons results in an uncertainty h/d in their initial tangential momentum. The total momentum of the emitted electrons, therefore, must be equal to at least $h/2d$; in consequence, the spot which would have a diameter dR/r if tangential emission at zero velocity were possible is enlarged by $\dfrac{4Ru_t}{u} = \dfrac{4R}{d}\dfrac{h}{2mu} = \dfrac{2R\lambda}{d}$.

The total diameter of the spot is $R\left(\dfrac{d}{r} + \dfrac{2\lambda}{d}\right)$ which is analogous to the result obtained earlier; it has a minimum when $d^2 = 2\lambda r$, with the intermediate result that the electrons which are emitted by a detail which lies at the limit of resolution must have an energy of about 10 eV. This estimate of the resolving power shows definitely that it is possible to resolve a pair of adjacent atoms, but only in exceptional cases, and it is quite remarkable that this limiting resolving power seems to have been achieved in the examination of a pair of oxygen atoms. The explanation is to be sought for in the extreme simplicity of the point-projection microscope, as a result of which it is insensitive to the numerous causes of instability such as mechanical vibrations or electromagnetic imperfections which are of primary importance in restricting the quality of the image in a microscope with lenses. To obtain a resolving power of 2 Å is, nevertheless, exceptional; 5 Å, however, which is common, is adequate to provide a spectacular analysis of the structure of molecules such as phthalocyanine (Müller, 1950, see Fig. 167b).

In conclusion, we mention that Müller (1951) has succeeded in obtaining images with ions extracted by field emission. By applying a positive high tension (20 kV) to the point, which is surrounded by hydrogen at a pressure

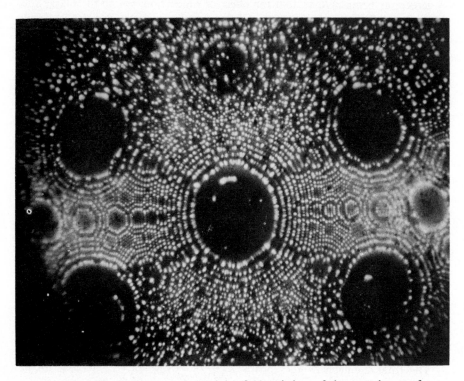

FIG. 167c. The ion image, obtained by field emission, of the central part of a tungsten point; the dark plane (011) is at the centre, and is surrounded by the four dark planes (112) of about 700 Å radius, annealed at 2000°. Helium pressure: 0·2 μ Hg; potential: 19,000 V; magnification: 1,300,000.
The "pearly" rings which surround the main planes represent the edges of lattice planes with small indices. Certain planes with high indices are fully resolved
(Photo: E. W. Müller).

of 5×10^{-3} mm Hg, the molecules which are attached to the tungsten surface become detached in the form of ions, and travel towards the screen. The image is less bright but finer (see Fig. 167c) and according to Müller, this process renders the individual steps of atoms visible. Too high an accelerating potential, however, produces confusion of the image. Müller explains this as the result of ionization of the hydrogen molecules before they arrive at the point.

A general review of field emission phenomena, in particular in connection with microscopy, has been written for the new *Encyclopædia of Physics*, **21**, (1956), 176–231, by R. H. Good and E. W. Müller. A detailed examination has also been made by Dyke and Dolan (1956), and more recently, a very complete review article has been published by Müller (1960).

CHAPTER 14

THE IMAGE-CONVERTER AND THE IMAGE-INTENSIFIER

14.1 THE BASIC PRINCIPLE

An *image-converter* transforms an invisible image which has been obtained by irradiating the object with infra-red or ultra-violet light into an image which is visible to the eye. The instrument which effects the same transformation of the image which has been formed with an objective which provides visible images is known as an *image-intensifier*, and in this case, it is the brilliance of the original image which has to be reinforced. Both types of instrument, however, work on the same principle: a suitable objective throws the optical image onto a light-sensitive layer, which emits electrons. The latter are powerfully accelerated, and focused, and produce the final image on a fluorescent screen or a photographic plate.

It was due initially to military requirements that these instruments have been developed to such a high degree; they are used to examine areas which are illuminated with infra-red transmitters at night or in fog, and to detect distant machinery which gives off infra-red radiation (artillery batteries and aircraft, for example). Physics too has benefitted from this progress, however; astronomers, for example, in their attempts to expand the limits of the visible universe, have been using "amplifiers" for fifteen or twenty years now, under the name of "electron telescopes", and it seems that the recent attempts to use these techniques in the microscope are likely to be successful. Finally, the need for image intensifiers with a very high gain has been felt since the invention of plastic fibre scintillators which emit photons when penetrated by fast ionizing particles; the search has been tenaciously pursued, therefore, both for new optical systems and for ways of perfecting existing systems, especially the photoelectric layers.

14.2 THE MAIN TYPES OF INSTRUMENT

14.2.1 The Image-converter with Uniform Field

The prototype of all these instruments is that of Holst, de Boer and Veenemans (1934)—shown in Fig. 168. It consists of an evacuated glass envelope, two of the faces of which are semi-transparent parallel metallic planes, a few millimetres apart, between which a potential difference of a few kilovolts is maintained. One face is covered with a photosensitive layer while the other carries the fluorescent screen.

The very intense electrostatic field enables a bright spot of small radius to be formed by each point which is emitting, so that the resolving power is acceptable. The radius of the spot can easily be calculated, and proves to be given by

$$r = 2d\sqrt{\frac{\Phi_0}{\Phi}},$$

in which Φ is the accelerating potential, Φ_0 is the mean energy of the electrons which are emitted making a grazing angle with the cathode, and d is

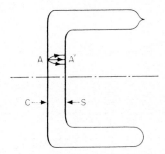

Fig. 168. The Holst converter.

the distance between the cathode and the screen. This tube, which was widely used during the last war because of the simplicity of its construction, is, however, only suitable for examining brightly illuminated objects. If the object is dim, the brilliance at the screen must be augmented by increasing the potential difference which is applied between the two planes; this produces parasitic emission, however, which results in an unwanted continuous background, as a high potential drop corresponds inevitably to a high field at the surface of the cathode, and hence to parasitic field emission.

14.2.2 Image-conversion with a Magnetic Field

To overcome this drawback, the anode and the cathode must be placed further apart, and this also reduces the risk of *poisoning* the photoelectric layer with the fluorescent sulphide. The image, however, must now be focused on the screen. In the first attempts, a combination of a homogeneous electrostatic field E which accelerated the particles and the homogeneous magnetic field, B, of a long solenoid which focused them was used (Fig. 169). The distance between the cathode and the screen can now be several centimetres. The electrons spiral round the lines of force of the magnetic field, intersecting them from time to time, and by varying B we can arrange that

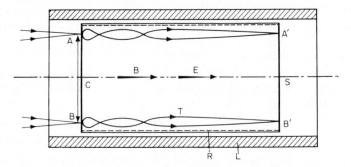

FIG. 169. A converter with homogeneous electrostatic and magnetic fields.

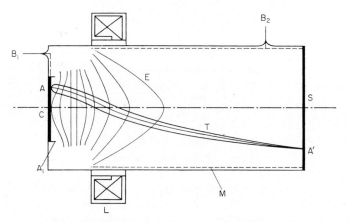

FIG. 170. A converter with a short magnetic lens.

one of the foci lies on the fluorescent screen. This system produces a clear image of an extended plane surface with unit magnification. In order to make the electrostatic field more uniform, and to avoid the perturbations which parasitic charges on the glass walls would produce, a conducting layer which has a high ohmic resistance is evaporated onto the latter. This system is still used in some image-forming tubes in television for tele-cinema use; the elementary pencils of electrons pass through a double scanning system, and fall on the cathode of a photo-multiplier (Farnsworth, 1934). As in Holst's system, the magnification is unity and the image is upright.

Figure 170 shows a variant of the same idea, in which a short shielded lens forms the image of a cathode C on a screen S; C is encircled by a guard-ring A_1, and the electrons are accelerated between C and S. This arrangement is used in super-emitron television tubes (McGee and Lubzinsky,

1939; Iams, Morton and Zworykin, 1939), in which the electron image is formed directly on the silver oxide mosaic. The anode in this case consists of a metal layer (M) on the outer walls of the tube, the accelerating potential is a few hundred volts, and the magnification is about 5.

In the devices described above, the magnetic field can be produced by permanent magnets, as it does not exceed a hundred gauss, although it does fill a volume of about 500 cm³—even so, the instrument is still heavy and cumbersome, and military requirements have led to the development of a light electrostatic instrument.

14.2.3 The Use of an Electrostatic System

(i) *Evolution and technology*

The earliest image-converters using an electrostatic electron optical system (see Zworykin and Morton, 1936) consisted of two tubes of the same diameter, each closed at one end by a plane disc —one of the latter supports the cathode C, the other the screen S (see Fig. 171). The slit between the two tubes acts as an accelerating lens, the magnification of which varies between

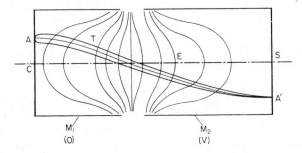

Fig. 171. An electrostatic two-cylinder converter.

0·2 and 2 according to the relative lengths of the tubes. It is, in short, a diode system, and lacks flexibility since the focus is fixed once and for all by the geometry; further, in common with all systems with a plane cathode, this system suffers from too much field curvature and distortion. The quality of the image has been considerably improved by Morton and Ramberg (1936) by using a curved cathode; the curvature cannot, however, be increased as far as the electron optics requires, as this would distort the image formed by the preceding glass optical system too acutely at the edges. During the last war, the result of extensive research into this problem was the development of a high quality electron optical system which was purely electrostatic (Morton and Flory, 1946; Schaffernicht, 1948; Eckart, 1954).

These devices are all sealed off. Special studies have been devoted to producing fluorescent screens which are capable of very high brightness—zinc sulphide activated with silver has received particular attention—and photosensitive layers with a high quantum efficiency (10 to 20 per cent) which can be used in the same envelope without either lowering the photo-electric sensitivity or reducing the secondary emission coefficient of the screen below unity, which would make its potential unstable. This inconvenience can be avoided by adding a layer of aluminium to the screen (§ 1.4). A special method for evacuating tubes of this kind is always used. The screen is formed first of all and baked, and the tube is pumped down before the somewhat complex photoelectric layer is introduced; then, without destroying the vacuum, two small evaporators are introduced through two lateral openings in the tube, and with these, first a lining layer of silver or antimony vapour is inserted, and then a photoelectric layer of caesium. These evaporators, which have flexible connexions, are afterwards retracted magnetically into lateral tubes, which are cut off with a blow-lamp. A different procedure has been suggested by Zacharov and Dowden (1960), which prevents, as far as possible, the electrodes and insulators in the tube from becoming polluted by caesium vapour. The cylindrical tube of the converter is divided into two parts by a flat glass disc; after forming the screen in one of the compartments, Sb and Cs are evaporated on the face of the disc which lies in the other compartment. The disc is then turned round *in vacuo*, so that the photocathode faces the optical system and the screen.

This technique has been highly developed in France, for another purpose, by the astronomers Lallemand and Duchesne, who have built an image-intensifier which is extremely sensitive to visible radiation: this is the *electron telescope*. In fact, it is an ideal photon receiver, in which the photons fall on a photoelectric layer which, unlike a photographic plate, has no threshold of sensitivity. Each electron which is emitted is accelerated to an energy at which it produces an impression on the photographic plate whereas in the case of very feeble illumination there would have been none.

(ii) *An infra-red converter with high-quality optical properties*

A modern infra-red converter is illustrated in Fig. 172 (Schaffernicht, 1948). It is a triode system in which the curved cathode C (of radius about 12 cm) and the metallized glass wall M which surrounds it constitute the first electrode, which is held at a potential V_2. The first anode A_1, which is about 6 cm from C and is held at a potential V_1, has an opening of 1 cm; the other anode A_2 which lies about 2 cm from A_1 and is connected to earth (together with the metallic layer of the fluorescent screen) has a 7 mm opening. The screen is 162 mm from C. The electrodes are made of very thin sheets of aluminium. This device is used with a potential V_1 of 18 to 20 kV and a potential difference $(V_1 - V_2)$ of 7 kV. In all instruments of this kind, the magnification G (which is always less than unity) varies with different designs between 0·7 and 0·1; this increases the image brightness which is multiplied by $1/G^2$ in any particular case and hence by a factor between 2 and 100. The image is focused by varying V_1.

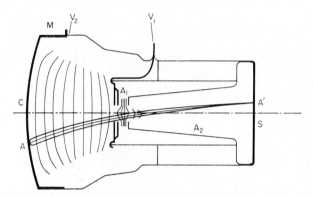

FIG. 172. An electrostatic three-electrode infra-red converter.

By using electrodes of more complicated shape, or by increasing their number to six, it has been found possible to make these converters work at potentials higher than 50 kV without increasing the field E at the sensitive layer. Post-acceleration is used in other types.

(iii) *An image-intensifier with a very high yield: the "electron telescope"*

Image-intensifiers very similar in design to those described above have been built commercially (Linden and Snell, 1957). A fine-mesh grid which is placed parallel and very close to the cathode plays at the same time the role of intermediate electrode and control grid; in its latter capacity, this grid can be used either to cut off the beam at will by applying a negative potential, or to provide pulsed operation. The image is again formed on a fluorescent screen.

Lallemand and Duchesne's "electron telescope" is quite a different affair. Its structure originally resembled that of a converter very closely, but successive improvements have gradually modified it (Duchesne, 1953a, 1961). Figure 173a represents a cross-section of the optical system of one of the latest models, which has a magnification 1/7. A supplementary electrode has been added, between the cathode and the first anode, and is held at 200 volts; the potential of A_1, however is 2000 volts, and the final potential lies between 25 kV and 50 kV. Figure 173b shows the outside appearance of the optical assembly, with the plate-container attached to the base. The whole instrument is surrounded by a glass envelope with high tension leads and numerous accessories. The cathode, which is all-enveloping, consists of a piece of copper, in the centre of which the metallized sensitive layer, 25 mm in diameter, is to be placed. Each caesium–antimony layer is kept in an extremely thoroughly evacuated thin glass envelope in a side arm until it is required (Lallemand and Duchesne, 1954). When a good vacuum has

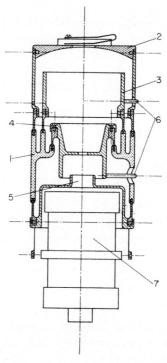

FIG. 173 a. Section of the optical system of a recent electron telescope. 1: glass insulator; 2: copper cathode; 3: first anode ($V = 200$ V); 4: second anode (2000 V); 5: third anode (20 to 50 kV); 6: high tension leads; 7: cassette of photographic plates.

been attained in the glass envelope of the instrument, and the electrodes and photographic plates have been outgassed, the small glass phial is broken and the layer spread in place, by moving magnetized parts from outside. The plates are cooled in liquid nitrogen to prevent the photocathode from being poisoned by the water vapour which they contain. The vacuum can be improved by using getters, and the most recent sensitive layers, which are themselves cooled to about $-60°$C, have a lifetime which is sufficiently long for the whole instrument to be detached from the pumping unit for several hours (Lallemand, Duchesne and Wlerick, 1960) and

FIG. 173 b. The optical unit of the electron telescope of Lallemand and Duchesne; the cassette of plates is visible in the lower part.

to be placed immediately behind the eyepiece of an astronomial telescope. With an overall operating potential of some 30 or 40 kV, the plates can be used in conditions where their sensitivity is maximum. Figure 173c shows the whole "telescope" which is mounted at the eyepiece of an astronomical telescope.

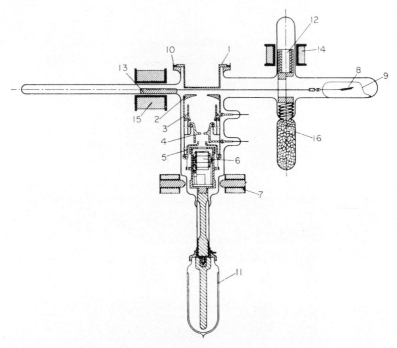

FIG. 173c. General diagram of the electron telescope installation. 1: (glass) entry window; 2, 3, 4, 5: the optical system; 6: plate-carrying drum (for 6 plates); 7: the magnetic system which controls the rotation of the plate-carrier; 8: photoelectric layer, placed inside its glass envelope 9; 10: joint; 11: cooling system for the plates (immersed in liquid nitrogen); 12: hammer for breaking the envelope 9, moved by means of the piece of steel 13, which is in turn attracted by the mobile magnet 15; 14: electro-magnet to set 15 in action; 16: radioactive carbon (now replaced by a titanium pump). The cathode is connected to a block of copper which is placed in a third lateral arm, and immersed in liquid nitrogen.

(iv) *Image-intensifiers with a thin window*

The solution to the problem of making the best possible use of the energy of the photoelectrons suggested by Lallemand and Duchesne is evidently by far the most satisfactory, the more so in that the resolving power of the instrument is a maximum in these conditions, as we shall see later. The life-time of the cathode is limited, however, and an exceptional degree of experimental skill is needed to set up and operate the instrument.

Attempts have been made to produce sealed-off tubes commercially, in which the yield remains very good but the photographic emulsion is outside the evacuated chamber in which the cathode is situated. The electrons fall on a fluorescent screen and the resulting photons make a direct impression on the photographic emulsion without any intermediate glass optical system which would inevitably collect only a small proportion of the light emitted by the screen. If an extreme loss of resolution is to be avoided, however, the fluorescent material and the emulsion must be practically in contact; this explains the need for extremely thin windows, the whole area of which must be of uniform thickness, and which must be vacuum-tight and capable of supporting atmospheric pressure. Mica walls of large diameter (2·5 to 5 cm) have been produced by Zacharov and Dowden (1960), which are capable of withstanding pressures of several atmospheres without fracture. A special arrangement enables us to apply the flexible photographic emulsion to the outer face of the window. According to Zacharov, the final yield of the system is not far from that of an image-intensifier, but the resolution is clearly worse.

Tubes of a slightly different design are used in astronomy (Hall, Ford and Baum, 1960), which contain an even thinner membrane (500 to 1000 Å), whose function is to separate the conversion region from a second less highly evacuated compartment in which the emulsion is situated. When the tube is in use, the emulsion and the membrane are in contact. The electrons are accelerated and pass through the membrane to make an impression on the photographic plate; the resolution is much better than in the preceding case, as the scattering produced by the layer of aluminium is very small.

Tubes of the same type have been built by Hiltner and Pesch (1960), who use thin foils of aluminium or silicon monoxide evaporated on a layer of plastic. This plastic is destroyed by heating to 350°C *in vacuo*, and the life of the cathode is on the average several weeks. A considerable improvement is obtained if aluminium oxide windows, which are robust and leak-tight, are employed.

14.3 PRINCIPAL CHARACTERISTICS

14.3.1 Sensitivity

The sensitivity of the whole instrument can be defined in terms of the minimum power emitted by a test object P_m which produces a perceptible image on the screen (or the photographic plate). A visible image is produced as soon as the eye can discern a difference of brightness or darkness between the object and the continuous background illumination, B_0, which is produced by parasitic currents. A photometric measurement of the luminous flux which arrives at the layer, and in terms of which P_m is defined, is made;

the sensitivity of the system is dependent, above all, upon the sensitivity of the photoelectric layer, and the efficiency of the fluorescent screen or photographic plate.

(i) *The sensitive layer*

The most sensitive photoelectric layers for making observations in the infra-red consist of caesium on oxidized silver; the maximum sensitivity is somewhere about $0.8\,\mu$, but acceptable images have been obtained with radiation of a few microns. The sensitivity can reach 20 or 40 μA lumen^{-1} (or about 10^{-2} A watt^{-1} in the near infra-red), with a quantum efficiency of about 3 per cent; in the visible region, the most common choices are layers of caesium–antimony, which are easy to prepare. Their sensitivity can reach 50 μA lumen^{-1}, with a quantum efficiency of 10 per cent while their maximum sensitivity is around $\lambda = 4,200$ Å. For special applications, however, far more sensitive layers are employed.

A layer of SbCsO gives 60 μA lumen^{-1} towards 4,500 Å, with a quantum efficiency of 45 per cent; in the ultraviolet, the efficiency remains very good, even in the far ultra-violet.

Recently, a layer consisting of antimony and the three alkali metals, sodium, potassium and caesium, has been developed (Sommer, 1955). This layer is capable of giving 150 μA lumen^{-1}, and the efficiency may reach 20 per cent. Its main use is in certain types of television camera (Gibbons, 1960). Once again, the maximum sensitivity is around 4,200 Å.

The sensitivity of the instrument is limited by the parasitic electrons emitted by the layer (the "dark current") which produce an unwanted illumination at the screen and reduce the contrast. One part of this current is due to thermal emission, and obeys the law $I_1 = A\,T^2\,e^{-b/T}$ while the other is due to field emission, $I_2 = C\,E^2\,e^{-d/E}$ (in which E is the electrostatic field). I_1 can reach 10^{-12} A cm^{-2} at normal temperatures. Because of this possibility, the layer is cooled in liquid nitrogen and the field E is kept very small (at most 1000 V cm^{-1}) in order to keep the parasitic currents as low as possible. Another source of parasitic electrons is field emission at the electrodes or the insulating walls. A very thorough examination of this phenomenon, which also produces a continuous background on the fluorescent screen even when no illumination at all is present, has been made by Essig (1960). Several precautions may be taken to reduce it to the maximum:

Strong fields must never be produced at any point of the structure, and bombardment of the insulators by the parasitic electrons produced at the electrodes is to be avoided as far as possible, as this produces very high fields at certain points on the insulators, at which the emission is in consequence reinforced;

Contamination by caesium vapour is to be avoided, as we have already mentioned.

Finally, the tube should, if possible, be cooled in liquid nitrogen, a technique similar to the one used to reduce the background noise of a photomultiplier.

In a commercial tube working at about 10 kV, the darkness current at the screen can be kept down to 10^{-15} A cm^{-2}, which corresponds to the threshold sensitivity of the eye, but this current increases very rapidly to 5×10^{-13} A cm^{-2} at 20 kV (Essig, 1960).

By taking this phenomenon into account, we can explain the way in which the sensitivity of an image-converter varies with the accelerating potential; the minimum luminous flux at which the image is just detectable falls at first as the potential is increased, passes through a minimum and then begins to increase. This increase is due to the reinforcement of the continuous background by the increasing darkness current.

(ii) *The sensitivity of the receiver*

The brightness, A, at a screen is proportional to the current density, j, which is incident upon it; if V_0 is the threshold potential, $B = B_0 j (V - V_0)^\alpha$ with $\alpha \simeq 2$ for aluminized screens near $V = 20$ kV. Precise photometric measurements have shown that the brightness is strictly proportional to the illumination of the cathode between 5 and 50 kV for reasonable values of the illumination. Equally, it depends upon the thickness of the screen, and passes through a sharp maximum when the thickness is equal to the depth to which the electrons penetrate (cf. § 12.2.6). It seems advantageous, therefore, to increase V to the maximum, but this, as we shall see, reduces the resolving power; an acceptable compromise must be made, taking into account the increasing dark current which we have already mentioned.

When the instrument records directly on a photographic emulsion, we work in the zone of maximum sensitivity of the plate, which corresponds to a potential V of 30 or 40 kV for the best emulsions. The sensitivity of modern emulsions has been examined both at normal temperatures and at the temperature of liquid nitrogen by Duchesne (1953 b). We are faced with two very different problems; either we try to obtain the image of a star with a considerably shorter exposure than would be needed for direct photography, or else we should like to record on the plate as much information as possible about those regions of the sky of which the brightness is very feeble, and of which we are largely ignorant. In the first case, the research is directed towards finding very fast emulsions, which have at the same time a sufficiently fine grain for the resolving power not to be diminished—at present, Crumière Spécial D or Ilford G 5 give the best results. The exposure can be reduced by 50 or 100 times. Figure 174:1 is a photograph of Saturn, taken with an exposure time of half a second.

In the second case, we try to reduce to the minimum the continuous background of the plate so that every possible useful detail appears, even if it is very poorly contrasted. If very fine grain nuclear plates are used

FIG. 174:1. Photograph of Saturn, taken with an "electron telescope" on an Ilford K 2 plate with an exposure of $^1/_2$ sec (Photo: Wlérick, Rösch and Dupré).

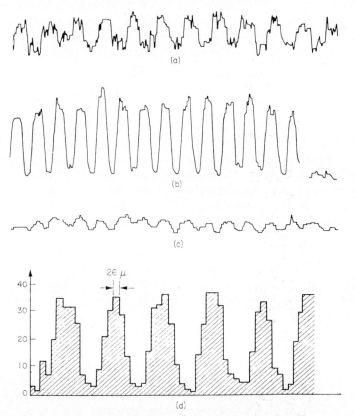

FIG. 174:2. The recording of a very low luminous flux by counting the traces of the photoelectrons. (a) Microphotometry of a grating, obtained by ordinary photography on a Kodak Rochester 103–0 plate; the period of the grating is 200μ, the exposure time, 100 sec. (b) Microphotometry of the same grating, obtained by electron photography on an Ilford G5 plate; electron magnification: 98; exposure: 1 sec. (c) Electron photography, as in (b), but with an exposure of $^1/_{10}$ sec; the grating is still distinguishable against the continuous background. (d) Detection by counting the traces corresponding to (c). The ordinate of each horizontal line is equal to the number of traces within a rectangle of $26 \times 170 \, \mu^2$ (its longer sides parallel to the lines of the grating).

(Ilford G 5 or Kodak NT 2A, for example), each electron gives a perfectly recognizable trace provided it has been sufficiently accelerated. If the number of photo-electrons is not too high, it is then possible to detect the image which is formed by an extremely small light flux, by counting the number of electron traces. This method is limited by the "veiling" (produced by silver grains which are developed even outside the traces) and by the electron "background noise" (which is produced by parasitic electrons emitted by the photocathode). By using recently prepared emulsions and short exposure-times to limit these disturbing effects, the method proves to be most effective: Lallemand, Duchesne *et al.* (1959) have found a gain of 10^4 over the direct recording of the photon flux on the plate, as Fig. 174:2 displays. (In each case, the most sensitive type of plate has been used, either to photons or to electrons as the case may be.)

14.3.2 Resolving Power

The resolving power is restricted by a number of factors:
the defects of the optical system;
the grain size of the screen (or plate);
the reduction of contrast which can occur when the phosphor is saturated by high local brilliance; with mobile objects, the contrast may be reduced by the "remanence" of the screen.
We shall examine only the first two of these possible factors.

(i) *The resolving power of the optical system*

The resolving power in the neighbourhood of the axis is calculated from the formula $d = k \dfrac{\Phi_e}{E_0}$ where $k \simeq 1$, and d is the diameter of the aberration disc, just as in all emission systems (cf. § 13.2.2). For a caesium cathode, used in the infra-red, we can take Φ_e to be 0·5 V; for $E_0 = 500$ V cm^{-1}, we should have $d = 10\ \mu$. Off the axis, d can be as large as 100 or 150 microns even when a curved cathode is used, as there is an effect due to residual field curvature and to the aberrations of the rays which are far from the axis. The average quality of the image can be improved by choosing the focus to be a certain distance from the centre (Fig. 175a). d can be reduced by increasing E, but we have to make a compromise between the mean resolving power which is wanted and the "darkness current" of the layer.

In converters with a homogeneous magnetic field, the resolving power is given by a formula of the type

$$\delta = k \frac{\Phi_0}{E} = k \frac{\Phi_0}{\Phi} L,$$

in which L is the length of the tube and Φ is the accelerating potential; the values of k quoted by different authors lie between 1·6 and 2, if we select for Φ_0 the potential which corresponds to photoelectrons of the maximum energy (for an Sb–Cs cathode, this is 2·3 V). The measured resolving power is some 10 μ for a potential of 5 kV (Wendt, 1955), over the whole of the screen.

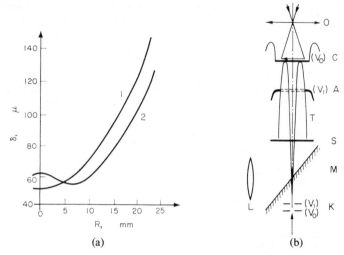

FIG. 175. (a) Resolving power δ, as a function of the radial distance R from the axis. (1) image focused on the axis; (2) focus at 7 mm from the centre.

(b) Semi-conductive mirror converter (Schaffernicht).

C: semi-conductive layer.   A: anode.
S: transparent fluorescent screen.   M: mirror.
K: electron source.   L: magnifying glass.

(ii) *The resolving power of the receiver*

Even in the most favourable cases, the grain of a screen obtained by sedimentation is still about one or two microns in diameter, so that the resolving power of screens of this kind cannot be less than 5 or 10 μ; the thickness, however, is equally relevant—it has been estimated that the resolving power is effectively equal to the depth to which the electrons penetrate into the layer, and hence to the thickness of the screen under the conditions of maximum brightness. At 20 kV, this depth is some 10 μ, while at 80 kV it is 100 μ. The resolving powers which are found at the screens of the infra-red converters described earlier are about 50 μ at the centre. This result can be improved by a factor of 3 or 4 by using screens of evaporated zinc oxide (Studer, Cusano and Young, 1951) and with these the resolution can reach 15 μ (the distances are to be understood as having been projected back into the object plane of the layer, taking into account the fact that the magnification is less than unity). It is clear, then, that at the centre it is the grain size of the screen which governs the resolving power of the instrument, while at the edges the latter is limited by the aberrations. In laboratory instruments, however, a curved screen can be provided, so that the resolving power is virtually constant over a region of about 30 mm diameter.

In recent instruments, where the electrons record directly on a photographic plate after passing through a thin window, the resolution is worse than that of the optical system as we must take the scattering produced by the window into account; a value of 30 Å can be reached, however, which is adequate in many applications.

If the image which is recorded by the emulsion is provided by the photons emitted by a screen placed against the thin window, the resolution is of the same order of magnitude (30 μ on the axis).

In the "electron telescope", where all these perturbing influences are removed, the resolving power is the same as that of the optical system itself, about 10 μ at the centre.

14.4 THE DEVELOPMENT AND THE FUTURE OF THESE INSTRUMENTS

14.4.1 Improvement of the Sensitivity

(i) *Converters using secondary emission*

A new type of image-intensifier tube was proposed by Sternglass in 1953, in which the secondary emission phenomenon was employed in cascade. The tube is an image-converter tube with homogeneous electric and magnetic fields (see Fig. 169), in which thin, planar, regularly spaced dynodes are placed perpendicular to the axis. Each dynode consists of a very thin metal foil (facing the incident electrons) on which a layer of an insulator with a very high secondary emission coefficient is deposited. The photoelectrons are accelerated towards the first dynode, and on passing through the metal and the insulator, produce secondary electrons which are in turn accelerated towards the second dynode; the same cycle recurs in cascade, with an amplification of the current at each stage. Finally, the electrons which emerge from the last dynode fall onto a fluorescent screen.

Considerably improved versions of this type of instrument have been described by Wachtel, Doughty and Anderson (1960), who have developed a technique for the very difficult task of making the dynodes; it has been possible to obtain secondary emission coefficients of 5 with films of potassium chloride, 500 Å thick, deposited on aluminium, and an accelerating potential of 4 or 5 kV. Layers of barium fluoride deposited on aluminium give the same yield, and have a longer life. With tubes of this kind, we can hope to obtain a resolution of 40 or 50 μ, with six dynodes 15 mm apart,

13   EO

a magnetic field of 400 gauss, inter-dynode potentials of 2·5 or 3 kV, and a
potential of 10 kV between the last dynode and the screen. In this way,
electron gains of 1500 have been obtained; gains in brilliance of 10^4 have
been measured (Anderson, 1960), using a layer with a sensitivity of
15 μA lumen^{-1}, an inter-dynode potential of 4 kV, and a potential drop
of 15 kV between the last dynode and the fluorescent screen (zinc sulphide
activated with silver). The limiting resolution is about 40 μ, inside a circle
25 mm in diameter.

Gains of about 10^5 have been measured in blue light (with a photocathode and a screen for which the spectral response is virtually identical)
by Wilcock, Emberson and Weekley (1960), using a tube containing five
dynodes, supporting a total potential drop of 35 kV between the cathode
and the screen. Once again, the resolution is about 40 μ.

(ii) *Converters in cascade*

An alternative suggestion for increasing the gain is to couple several
converters together in cascade. The best solution, due to Schaffernicht
(1948) consists of setting the screen of the first converter close to the photosensitive layer of the second, to reduce the loss of resolution as much as
possible. Practical versions of two- and three-stage tubes have been described by Stoudenheimer (1960). Each stage is a converter with a purely
electrostatic electron optical system (see Fig. 175b) and unit magnification.

In the two-stage tubes made by RCA, the first photocathode is a layer
of Sb–Na–K–Cs and the second, of Sb–Cs. The maximum sensitivity of
both of these two layers is in the blue region, around 5000 Å and their
characteristics satisfactorily match those of a screen of zinc sulphide activated with silver. The two stages are separated by a "sandwich", which
consists of a thin glass sheet (about 0·1 mm thick) on one side of which is the
screen and on the other, the Sb–Cs cathode. Although less sensitive, this
type of cathode is preferred to the "three-alkali" cathode, being easier to
prepare. More recently, three-stage tubes of a similar design have been
produced (Stoudenheimer, Moor and Palmer, 1960) with a potential of
10 kV at the first stage and 20 kV on each of the two succeeding stages.
At the maximum sensitivity, which occurs at about 4200 Å, the radiant-flux
conversion gain—the ratio of the number of photons which emerge to the
number of photons which are incident upon the photocathode—is given by

$$G_{th} = S_1 \, S_2 \, S_3 \, V_1 \, V_2 \, V_3 \, E_1 \, E_2 \, E_3 \, K^2$$

in which S_1, S_2 and S_3 denote the sensitivities of the three layers respectively,
in amp watt^{-1}, $V_1 \, V_2$ and V_3 are the accelerating potentials, $E_1 \, E_2$ and E_3
are the efficiences of the three screens and K is an adaptation factor, with
the aid of which the relative sensitivities of the adjacent photocathodes and
screens can be taken into account.

With

$$\begin{cases} S_1 = S_2 = S_3 = 0.06 \text{ A watt}^{-1}, \\ V_1 = 10,000 \text{ V}, \quad V_2 = V_3 = 20,000 \text{ V}, \\ E_1 = E_2 = E_3 = 0.1, \\ K = 0.9, \end{cases}$$

we find $G_{th} = 7 \times 10^5$. In practice, gains of 10^5 have been obtained with some experimental tubes.

The resolution of the image-converter is of the order of 40 μ in the neighbourhood of the axis, and falls to 60 μ, 10 mm away from the axis. This decrease is due to the field curvature of the imaging field in the second stage. To counteract this, a photocathode curved in the opposite sense to the image which is formed upon it by the first optical system would have to be used, but this would worsen the aberrations of the first stage: we are forced to compromise; a further loss of resolution arises from the passage of the photons through the support of the intermediate screen.

(iii) *The channelled image intensifier*

For applications in which an enormous gain in brilliance is required rather than a very high resolution, special amplifiers have been developed in several laboratories, in which the optical system is wholly different from those described above. The cathode is divided into small units, and the electrons emitted by each unit are accelerated towards a fluorescent screen along channels formed by the successive dynodes which amplify the current by secondary emission exactly as in the photomultiplier. The resolution is clearly fixed by the transverse dimensions of the individual channels, and depends neither upon the number of stages nor upon the velocity distribution of the electrons which emerge from the cathode. The dynodes, which are held at progressively higher potentials, are placed one after the other in such a way that all the secondary electrons which are emitted by

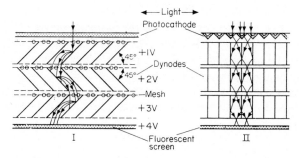

FIG. 176 a. Channelled image-intensifier (McGee). (I) The "venetian blind" type. (II) The model with connecting cylindrical tubes.

any one dynode subsequently fall onto its successor. One possible structure, which has been proposed by McGee and his collaborators (1960), is shown in Fig.176a, and another due to Burns and Neumann (1960a, b) in Fig.176b; some typical trajectories have been included.

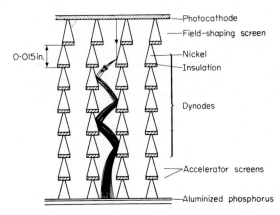

FIG. 176 b. Another type of channelled image-intensifier with successive dynodes (Wachtel, Doughty and Anderson). Some typical electron trajectories are shown. The period of the structure is $^{15}/_{1000}$ of an inch, which explains the considerable technological difficulties encountered in the construction.

The dynode surfaces are composed of evaporated layers of an Ag–Mg alloy, which has a secondary emission coefficient of 5 or 6 for an accelerating potential of 200 V. Work is at present in progress on large scale models, with the object of measuring the electron gain of the different channels. The hardest problem which has still to be solved seems to be the insulation between the stages; poor insulation produces micro-discharges, which result in an intense continuous background at the final screen.

(iv) *The use of feedback in the image-intensifier*

A great deal of nuclear research—in the domain of plastic fibre scintillators, in particular—requires high gain amplifiers. We should like to be able to detect the passage of ionizing particles which produce only a very few photons in their passage through a unit of the counter. If we define the quantum efficiency, R, of a converter as the ratio of the number of photons emitted to the number of photons incident, present-day single-stage tubes can provide

$$20 < R < 50,$$

whereas the amplification which is required here is of the order of

$$10^5 < R < 10^8.$$

Some authors have suggested that the image formed on the screen should be fed back to the cathode by an appropriate glass optical system in such a way that this image coincides with the original image. The system must use pulsed operation, so that the final gain in the brilliance can be controlled. A discussion of this problem and of the various different designs which have been proposed is to be found in an article by Perl and Jones (1960).

(v) *The use of "memory" tubes*

The optical image may be formed upon a photosensitive cathode, which produces an electron image on a layer of insulator with a high secondary emission coefficient; alternatively, it may be formed, either directly or after amplification by an image-converter provided with a fluorescent screen, on a photoconducting layer. In the first case, the electrons, and in the second, the photons, lead to the appearance of a "potential image" at the surface of the receptive layers. This image is then scanned by a beam of slow electrons, which erases the differences of potential across the layer; an electrical signal which is proportional to the potential difference at any point is obtained on a special electrode which is in contact with one face of the layer; this signal is subsequently transmitted to an ordinary television tube, whose scanning is synchronized with that of the beam which "reads" the image. The beam in the television tube is modulated by the signal, and on the screen of this tube we obtain the final image, improved, however, by a marked increase in brilliance.

This class of image-converters closely resembles tubes of the type which are used as television cameras, and further, allows us to benefit from the long integration time of the sensitive receptive layer, before actually reading the potential image (Beurle and Slark, 1960; Heimann, 1960).

14.4.2 Application to Microscopy

In ordinary light microscopy, converters can be used to examine the infra-red and ultraviolet. In electron microscopy, the resolution in the image plane is often no better than $50\,\mu$, for a magnification of 50,000 (which corresponds to 100 Å at the object) and amplifiers can be useful to reinforce the image so that beams of very weak intensity can be used. It is in ion microscopy, where the lack of brightness at the screen is the main difficulty, that an image amplifier would be most valuable.

One solution would be to replace the screen of the microscope with a semi-transparent fluorescent screen painted onto the glass wall of the converter, on the other side of which would be the light-sensitive layer. This layer must be kept sealed off in its own envelope, as the vacuum in the microscope is inadequate to ensure that it has a reasonable lifetime. With such an arrangement, however, the resolution of the whole device will be considerably reduced by the diffusion of light in the glass wall of the converter. The thickness of the glass between the layers cannot be reduced below a few tenths of a millimetre, as it has to be capable of supporting atmospheric pressure; on the light-sensitive layer, it will be an emissive spot, a few tenths of a millimetre in diameter, which corresponds to a point on the electron or ion image.

Three ways of reinforcing the image without loss of resolution have been tried.

The electron (or ion) image which is obtained on the screen of the microscope is recorded with a very sensitive television camera, which transmits it, after suitable amplification, to the screen of a television receiver. With the aid of special circuitry, the contrast can be considerably reinforced by this procedure, and any troublesome continuous background can be suppressed.

An image-converter using secondary emission and deprived of its photosensitive layer can be employed. This procedure was suggested by Möllenstedt (1953) and Le Poole (1954), and has recently been tried out experimentally by Castaing, Jouffrey and Slodzian (1960), as a means of reinforcing ion images.

The principle is as follows—after passing through the objective L_1 at 40 kV, the ion beam is slowed down to 10 kV, and the first low-magnification image ($G_1 \simeq 100$) is formed on a perfectly polished metal plate (stainless steel, for example), which acts as the cathode of an immersion objective. The secondary electrons which are emitted are accelerated in the opposite direction and are focused by the immersion objective on a fluorescent screen (see Fig. 177), through which the primary ions have been able to pass by means of a small hole pierced close to the focus of L_1; the final magnification can reach $G_2 \simeq 10,000$. If we suppose the secondary emission coefficient equal to unity, the gain in brightness between a direct ion image

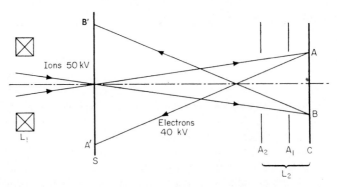

FIG. 177. Diagram of an image-intensifier for ion microscopy.
L_1: electrostatic objective. AB: ion image.
L_2: immersion objective. $A'B'$: secondary electron image.
C: polished metal plate. S: screen.

with $G = 10{,}000$, and the image obtained in this way with the same magnification is very much greater than unity, as the plates are considerably
more sensitive to electrons than they are to ions. The first attempts in this
direction have led to gains of the order of 100, with an ion microscope
working with Li_1^+. Further, the cathode of the immersion objective can be
made from a metal which has a very high secondary emission coefficient
when bombarded with ions.

Finally, a new device for intensifying the image has recently been built
by Haine, Ennos and Einstein (1960). This device makes use of the conductivity which is induced in a thin film of selenium by fast electrons (50 kV).

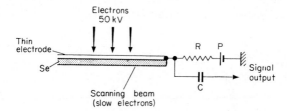

FIG. 178. Image-converter with a semi-conductor element, for electron microscopy
(Haine).

Figure 178 shows the principle: an amorphous layer of selenium is deposited
onto a metallized plastic layer which acts as signal electrode; the latter
is connected both to a source of polarization and to an electronic amplifier
which is in turn connected to the modulation electrode of an ordinary
television tube. The local conductivity of the selenium layer varies with
the density of the incident beam, although only in the direction normal
to the layer. These variations are "read" by a second electron beam, this
time of very slow electrons, which scans the lower surface of the selenium
layer. In the absence of any "illumination", the potential of the lower
surface is uniform and equal to the potential of the cathode of the "reading"
gun, by virtue of the phenomenon of secondary emission with a coefficient d
less than unity. Between two sweeps of the beam, when the layer is "lit" by
the 50 kV electrons, the charges on the lower face flow towards the polarized
electrode, and create local variations of the surface potential. The "reading"
beam provides charges which restore a uniform potential, which results
in a variation of the equivalent potential at the signal electrode, an effect
due to the capacitance of the selenium layer. This variation, after suitable
amplification, is used to modulate the intensity of the electron gun of a
television tube.

The resolution in the plane of the electron image is about $80 \, \mu$, but this
can be improved to $30 \, \mu$, which is in practice adequate for high magnification (for $G = 10^5$, we should have a resolution of 3 Å at the object).

Electron densities of 10^{-12} A cm^{-2} at 50 kV, for which the image would be barely perceptible even on a very good fluorescent screen, produce acceptable images on a television tube with this procedure.

14.5 AMPLIFICATION OF THE BRILLIANCE OF
RADIOSCOPIC IMAGES

In one recent design, the usual arrangement is used, but the screen which is fluorescent to X-rays is placed within an evacuated envelope on a very thin wafer of glass on the opposite side of which is the light sensitive layer. In another instrument (Eckart, 1954), an insulating layer only a few tenths of a micron thick separates the screen from the layer.

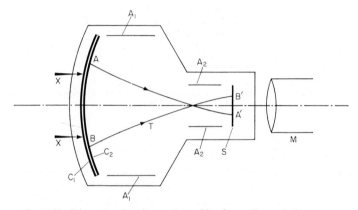

FIG. 179. Diagram of an image-intensifier for radioscopic images.
A_1, A_2: electrodes.
S : fluorescent screen.
C_1 : fluorescent layer, sensitive to X-rays.
T : electron trajectory.
M : microscope for examining the image.

In connection with this, we must also mention the design of Teves and Tol (1952; Fig. 179), which has the following characteristics—the operating potential is 25 kV, the electron magnification is 1/9, and the gain in brilliance is about 1000. A small microscope is attached to the screen and provides a final magnification of unity. The surface examined, however, is some 25 cm^2 in area, and only technological problems at present prevent this area from being doubled or tripled.

The compactness of this instrument allows us to use it in conjunction with a television camera to retransmit radioscopic images which are formed in a region which is dangerous for an observer (physico-chemical treatment of radio-active bodies, for example, can be examined from a distance). Tubes of this type are now produced commercially by several firms.

A slightly different model has been constructed by Timofeev and Sorokina (1960). Inside a vacuum chamber of the type illustrated in Fig. 179, the convex portion is covered with a thin film of aluminium, transparent to X-rays, on which the fluorescent material is deposited (a few tenths of a millimetre to obtain the maximum brightness).

The caesium–antimony photosensitive layer is evaporated directly on this material, facing the anode. The grid is connected to the cathode; the anode is a tube with an entry stop, behind which the fluorescent screen is placed. The shape of the electrodes is such that the potential distribution $\Phi(r, z)$ is of the form

$$\Phi(r, z) = \Phi_0(2z^2 - r^2 + 2pz).$$

In the useful region, the equipotential surfaces are hyperboloids of revolution; the trajectories perform one semi-oscillation in a radial plane between the cathode and the final screen. The magnification depends upon the model, and is either 1/6 or 1/9. The resolution is limited above all by the thickness of the fluorescent material at which the primary image produced by the X-rays is formed; in the cathode plane, the resolution is of the order of 0·1 or 0·2 mm. The increase of brightness between the primary image and the final image produced by the electrons is of the order of 1000 for $\Phi = 25$ kV and $G = 1/6$. In the model for which $G = 1/9$, gains of 3000 have been attained. In some models, the useful area of this cathode is 10 cm in diameter.

CHAPTER 15

THE ELECTRON MICROSCOPE; STRUCTURE AND SPECIAL FEATURES

15.1 THE NOTIONS OF RESOLVING POWER AND USEFUL MAGNIFICATION

The electron microscope provides a highly magnified image of a small object which is being irradiated with electrons. The importance of this type of microscope lies in the fact that by using electrons, the effective magnification G can be 100,000 times or more, whereas with light, it is only with difficulty that G can be increased beyond 1500 times. Simple though this statement is, it is easy neither to analyse this fundamental property in detail, nor to give a precise definition of it, as we shall see in Chapter 16.

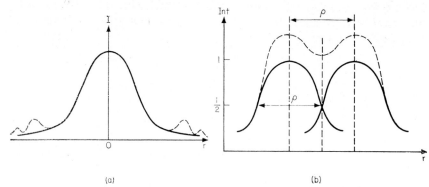

(a)   (b)

FIG. 180. (a) The form of the curve which gives the intensity at the image spot as a function of the distance from the centre. (b) The limiting position for the images of two points which can be "separated". The resultant intensity is represented by the dotted curve.

In preparation for this final goal, we shall tabulate the various factors which affect the magnification. In this way, we shall obtain criteria from which we can judge the essential quality of the instrument, understand the stages in the historical evolution of its construction, and compare the different variants which exist at present.

Like all optical instruments, the microscope has intrinsic imperfections. The image of a point is not a point but a small spot in which the brightness decreases outwards from the centre (Fig. 180a) according to a law which depends on the operating conditions. When diffraction phenomena preponderate, a set of small concentric rings is produced about the central maximum, with progressively decreasing intensities. As it is only the central peak which is of interest, however, the distribution of (luminous or electron) intensity can in general be represented to a reasonable approximation by a bell-shaped curve; often, the Gaussian error function is selected to describe this curve analytically, when a definite model is required for a calculation.

The limit of resolution, or simply "resolution", corresponds to the minimum distance between two maxima of the light distribution (or image points), at which the two can still just be separated. This distance is always referred back to object space, and is, therefore, the quotient of the real distance observed at the image and the magnification of the latter. The expression "resolving power" (*pouvoir séparateur*) which is frequently used for this minimum distance is not a particularly fortunate choice, as the resolving power increases when this minimum distance decreases. Let us consider the images of two bright points very close together (Fig. 180b) for which there is no possibility of separating the respective luminous intensity curve. The curve which represents the resultant intensity at each point has a slight depression which becomes less and less perceptible as the two points are brought closer and closer together. We decide *arbitrarily* to call two images "resolved" (or separately discernible) if the distance between the two maxima is equal to the width at the point at which the two bell-shaped curves have fallen to half their maximum height (the width, that is, above which the intensity is more than half the maximum intensity). This clearly does not exclude the possibility that the minimum might still not be obscured when the two curves are slightly closer together. The relative depth of the minimum does, however, decrease very rapidly as the two images approach each other, and in practice it is very difficult to resolve two images which are separated by a distance which is less than their half-width. This, therefore, is an excellent convention—it is almost equivalent to the more well-known definition which is in general the one which is adopted in optics when only diffraction is present (cf. Bruhat and Kastler, *Optique*, Masson, Paris, pp. 233 and 239).

The image formed by an (optical or electron) microscope has a certain *direct magnification*. Whether the image can or cannot be further magnified depends upon its quality; this additional magnification would be effected with a photographic enlarger, and would give the image a *useful magnification* which is determined by the resolving powers both of the instrument and of the human eye—this magnification has to be adequate to magnify the smallest details in the object to a size at which the eye can discern them.

If ϱ_{eye} and ϱ are the resolutions of the eye and the microscope respectively, we have

$$G_u = \frac{\varrho_{eye}}{\varrho}$$

(ϱ_{eye} corresponds to an object placed at the least distance for distinct vision, about 20 cm from the eye). There are many cases, however, in which it is convenient to work with double the magnification which we have just defined. This magnification is adopted to avoid straining the eye by using it continually at the limit of its capability—in this way, the final image can be scrutinized without fatigue.

In the majority of research with the microscope, the image is first of all directly recorded in the apparatus, and this first negative is then magnified by the usual photographic procedure which produces the final positive print; it is this overall final magnification which has to be at least equal to G_u. The limit of resolution with the naked eye is $\varrho_{eye} \simeq 0\cdot1$ mm, which gives the following useful magnifications for various types of microscope:

Type of microscope	ϱ	G_u
Optical	$0\cdot2\,\mu$	500
Ultra-violet	$0\cdot12\,\mu$	800
Electrostatic	$1\ \text{m}\mu = 10\ \text{Å}$	100,000
Magnetic	$0\cdot5\ \text{m}\mu = 5\ \text{Å}$	200,000

Using a magnification higher than G_u would not reveal any further detail; two points which are just resolved with a magnification of 100,000 will simply be further apart if the magnification is increased, but no fresh details will appear between them. The image will be larger but more blurred.

15.2 A COMPARISON OF THE ELECTRON AND
OPTICAL MICROSCOPES

The use of glass objectives with wider and wider apertures, which are continually being more satisfactorily corrected, might suggest that the increase of the magnification would lead to an unlimited increase in the resolving power of instruments of this type. It is soon clear, however, that however perfect the instrument may be as far as geometrical optics is concerned, it is never capable of forming distinct images of two points on the object which are very close together; biological specimens are rich enough in details for this to be verified by simple observation.

It was Abbe who, more than eighty years ago, gave the first explanation of the origin of this limit, and provided a quantitative theory which was soon to be confirmed independently by the original work of Lord Rayleigh. He showed that the limit of resolution of the microscopes in use at the time was restricted not by the degree of optical or mechanical perfection of the instrument but by the phenomenon of diffraction which is inherent in the nature of light, and which is due to its wave-like character; the wavelength of the light determines the order of magnitude of the smallest detail which will be visible. No objective, however perfect it may be from the point of view of geometrical optics, can resolve details which are separated by less than about half the wavelength of the radiation in the medium surrounding the object. The results, therefore, ought to be improved in proportion as the frequency of the radiation is higher and the refractive index of the medium surrounding the object is larger.

For visible light, the wavelength λ of the colour to which the eye is most sensitive, yellow, is about $0.5\,\mu$. The liquids in which the object is immersed have a refractive index which does not exceed 1.6. This shows that the limiting resolving power of the optical microscope is in the region of $0.2\,\mu$, as we have already stated. This value had been attained before the end of the nineteenth century, and the only improvement which has been made since, before the advent of electron microscopy, is the use of ultraviolet light and the optical properties of quartz, which reduces the wavelength λ to about $0.25\,\mu$ and results in a resolving power of about $0.12\,\mu$. It is not possible to use electromagnetic waves of shorter wavelength, as no refringent media exist with which the rays could be suitably refracted, and the latter would moreover be difficult to produce. Kirkpatrick and Baez (1950) have tried to use X-rays, and the resulting technique is interesting, but no real microscopes have as yet been built successfully, and magnifications exceeding 100 have not been attained.

The work of L. de Broglie has established that a wave is associated with every moving particle, and that both the frequency and the wavelength which characterize this wave depend upon the energy of the particle. The wavelength λ which he attributed to the waves associated with the particle is given by

$$\lambda = \frac{h}{p}, \qquad p = \frac{m\,v}{\sqrt{1 - \left(\dfrac{v}{c}\right)^2}}.$$

If the velocity is low, $(v/c)^2$ can be neglected in comparison with unity, and we obtain

$$\lambda = \frac{h}{m\,v}.$$

If the particles are electrons with charge e, accelerated through a potential drop V, we have $e\,V = \frac{1}{2}\,m\,v^2$, so that

$$\lambda\,(\text{Ångströms}) = \frac{12.25}{\sqrt{V\,(\text{Volts})}}, \qquad (\text{and } 10\,\text{Å} = 1\,\text{m}\mu).$$

In Fig. 181, the variation of the wavelength of the waves associated with electrons and with various ions is shown as a function of the accelerating potential. Electrons accelerated through a potential of 60,000 V, for example, have a wavelength of 0·05 Å, which is 100,000 times shorter than the wavelength of blue light. Electron radiation is immediately advantageous in microscopy, therefore, as its wavelength is very short and the diffraction phenomena which are proportional to λ are extremely small. A very high resolving power is to be expected with an electron instrument.

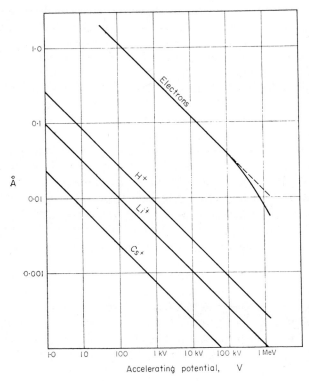

FIG. 181. The wavelength of the associated wave, plotted as a function of the accelerating potential for various ions and for electrons.

At first sight, an electron microscope ought to be much more "magnifying" than a light microscope. If this advantage is to be achieved, we must first find a way of building electron lenses as perfect as the corresponding glass lenses as far as the correction of aberrations is concerned; this, unfortunately, is an extremely difficult problem to solve (cf. Chapter 7), and at present, the geometrical qualities of glass lenses have not been equalled with electron optical lenses. We are reduced, therefore, to using beams with very small numerical apertures in the electron microscope, of the order of 10^{-3} (instead of unity, which is often attained in optics). This small aperture produces, amongst other defects, diffraction effects which vary inversely with the aperture. The possibilities of improving the resolving power which are implicit in the shortness of the wavelength of the waves which are associated with electrons are far from being fully realized. The quality of electron optical images is only 100 times better than that of the best light image, as a result of the defects of electron lenses.

In the laboratory, under ideal conditions, a resolving power of 6 or 8 Å has recently been obtained, but only by choosing a particularly favourable type of object. This is a record, and normally, 15 or 20 Å is reckoned to be highly satisfactory. In the large majority of applications, however, a resolution of 50 or 100 Å is sufficient, which corresponds to a useful magnification between 20,000 and 40,000. Most of the microscopes available commercially are of high enough quality to enable the average experimental scientist to obtain photographs of this quality habitually.

15.3 THE GENERAL PROPERTIES OF THE INSTRUMENT

To each of the two types of electron lens—the electrostatic lens and the magnetic lens—corresponds a different type of microscope; these two types have retained, despite the similarity between their structures and their quality, a pronounced individuality. In France, the microscopes which have been constructed have, until recently when a magnetic microscope has appeared, all been electrostatic; in England and the United States, however, magnetic microscopes have been built almost exclusively. In Germany, on the other hand, the two versions have never ceased to be developed side by side since the inception of the subject, and their progress has been parallel. In conclusion, we should mention a Swiss microscope in which a combination of both kinds of lens is used, and point out the fact that Russia and Japan have mass-produced both types since the war.

15.3.1 The Analogy with the Structure of the Light Microscope

Although the details of the construction differ considerably from one instrument to another, it is quite clear that in all the variants the structure of the electron microscope is analogous to that of the light microscope; this is illustrated in Fig. 182, in which the paths of the rays and the positions of the images are shown in both types of instrument.

The "projection" light microscope comprises a condenser which concentrates the light from a powerful lamp onto the object to illuminate it, an objective which forms a highly magnified image of the object, and a projective lens which magnifies this first image a second time and projects it onto either a photographic plate or a screen.

The same arrangement is to be found in the electron microscope. The object is no longer illuminated by light, but by a beam of electrons which

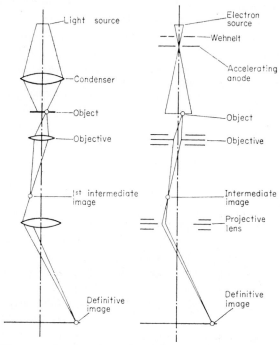

FIG. 182. Diagrams showing the points of comparison between an electron microscope and a projection glass-optical microscope.

emerges from a "gun", similar to the gun of a cathode ray oscillograph. This electron beam is directed onto the preparation, either directly or after passing through a magnetic or electrostatic condenser which plays the same role as the "second lens" in a cathode ray oscillograph. The electrons then pass through an objective, a projective lens, and finally form a magnified image either upon a screen for direct observation or upon a photographic plate or film. The only complication, which will be new to the person used to an optical microscope, is that the whole length of the instrument has to be evacuated (to about 10^{-4} to 10^{-5} mm Hg).

The focusing, which was for a long time obtained by moving the object about within the vacuum, is nowadays produced by modifying the focal length of the objective electrically.

15.3.2 Depth of Field

This similarity of structure does not prevent wide disparities from appearing between the characteristic magnitudes in the optical microscope and its electron optical junior; these stem from the fundamental difference between the numerical apertures of the two objectives, unity for a glass lens, 10^{-3} for an electron lens.

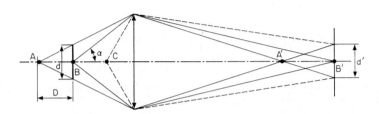

FIG. 183. The depth of field (to simplify the drawing, d and D have been drawn on a larger scale than the rest of the diagram).

We consider a lens (see Fig. 183) which forms the image of a point B on a screen at the point B'; another point A, distance D from B, will be imaged at a point A', and on the screen we shall see a blurred circle of diameter d'. We refer the length of this diameter back to the original object plane by considering the circle of diameter d' as the image of a circle of diameter d lying in the plane of B. We shall say that the image of A is "well focused" on the screen provided d' is less than the "width" of the diffraction disc which surrounds the image point B', or referring back to the object plane, $d \leq \varrho$ (in which ϱ is the resolution). As the distance D is small, together with the angle α in the case of the electron microscope, we find $\dfrac{d}{2} = \alpha D$. A glance at Fig. 183 shows that there is another point C, virtually symmetrical to A with respect to B, which is also "well focused" onto the screen; the *depth of field*, therefore, is given by $p = 2D$, and so

$$p = \frac{d}{\alpha} \simeq \frac{\varrho}{\alpha}.$$

In normal working conditions, $\alpha \simeq 10^{-3}$ and $\varrho = 30$ Å, we find that the depth of field is close to 3μ. If a poorer resolution is acceptable, however, the depth of field is considerably greater, and can be as high as a fraction of a millimetre when the resolution is the same as that of an optical microscope. This is an extremely large value, as can be seen by comparing it with the depth of field of an optical microscope which is in the neighbourhood of 0.2μ, and hence of the same order of magnitude as the limit of resolution;† in the electron microscope, the depth of field is 1000 times higher than the limit of resolution. This depth of field is enormous, also, if we compare it with the "field of view", the portion of the object which is seen at the image—for a direct magnification of 20,000, the latter is in the region of 5 μ^2.

It should be realized, however, that to be able to increase the depth of field is not always a particularly important advantage in electron microscopy. Indeed, the restricted depth of field of the optical microscope is often valuable, since for thick objects with very tiny details (which are often encountered in biology), different cross-sections of the object can be examined by varying the focal length; the image which is obtained, therefore, represents a section which is thinner than the original specimen, while the details of the other parallel sections only appear diffusely at the image. We can, therefore, build up the structure of the image in the third dimension by examining a number of different cross-sections. In electron microscopy, such a procedure is impossible; the whole object—which has of course to be very thin in order to be transparent to electrons—comes into focus simultaneously as soon as any particular detail is in focus. It is most difficult, especially with thick biological sections, to localize the relative positions of the various details in space.

The large depth of field of the electron microscope is, from another point of view, an advantage, since focusing, which would be an extremely critical operation if the aperture of the instrument were not so small, is made considerably easier; electron microscopists are already faced with enough difficulties in focusing correctly, which are due to numerous secondary causes which will be discussed later.

15.3.3 Stereographic Reconstruction

Another consequence of the large depth of field is the possibility of obtaining a "stereographic reconstruction" of the object—this entails taking two photographs of the same object after tilting the object plane 5 or 10 degrees each way from the axis. If these two images are examined through an ordinary stereoscope, the effect of stereoscopic relief is striking. Alternatively, the twin photographs can be projected onto a single screen using two projecting beams polarized at right angles, and the composite projected image examined with polarizing spectacles which filter off the component which is needed to reconstitute the relief for each eye.

---

† In a glass optical microscope, the depth of field can of course be increased by reducing the aperture α, but only at the expense of the resolution. Good resolution and outstanding depth of field are, therefore, incompatible properties. Generally speaking, a compromise must be made; an optical microscope is difficult to use at its optimum resolution, as various details of the object which lie in different planes are seen simultaneously, and they cannot all be clear at the same time. In practice, the best results are often obtained when the depth of field is between twice and five times the resolution.

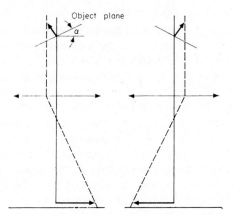

FIG. 184. The method of obtaining stereoscopic images by rotating the object plane between two successive photographs.

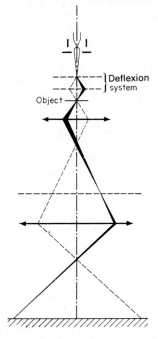

FIG. 185. A simple means of obtaining stereoscopic images. The illuminating beam is slanted between two successive photographs.

Two different methods are in use. In one of these, which provides excellent images, the object plane is differently orientated for each of the two photographs. For this, we can turn the object about an axis perpendicular to the beam (see Fig. 184) and it is the object-support of this kind, which has been described by von Ardenne (1940c) and Marton (1945), which is usually employed. The other method, which is simpler, was first recommended by Marton and Schiff (1941); the object is fixed, but in the neighbourhood of the object, the direction of the optic axis is varied. This can be done by deflecting the beam in such a way that it passes obliquely through the object (Fig. 185), with the aid of a magnetic or electrostatic field; the resolution which is obtained with this method is less high as the beam remains inclined with respect to the axis of the objective. As a result of the simplicity of its construction, however, Kinder (1946) has been able to obtain the two stereoscopic images simultaneously on the two halves of a fluorescent screen—the effect of relief can then be reconstructed in the microscope itself where it will be directly visible. So far, however, the total magnification is only 100 in this procedure.

Once the two stereoscopic photographs have been taken, the form of the object in space can be reconstituted with the aid of instruments similar to those which are used to determine altitude from stereoscopic aerial photographs (Eitel and Gotthardt, 1940, 1942; Heidenreich and Matheson, 1944; and Helmcke, 1954). The separation into layers can be determined to within about ± 50 Å; so far, the applications of this method have been rather limited as it can usually be replaced by the shadow method which is considerably easier to use.

15.3.4 Depth of Focus at the Image Plane

Suppose that the geometrical image A' of a point A (Fig. 186) is produced at a fluorescent screen or a photographic plate, the structure of which permits the actual resolution to be d (and in a good instrument, d is clearly smaller than the diameter of the diffuse spot which surrounds the image point A'). Displacing the screen along the axis through a distance x spreads the image out by a further amount, $2x\alpha'$. Provided that the spreading out which is produced by this adjustment is definitely less than d, a dis-

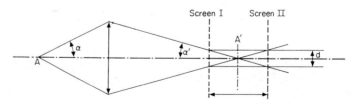

FIG. 186. The depth of focus in an electron microscope.

placement of this kind will have no effect on the resolution which is observed. The screen, therefore, can be moved between the two positions I and II without deterioration of the image clarity; the "depth of focus" l, the distance between I and II that is, is given by:

$$2 \frac{l}{2} \alpha' = \frac{l\alpha}{G} = d,$$

or

$$l = \frac{dG}{\alpha},$$

in which G is the magnification of the image. For $d \sim 10^{-3}$ cm (photographic plate), $\alpha \sim 10^{-4}$ and $G = 10^4$, we find that $l = 10^4$ cm; the depth of focus, therefore, is effectively infinite.

This explains why it is not at all important that the photographic plate should be situated in exactly the same plane as the screen, and why, on the contrary, it is often placed some distance from the screen upon which the image is focused. This fact can be used to simplify considerably the way in which one is replaced by the other, and in the old commercial Philips instrument, great advantage was taken of this possibility by placing the camera quite a long way in front of the screen, where the image which is photographed is one-quarter of the size of the image at the focal point (Le Poole, 1947; van Dorsten, 1948).

Another consequence of this great depth of focus is that the convergence of the projective lens can be varied over a wide range to alter the magnification, without the final image ceasing to be clear.

15.4 EXAMINING AND RECORDING THE IMAGE

15.4.1 The Screen

(i) *Characteristics*

Electron images are transformed into light images by a screen covered with some fluorescent substance. In the early stages of this subject, the photographic image was obtained by photographing the screen, but this method is no longer used except in very exceptional cases such as the recording of cinematographic images (Gauzit, 1954b; and Arnal, 1956). In all modern instruments, the electrons act directly on the photographic layer—this means that the process can be much faster, the resolution higher, and the gradations of contrast can be far wider.

A fluorescent screen, whether it be examined by reflexion or by transmission (Deubner and Hieber, 1954) is characterized by its luminous yield and by its own resolving power. The substances which are habitually employed are zinc sulphide and cadmium–zinc sulphide, activated by traces of various kinds of impurity in such a way that they fluoresce greenish-yellow, the colour to which the eye is most sensitive. More details about fluorescent screens are to be found in § 1.4, and in Chapter 12 (§ 12.2.5); here we shall simply give a few characteristics which are peculiar to microscope screens. The optimum thickness of the fluorescent layer has been studied systematically by von Borries (1948). Between 50 and 100 kV, he found that the best results were obtained with 44 mg cm^{-2} of fluorescent material (or a layer of about 100 μ thick). In general, the minimum intensity which must be incident on a good screen, observed directly, is about 5.10^{-11} A cm^{-2} between 40 and 80 kV. In these conditions, the eye perceives a sufficiently bright image for the observer to adjust the focusing precisely. In this operation, the resolution of the screen is an important quality. The resolution of the image which is seen on the screen depends both upon the size of the grains of the fluorescent material and upon the scattering of the beam within the layer; in particular, as the penetration increases, the resolution decreases as the potential is raised. The normal resolution for a screen is 70 or 100 μ (Hinderer, 1942). The grain size in the finished screen, which is composed of a fine powder, is very variable and depends upon the way in which the fluorescent layer is prepared; the method of doing this must be chosen rather carefully so that the finest particles lie on the surface.

(ii) *Construction*

One method of preparing the screen is by sedimentation (in alcohol, with silicate of soda as binding material; or in acetone, with 0·1 per cent of collodion); alternatively, a solution of cellulose acetate thoroughly mixed with the fluorescent material can be poured over a glass plate, and after the solvent has evaporated, a thin cuticle remains which is stripped from the glass surface and stuck on the metal support of the screen. The surface which had been in contact with the glass is the surface which is observed. Screens which are very brilliant because they contain almost no binding material (this is essential when the electrons are replaced by ions—see Chapter 18), and which have a high resolution, can be obtained by sprinkling the metal support of the screen with the fluorescent material, after first spreading a thin film of oil over the support. A glass plate is then pressed against the outer surface of the fluorescent layer, thus making it perfectly smooth.

When the electron optical system is to be used at low magnification, the image has to be examined with the aid of a high-magnification light microscope in order to focus the system correctly, but we are then faced with two difficulties. The grain-size of the screen limits the resolution of the images, and the depth to which the electrons penetrate the phosphor becomes important. The numerical aperture of the light microscope has to be reduced until its depth of focus is of the same order of magnitude as this penetration depth, and screens free of any structure must be employed. Attempts to produce the latter have been made in which either certain monocrystals are used (von Ardenne, 1939), or certain kinds of uranium glass, or special zinc oxide screens (Kopp and Möllenstedt, 1946, 1947); crystals, which are perfect as far as the absence of granular structure is concerned, have the disadvantage of being rather insensitive (a hundred or a thousand times less sensitive than ordinary screens).

A technique is available with which screens free of granular structure can be obtained in the form of a thin layer, by evaporation of zinc sulphide followed by a suitable annealing (Feldman and O'Hara, 1957). These thin screens, which are grain-free and are about as thick as the penetration depth of electrons accelerated to a few keV, can thus be examined at their brightest with an objective of large angular aperture; this has recently enabled microscope images to be examined at very low accelerating potentials (Nixon, 1958).

15.4.2 The Photographic Plate (see also § 1.5)

(i) *Plates and films*

The camera is to be found beneath the screen and is situated like the latter in the vacuum; the shutter is often simply the screen itself. Non-sensitized emulsions are used, mounted either on glass or on film, although the latter has the drawback of containing certain volatile substances which means that the pumping-time is rather long; emulsions spread on film, however, seem to be faster to electrons than the corresponding plates. Certain films have the additional disadvantage of being too good insulators—parasitic deformations are then seen in the image which is recorded, caused by charges which have been unable to flow away (Couchet, Gauzit, and Septier, 1951a and 1951b).

(ii) *Choice of exposure time*

The exposure time is usually estimated by the operator on the basis of his visual impression of the brilliance of the screen, but just as in ordinary photography, a photometer is a most useful accessory. An equal intensity photometer can be constructed with the aid of an auxiliary electric light bulb of which we can vary the brilliance; the light which is emitted is filtered by a coloured plate and produces a bright spot of the same intensity beside the image (van Dorsten, 1950). The current which falls on the screen can also be measured, in terms of the potential difference which it produces across a large resistance of the order of tens of megohms which is connected between the screen and earth (Hamm, 1951). A more accurate and robust instrument, which is, however, more expensive, consists of an exterior photomultiplier which is placed against the glass window (Cuckow, 1946; Bishop, 1950; Frei and Hirsfeld, 1951; Gauzit, 1954b; and Tristram, 1956).

Long exposures are to be avoided, as the probability that the image will be displaced during the exposure (by an accidental vibration, or by mechanical or electrical drift) is too high; very short exposures, too, are not advisable with the microscopes commercially available as the shutters are too coarse and produce vibrations which continue during part of the exposure. The best exposure time is about three or four seconds. Extremely good results can, however, be obtained with an exposure of a second or less provided that an "electrostatic shutter" is installed, which diverts the beam except during exposures, or alternatively, an electromagnet which acts on a very light shutter under the projective.

(iii) *Sensitometry*

Photographic emulsions are characterized by their sensitivity, their contrast and their resolution. Their behaviour in the presence of electrons is, generally speaking, analogous to their behaviour in the presence of light; in the electron case, the properties of emulsions vary in the same sense but over a considerably more reduced domain than in the optical situation. The photographic plate is extremely sensitive to the arrival of electrons (von Borries, 1942, 1943, 1944); whereas between 10 and 100 incident photons are necessary before a grain of bromide can be developed, a single electron can render one or even several grains ready to be developed. The energy yield (which is defined as the number of watt cm^{-2} which provide a given intensity) is, however, about twenty times weaker for electrons than for light. This explains the failure of the reinforcing fluorescent screen technique, which had been intended to increase the sensitivity of the photographic emulsion to electrons (von Borries, 1942; Baker, Ramberg and Hillier, 1942, 1943). For the same reason, neither a colouring material to increase the sensitivity nor a material which reduces the sensitivity has any effect on the speed of the emulsion (Nissen, 1944).

When the high tension is increased, the sensitivity of a solution is reduced; the optimum operating potential is in the neighbourhood of 50 kV. Recent attempts at very low tension microscopy (see § 19.2.8) have reawakened research into this question, however, with the object of finding an emulsion on the commercial market which still has a good sensitivity at 10 kV. Van Dorsten and Premsela (1960) recommend the Gevaert 9 E 56 films which only require an increase in current density of 1·6 times to obtain the same degree of blackening when we pass from 52 to 13 kV, whereas the films which are in everyday use in microscopy would require the intensity to be increased by ten to a hundred times.

The forms of the curves which describe the photographic density of photographic emulsions in the presence of electrons resemble those which are produced by light; the base, however, is relatively longer, and has a weaker curvature, so that the straight line portion is only reached at the density 1·5 (whereas for light the value is 0·5). From Fig. 187, we can see that the rectilinear part of the photographic density curve corresponds to charge densities between 10^{-11} and 5×10^{-9} C cm^{-2}.

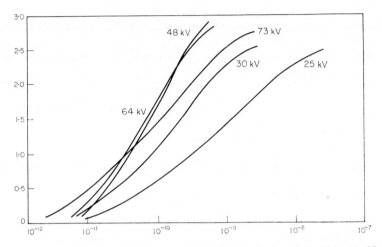

Fig. 187. Curves showing the way in which the blackening of the photographic plate varies with the incident charge density (in C cm^{-2}).

Just as in the case of light, slower plates have a better resolution and a higher contrast, or what is equivalent, a more vertical rectilinear portion. Since the "shadow technique" has been introduced, however, the contrast seems to be adequate, and the present tendency is to protect the object from being heated by reducing the illumination and using faster and faster emulsions which are, of course, less contrasty.

The resolving power, which varies widely between one type of emulsion and another, is difficult to measure. Von Ardenne (1939) found that the resolution of the best ordinary emulsion was 10 μ; von Borries (1944) after a precise and extensive examination, concluded that the result should be about 30 μ, and this estimate is generally accepted to-day. Certain special emulsions have, of course, a much finer grain; for example, Kopp and Möllenstedt (1946) considered certain plates of the Schumann type, and found a resolving power of one or two microns, but the sensitivity is far below that of normal emulsions. The price of this type of plate is high, and manipulation is difficult, so that despite the better contrast, it is only rarely employed.

A very thorough survey of the response of photographic emulsions to electrons has just been prepared by Valentine (1966). The reader will find much helpful information about the most recent emulsions in this article, as well as practical advice about developing films and plates.

15.4.3 The Choice of Direct Magnification

(i) *The influence of the grain of the plate*

The grain of the plate affects the magnification of the image at the fluorescent screen or on the photographic negative; as the microphotographs are always enlarged photographically, but not beyond about twenty times because of the grain size of the emulsion, the total magnification which can be attained is fixed at about twenty times the direct magnification. In order to exploit the qualities of the objective as fully as possible, an electron microscope should have a direct magnification of the order of one twentieth of its useful magnification.

The lower limit is given by

$$G_{\min} = d/\varrho,$$

in which d and ϱ are the resolutions of the plate and the instrument respectively (von Ardenne, 1939). The upper limit cannot be so simply defined, as many different emulsions with grains of widely different sizes are available, which can support enlargement between about five and twenty times (or even 100 times for certain emulsions which have been developed recently). As fine-grain emulsions are slower than ones with coarse grains, we must resign ourselves to a compromise, as a short exposure avoids many of the sources of blurring which are due to slow drifts or instabilities of the potentials, or the lens currents.

To take a photograph of which the resolution is to be 20 Å, for example, with an emulsion with a resolving power of $50\,\mu$, the electron magnification has to be at least 25,000. Experience shows that the best results are obtained by taking the photograph at the smallest electron magnification which is compatible with the resolution of the emulsion, and then exploiting to the maximum the optical properties of photographic enlargement.

(ii) *Heating the object*

Among other advantages, this rule means that the object is treated as gently as possible—given the minimum current density which is admissible at the image $I'_{\min}$ (which is of the order of 10^{-11} A cm^{-2} for an exposure of 10 sec) we can calculate the minimum current I_{min} which passes through the object; we find

$$I_{min} = I'_{min} \left(\frac{d}{\varrho}\right)^2.$$

In the example above, I_{min} is of the order of 6×10^{-5} A cm^{-2}.[†] We see, therefore, that the value of using a plate with a high resolution resides in the fact that for a given exposure the intensity to which the object is subjected will vary with the square of the resolution associated with the grain, provided we can assume that the sensitivity remains constant.

(iii) *A measure of the quality of emulsions*

Since fine-grain plates have a lower sensitivity, the advantage which we have just mentioned is considerably diminished by the fact that I_{min} is in-

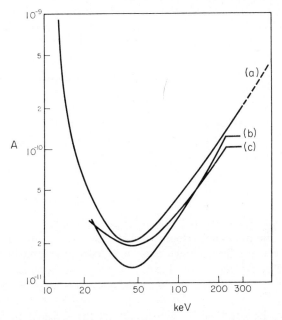

FIG. 188. The dependence of the *Bildpunktarbeit* upon the accelerating potential, for various emulsions (B. von Borries). (a) Agfa Normal. (b) Agfa Mikro. (c) Agfa Kontrast. (Developer—Agfa Medional 85 g 1^{-1}, 18°, 8 min.)

[†] At this intensity, a considerable power is being concentrated on the object. At a potential of 50 kV, the incident power is 300 watts cm^{-2}, whereas the earth only receives 0·1 watts cm^{-2} from the sun. Von Ardenne (1939) compares the current I_{min} with the critical electron density, 3×10^{-6} C cm^{-2} at which living matter is just killed by the effect of ionization; his conclusion is that at present it is not possible to observe living matter in an electron microscope working at a very high magnification, in the potential range 50–100 kV.

creased. We ought, therefore, to characterize the quality of an emulsion by some index in which speed and fineness are weighed against each other. Baker, Ramberg and Hillier (1942, 1943) and von Borries (1944) have taken as a measure of the quality the coefficient A which is defined by

$$A = U q_{0.5} \cdot d^2,$$

and which they call the *Bildpunktarbeit* (image-point energy); d is the least distance resoluble at the plate, U is the accelerating potential, and $q_{0.5}$ is the charge density per unit of surface which darkens the plate to an optical density of $d = 0.5$. The product of U and $q_{0.5}$ is a measure of the energy which must be incident upon unit area of the surface of the plate in the form of flux of electrons to produce the stated photographic density, so that A is an energy which explains the name which von Borries, to whom this definition is due, gave it.

This measure does indeed take the speed of the emulsion into account, since $q_{0.5}$ is proportional to the exposure for a given current density at the image plane. If we introduce this final current density $i_{0.5}$ at the plane in which the plate lies, A takes the following, more suggestive form which emphasizes the role of the exposure time

$$A = U i_{0.5} t d^2.$$

The smaller the coefficient A, the better is the emulsion. Figure 188 shows von Borries' results, which describe the behaviour of A as the potential is varied for various emulsions.

(iv) *The role of the eye—the use of a magnifying-glass for focusing*

The choice of the direct magnification is also affected by the fact that the image has to be inspected visually in order to focus it onto the fluorescent screen, and the smallest details have to be capable of being separated by the eye. The resolution of the eye in normal conditions is about 0·1 mm, so that if two details 20 Å apart are to be distinguished, the electron magnification must be at least 50,000 which is far too high for the photographic conditions to be good, since the intensity at the screen falls as the square of the magnification. For this reason, a compromise between the exigencies of eye and plate is often adopted, in which the adjustment is made and the photograph taken at some intermediate magnification between 25,000 and 30,000, say.

There is, however, an optical way out of this dilemma. At a direct magnification of about 10,000, the brilliance at the fluorescent screen is quite adequate to prevent the eye from being too fatigued, and to allow good photographs to be taken, but this is not a high enough magnification for details near the limit of resolution of the instrument to be distinguished—

the contrast is too slight and the brightness is still low enough to diminish the resolving power of the naked eye.† We can, therefore, simply use a small microscope which magnifies the fluorescent image five or ten times; little brilliance is lost, and a great deal of resolving power is gained. If the objective has a very wide aperture, the image intensity and the intensity at the object are equal—for this, the pupil of the eye must be wholly covered by the exit pupil of the eyepiece. The only disadvantage of this artifice is that the field of view at the fluorescent screen is reduced to less than two centimetres, but the overall magnification can easily be of the order of 100,000 times and in research in which a high resolving power is necessary, this method is frequently employed. Finally, we should point out that the method can only be used if the grain-size of the fluorescent screen is less than $0·1/g$ in which g is the magnification of the eyepiece. To satisfy this condition, the fluorescent powder must be specially prepared, or else the smaller grains of an ordinary powder must be selected by sedimentation; pounding a coarse powder does not produce a good screen as the sensitivity of the grains is considerably reduced by so rough a fracture, and no subsequent reheating can restore it to its previous value. The reader interested in this point can usefully consult a detailed study by von Borries (1948).

15.5 THE IMPORTANCE OF THE PROJECTIVE LENS

15.5.1 The Role of the Projective

The purpose of this lens is to provide a direct magnification large enough for the resolution of the objective to be exploited to the full in an instrument which is short and hence mechanically very stable; the overall resolution, therefore, is determined by neither the photographic plate nor the fluorescent screen but by the objective. For this reason, it was possible to dispense with a projective lens in the early electron microscopes the resolution of which was not very high. Attempts were made to retain this simple structure even when the optical properties became more refined, and screens and photographic emulsions with a very fine grain were tried (with a resolution of between 4 and 12 μ); in this manner, Kopp and Möllenstedt (1946, 1947) were able to obtain images with a useful magnification of 30,000 by invoking all the resources of modern screen and plate techniques. This is a fine achievement, but difficult and costly, and recent work by Möllenstedt has shown that fine-grain plates do not give as good a performance with electrons as their resolving power with light would lead us to hope. Parasitic effects due to the imperfect conduction of the electrons are the cause.

In the electron microscopes which are available commercially, the inter-

† The reader is referred to § 14.4.2 for details of the use of image-intensifiers.

mediate image which is formed by the objective is in general magnified by one or even two successive projective lenses. The advantage of introducing two successive stages is partly that the length of the structure is reduced to the minimum (for rigidity) but also because various auxiliary functions of the instrument become simpler, as we shall see later. The "object" for the projection system is the image formed by the objective, and we choose the position of this image to be near the focus of the projector—that this is the "classical focus" (§ 4.3.4), we must state explicitly, as this objective image is a virtual object so far as the projective lens is concerned.

The path of a pencil of electrons through the projective is quite different from the trajectory in the objective. In the former, the image of an object with a relatively large surface area has to be formed with pencils of electrons with an extremely slender angular aperture—that of the objective divided by its magnification—in the neighbourhood of 10^{-5}. The object, which is virtual, can, however, be placed well within the electric or magnetic field of the lens so that with the same design, the projective can be given a considerably shorter focal length than the objective. The projective, therefore, has to give a good image of a considerable area (from a fraction of a mm^2 up to 1 mm^2) in such a way as to fill the screen on which it is observed for an extended range of magnifications. The spherical aberration no longer plays a direct role, as the pencils are very fine, and the aberration which is important is the distortion. It has the effect of magnifying the peripheral region of the intermediate image more than the central part, so that the final image displays "cushion" distortion in these conditions (§ 7.2.2) which is particularly pronounced as the magnification is low. It is this defect which provides a lower limit to the possible magnification, and which decides the minimum value which is acceptable for a given projective. To work at a smaller magnification, a projective with, for example, electrodes or pole-pieces of a larger diameter would have to be used, but this would increase the minimum focal length and reduce the maximum magnification. Only by building a more complicated system can we avoid the necessity of making a compromise between these two contradictory conditions; this entails placing a second projective lens after the first (§ 15.5.2). With a single projective, the magnification can be varied in the ratio 1 : 10. The magnification of the projective might be adjustable between about 25 and 250, for example, which would give a total magnification between 2000 and 20,000. As we have explained earlier (§ 15.4.2), the convergence of the projective can be varied over wide limits without affecting the resolution of the final image in consequence of the extreme depth of field of the objective.

We may equally well use a double projector (Rang and Weitsch, 1956), the electrostatic version of which contains four electrodes. This is in fact simply an electrostatic lens with two outer electrodes at earth potential, and two independently excited central electrodes. It is less bulky than two separate lenses placed side by side, but it is nevertheless still possible to obtain a very wide range of magnifications; if the two central electrodes are at the same potential, the focal length is of the order of 20 mm when the potential is three-quarters of the cathode potential, whereas if one electrode alone is excited and the other earthed, the focal length is $f_1 = 5$ mm in one case, $f_2 = 2\cdot2$ mm in the other. For the same excitation potential, the magnifications of each of these three possible arrangements are in the ratio $1 : 3 : 10$, provided the positions of the principal planes are taken into account.

15.5.2 The Use of Two Projective Lenses

(i) *The advantages of this combination*

Many modern instruments have two projective lenses (see Fig. 189), an arrangement which was first proposed by Marton (1944, 1945). This combination, which is more compact and considerably more flexible, is advantageous both mechanically and optically. In particular, the direct magnification can be varied continuously over an extremely wide range (from less than 1000 to more than 50,000) without either the object having to be shifted or the lens replaced. In addition, the diffraction pattern which corresponds to a zone which has been examined microscopically can be obtained without altering the position of the object; a simple variation of the excitation of a lens is all that is necessary to produce on the screen the diagram which corresponds to the part of the object which had been examined previously (see Chapter 21). In this way we can select some detail of the crystalline form and analyse it without error. The mechanical advantage of using two projectives should be underlined as well. For the same direct magnification, the three-stage instrument is far less cumbersome, and can therefore be made far more rigid and free from the aberrations which mechanical vibrations would produce.

(ii) *The path of the rays and the production of the magnification*

In practice, we should like above all to reduce the distortion to the minimum, and to make focusing as easy as possible. Experiment shows that the best results are obtained when it is the potential of the first projective that is varied, while the second has a constant convergence which corresponds to a magnification of about 100; in these conditions, the field in the second lens, and therefore the distortion which it produces, are small. The intermediate projective has large openings to reduce distortion and quite a large minimum focal length, in general about two centimetres. It may produce a radial diminution of the magnification, which leads to "barrel" distortion—Hillier (1946) has considered this case in detail.

(iii) *Some practical examples*

In the first case which we shall consider (see Fig. 189a), the two projective lenses P_1 and P_2 function identically, progressively magnifying the image, and providing an overall magnification which is the product of the magnifications of each of the two separate stages. When the convergence of P_1 is reduced to diminish the magnification, the image plane I_2 remains stationary while the object plane I_1 moves progressively towards the objective O; the convergence of the latter will have to be slightly increased to

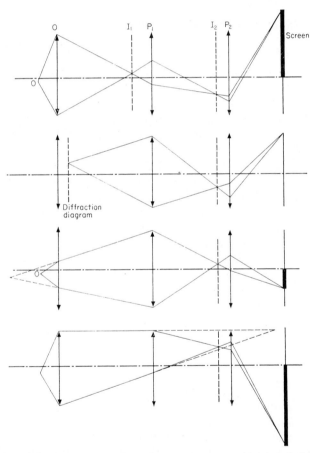

FIG. 189. The various ways of using a microscope with two projective lenses (from top to bottom): (a) Normal use, at high magnification. (b) Use as a diffraction device, the two projectives giving a magnified image of the diffraction pattern which is formed in the focal plane of the objective. (c) Use at low magnification. (d) Extreme case—the instrument behaves like an instrument with a single projective.

keep the image in focus when we move towards the smaller values of the magnification, although there is a certain latitude in the focusing. The minimum value of the magnification (which is of the order of 3000 in the normal types of instrument) is fixed by the fact that the convergence of the objective cannot be increased beyond a certain point.

It is, however, no problem to continue decreasing the convergence of the first projective P_1, and there comes a moment (see Fig. 189b) at which the focal plane on the object side of this lens coincides with the focal plane on the image side of the objective, the plane in which the objective focuses the "diffraction pattern" of the object at infinity. P_1 in this case works at a magnification of about unity, so that on the final screen, we find an image of the diffraction pattern magnified about a hundred times by the second projective. This diffraction arrangement was first employed by Boersch (1936); in this situation, as van Dorsten *et al.* (1950) have pointed out, an image is formed in the focal plane on the image side of the first projective P_1 at magnification f_1/f (f_1, which is of the order of a few centimetres, is the focal length of P_1 while f, which is of the order of a few millimetres, is the focal length of the objective). By placing a diaphragm in this plane, of which both the size of the opening and the position can be varied, we can isolate a small region of the object (a crystallite, for example) the diffraction pattern of which we should like to examine; the area of this zone can be less than a square micron if f_1/f is large enough (50, for example) and the diaphragm is small (50 μ).

If the convergence of P_1 is reduced still further, the focal plane on the object side of this lens passes beyond the objective. It is then possible to obtain an image with a very small magnification (about 500 times), if the power of the objective is reduced simultaneously, as Fig. 189c shows. The reason for this is that as far as P_1 is concerned, the electrons appear to have originated in the (virtual) image of the original object which is formed by the objective and lies "upstream" from the objective; the distance of this image increases in proportion as the convergences of P_1 and of the objective decrease. We can go on reducing the convergence of P_1, and if the latter is to produce an image at I_2, the objective must produce an image *beyond* I_2 (see Fig. 189d). In the limit, P_1 will cease to have any effect, its convergence will be zero, and the microscope will function as a two-stage instrument with a magnification of about 5000.

In the course of the series of adjustments described above, in which the adjustment of the second projective is left untouched, we have seen that the magnification begins at a very high value, of about 50,000, falls at first to nearly zero (this is the region which is useful for diffraction) and increases again to a mean value of about 5000 for example. Only two regions, on each side of the diffraction zone, cannot be used as a result of distortion or the value of the field. In case (d), the beams which form the intermediate image do not intersect in a crossover, so that the barrel distortion which results can compensate the cushion distortion of the second projective. Detailed analyses of all these focusing procedures have been published by Le Poole (1947) and by Challice (1950), who illustrate them with numerous numerical examples. Several modern instruments work in the way just described, for example the Metropolitan Vickers EM 3 (Haine, Page and Garfitt, 1950), the RCA type EMU (Simard, Burton and Barnes, 1945) and the CSF type M 9 (Grivet and Regenstreif, 1950).

Sometimes, still more complex systems are employed; certain instruments have three projective lenses, although only two are ever simultaneously in use (Le Poole, 1947; van Dorsten et al., 1950) while in other designs, certain lenses can be retracted mechanically while the instrument is in use so that their electrodes do not restrict the field (the Trub–Tauber Microscope, type KM 4 is an example).

15.5.3 The Projective Lens Stop

A field stop is usually placed within or beneath the projective; its purpose is to collect any electrons which have been reflected by the walls of the tube which supports the lenses, or scattered by the gas in the instrument (Hillier, 1950). In the electrostatic microscope, the stop has a supplementary function; it is placed just above the projective, in the focal plane on the object side, and it helps to eliminate rays which are following transgaussian trajectories far from the axis and which would converge onto the centre of the image where they would form a transgaussian star. The diameter of this diaphragm, which is about 1 mm, must be determined with care so that the field is not purposelessly reduced at small values of the magnification, but the transgaussian rays are interrupted.

15.6 MEASUREMENT OF THE CHARACTERISTICS OF A MICROSCOPE

15.6.1 Measurement of the Magnification

This is a fundamental operation with an electron microscope, though it is difficult and often neglected. In fact, various factors introduce a degree of uncertainty into the process of calibration, which is rather a long operation, so that the exact value of the magnification is often not known to within better than 30 per cent. First of all, the distortion, which is equivalent to an increase of magnification as we move from the centre towards the edge of the image, must be taken into account. Further, the position of the object with respect to the focal plane of the objective is not absolutely fixed, so that the focal length of the latter must vary to keep the image in focus. In a magnetic microscope, both the magnetic hysteresis of the lenses and variations in the high tension can be sources of error. Generally speaking, and contrary to the opinion of certain authors, it seems that the value of the magnification can only be determined to within ± 10 per cent (or ± 5 per cent if the microscope is electrostatic).

We now discuss the various methods which have been used to measure the magnification.

The magnifications of the objective and the projective (this method is suitable only for instruments with a single projective) are measured separately in two stages, and the total magnification is given by their product. Von Borries and Ruska (1940) examined holes with diameters sufficiently large to be measured accurately with an ordinary microscope.

The object can be laterally displaced through a known distance (measured with the aid of a micrometer, or by interferometry) and the magnification calculated from the corresponding shift of the image.

The electron image can be compared with an optical image; the useful magnification of an ordinary microscope is restricted to about 1000 times, however, so that a particle must be at least 25 μ in diameter if its dimensions are to be determined accurately. This is a prohibitively large diameter for electron microscopy, as the field is too small and there will be distortion.

In the most accurate method, the one which is in general use, an object of which the dimensions can be measured by some other method and related to the usual units of length is examined through the electron microscope. Spheres of various materials can be used (glass, silica, latex) which are uniform in shape and calibrated by centrifuging, by interferometry or by the scattering of X-rays. Backus and William (1949) and Gerould (1950) have found that the polystyrene latex marketed by the Dow Chemical

FIG. 190. The image of a replica of a diffracting lattice, with which the magnification in an electron microscope can be measured.

Company consists of highly regular spherical particles (2500 $\pm$ 25 Å in diameter). This method has, however, been the subject of a certain amount of dispute; Yudowitch (1951) and Danielson *et al.* (1952) have measured the diameter of the same particles with the aid of X-rays, and found that the diameters were 2740 Å $\pm$ 1 per cent and 2687 $\pm$ 0·5 per cent while Kern (1950) measured the diameter—electron microscopically—to be more than 3000 Å, when the preparation is "shadowed".

Another solution is to examine the replica of a grating (with at least 5000 lines cm^{-1}; see Fig. 190) the mesh length of which can be determined optically with a precision which is for the most part adequate. This replica must not be in an organic glaze, which is deformed when it is stripped off; instead, a layer of silica or alumina is necessary. The accuracy of the method (Rouzé and Watson, 1953) is limited at once by the fact that the spectroscopic measurement gives only an average value of the mesh-length, and not the local value which is observed in the microscope, and by the difficulty of measuring the exact distance between two corresponding points on two lines of the grating (the grooves may not be identical, for example, or the lines too wide). The most accurate method at present seems to be the one due to Farrant and Hodge (1948), who examine a glass fibre about 5 μ in diameter. This latter is measured interferometrically to a very high accuracy.

15.6.2 Measurement of the Resolving Power

(i) *The practical determination of the resolving power*

The theoretical definition of resolving power is based on the curve which describes the luminous intensity across the image of a bright point. Such images are only very rarely encountered in electron microscopy, however, except perhaps in "dark field" observation (see Chapter 19). No object is known with holes sufficiently close together to play a role analogous to that of diatoms in light optics. The finest details which are to be seen in electron optics are the images of small particles deposited on a transparent film (Fig. 191). The criterion which is used to define resolving power can be extended to this case, but the possibility of "resolving" the images of two small particles depends not only upon the distance between them but upon their individual opacity as well, that is, upon their scattering power. The empirical resolution and the scattering power are intimately related therefore, so that this is a most uncertain method of measuring the actual resolving power.

Great patience is needed in practice, to choose the zone on the image in which the particles with the smallest dimensions have collected together. Metal particles are usually employed, and obtained from a colloidal suspension of gold, silver or platinum (in an aqueous solution of Argyrol at $^1/_{1000}$ for example). Alternatively, a metal can be evaporated on the support *in vacuo* to produce, for example, a film of chrome, gold or platinum 5 or 10 Å thick. Von Ardenne (1941) suggested that the arrangement shown in Fig. 192 might well be used; in this, a razor-blade is placed 0·1 mm from the object-support, so that when the metal is evaporated onto the support, a transition zone 0·5 μ thick is formed. The supporting film must be as thin as possible and of the lowest practicable atomic weight. The electron intensity too must be as low as possible so that both recrystallization and the growth of the particles by contamination are prevented.

Figure 191 is a photograph of an aggregate of fine particles of an alloy of gold with palladium. The smallest particles which are distinguishable on this photograph are slightly less than 10 Å in size; the photograph was taken by Haine in his laboratory at Aldermaston with a Metrovick microscope, the resolving power of which he has coaxed down to 5 Å at the optimum operating conditions. With a similar type of object (an evaporated platinum–iridium film), Weichan (1966) has been able to discern separated grains, 4 Å in size, using a microscope with accelerating voltage 80 kV, an electron beam incident with angular aperture 10^{-3} and a contrast aperture of 10^{-2}.

Haine (1954) has studied the possibilities of using objects of this kind in practice to determine resolving power. The difficulties are considerable, for if the resolving power to be measured is of the order of 5 Å, the object which we have to use must be of atomic dimensions. Haine suggested that this problem could be avoided by using the Fresnel diffraction fringes which are also used to detect astigmatism (§ 16.4). This was not, fundamentally, a new suggestion, but it was based on original reasoning and a detailed study and suggests convincingly that this technique leads effectively to a correct estimate of these extreme resolving powers. The measurement is indirect, and the calculation of the resolving power entails introducing the spherical aberration constant, which has to be measured separately; nowadays, however, the values of the latter are well known.

A certain amount of research has gone into finding a different definition of resolving power which could be measured practically and which would be independent of the nature of the object. So far, however, no suitable test object has been discovered, and the image-quality which results almost always depends as much upon the nature of the specimen and the competency with which it has been prepared, as the quality of the instrument itself. A particular scrutiny of this point has been published by the American Electron Microscopical Society (1946), in which the various possible definitions are examined, but no method could be recommended in particular for its precision as a result of this analysis.

As we shall see later, a pseudo-image of the lattice planes can be obtained with a periodic object (a monocrystalline layer, for example) which

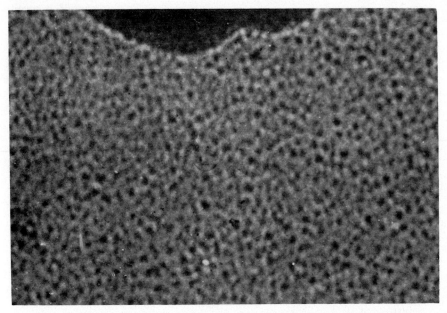

FIG. 191. The image of the grains of a gold–palladium alloy deposited onto a collodion film. $G = 1,000,000$ (Photo: Haine and Mulvey).

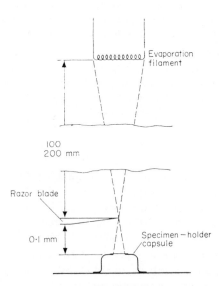

FIG. 192. The preparation of a test object by von Ardenne's method for measuring resolving power.

may, under certain conditions, have a higher resolution than we could anticipate from two neighbouring points or grains; the nature of the test object must therefore always be specified. Using an instrument with a practical resolution between neighbouring points of about 5 Å, it has proved possible to form images of the (200) lattice planes of gold, for which $d = 2\cdot04$ Å (Komoda and Otsuki, 1964), of the (220) planes of gold, $d = 1\cdot44$ Å (Komoda, 1966) and of the (220) planes of copper, $d = 1\cdot27$ Å (Watanabe *et al.*, 1966).

(ii) *Estimation of the quality*

This empirical method was proposed by Hillier (1945). A negative is enlarged to the maximum when a photograph is being printed, up to the point at which the details become blurred. This gives an experimental measure of the useful magnification and hence of the resolution, using the relation

$$\varrho = \frac{\varrho_{eye}}{G_u}.$$

This method, useful though it is for rapidly comparing several negatives produced with the same instrument or with different instruments, is not very accurate when it is an absolute value that is required; it depends on such subjective effects as the keenness of the eye and above all on the photographic contrast. The contrast plays a most important role in this way of estimating the image "quality", as the image of a "shaded" object always seems considerably better than the image of the same object in its normal state, even though the latter may often contain far more detail. If the estimate is to lean on the "clarity" of the detail (a more or less round particle, or a crystal boundary for example), a careful printing of the photograph onto "hard" paper can also improve the resolution artificially (by a factor of up to two), while in reality much of the detail of the negative is being lost. Nevertheless, a trained observer can very easily judge the degree to which his instrument has been well adjusted by examining a few photographs of a "test-object" to which he is used and the magnification of which is known to him.

(iii) *Typical experimental results*

Figure 193 shows the various values of the resolution which have been obtained since the creation of the electron microscope; a factor of about three separates the values which correspond to the electrostatic microscope from those which correspond to the magnetic microscope, a difference which we pointed out at the beginning of this chapter. It does seem, however, that the quality is tending asymptotically towards a limit, despite the increasing numbers of studies which are being devoted to the instrument.

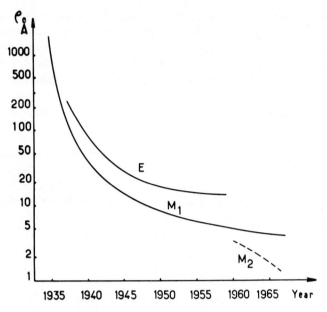

FIG. 193. Chronological chart, giving a general idea of the progress made in electron microscopy over the last thirty years. E: electrostatic microscope; M_1: magnetic microscope, non-periodic specimens; M_2 magnetic microscope, periodic specimens.

The best results which have been published at the present time are the following:

With the electrostatic microscope:

Boersch (1944)   28 Å
Grivet and Regenstreif (1950); with a Bertein stigmator 20 Å

and with the magnetic microscope (with a stigmator):

Hillier and Ramberg (1947)   10 Å
Leisegang (1954)   6–8 Å
Haine and Mulvey (1954)   6–8 Å
Engel, Koppen and Wolff (1962)   4 Å
  (non-periodic specimen)
Komoda and Otsuki (1964)   2 Å
  (periodic specimen)
Watanabe, Shinagawa and Shirota (1966)   1·3 Å.
  (periodic specimen)

CHAPTER 16

THE FACTORS WHICH GOVERN THE RESOLVING POWER IN ELECTRON MICROSCOPY

IN PARTICLE microscopy as in optical microscopy, the defects which limit the fineness of the image can be classified into three categories, according to the degree to which their presence is unavoidable. The first of these categories which we shall examine is diffraction, the influence of which is, even to-day, so very difficult to eliminate (Gabor, 1949 and 1951) that we must consider it to be inevitable. The final resolving power is determined by the diffraction, which is a characteristic of the radiation that is used in the microscope.

The theoretical limit of resolution of a microscope is determined by the combined effect of diffraction and spherical aberration. Finally, the practical resolution of any instrument also depends on the supplementary aberrations that arise from constructional defects, or the observer himself, or from the mechanism of interaction between the beam and the object to be observed. We shall consider each of these points in succession. Much complementary material will be found in a very searching recent article by Ruska (1966).

16.1 THE FINAL RESOLVING POWER; DIFFRACTION ALONE

16.1.1 The Abbe Formula

If we suppose that the lenses are perfect in the sense that they have no aberrations, which is quite possible in light optics, the resolving power is fixed by the wavelength of the radiation employed and by the aperture of the instrument. Only diffraction of the beam at the openings in the stops provides a limit to the quality of the image. This ideal resolving power is given by the general formula due to Abbe, a derivation of which is to be found in the book by Louis de Broglie (1950), which states that

$$\varrho = \frac{0 \cdot 61 \lambda}{n \sin \alpha},$$

in which λ is the wavelength of the light which is being used, n is the refractive index of the medium which surrounds the object, and α is the semi-angle of the cone defined by the useful rays which emerge from some point on the object. If the optical system which is employed allows the angular aperture to be high ($\alpha \simeq 70°$), the formula becomes $\varrho \simeq 0.61 \dfrac{\lambda}{n}$, or roughly $\lambda/2$.

16.1.2 The Optical Microscope

In an optical microscope, therefore, when the objective is corrected and of wide aperture, and the rays are immersed in a liquid the refractive index of which is close to that of the face of the objective at which they enter ($n \simeq 1.5$), the final resolving power which is obtained with $\lambda = 0.5\,\mu$ is 2000 Å which corresponds to a useful magnification of about 500. If the ultra-violet radiation provided by a mercury arc is used, the ideal resolving power is about 1000 Å, and the useful magnification about 1000. The term $\sin\alpha$ can never be larger than unity, and it is virtually impossible to find a liquid with a refractive index larger than that of cedar oil, for which $n = 1.6$. There is, therefore, an upper limit to the useful magnification attainable.

16.1.3 Attempts to Use X-rays

It is tempting to try to go down to the wavelengths of X-rays. Two different ways of exploiting this idea have been followed, and interesting results have already been obtained; we shall outline the essentials of these.

The first idea was to try and construct a very fine X-ray source, and then to project a highly magnified "X-ray shadow" of the object which was being studied onto a fine-grain photographic plate. The finer the source and hence the smaller the penumbra, the sharper is the image. The whole problem, therefore, is to construct a suitable source; recently, the extremely tiny spot which is obtained when the gun replaces the photographic plate in an electron microscope has been used (Cosslett and Nixon, 1952, 1953; Cosslett, 1954). It is thus the smallest possible image of the tip of the filament which is employed. This is not an easy exercise, and the operation of the electron microscope in reverse is described by exactly the same theory as the normal electron microscope. The anti-cathode, which is a very thin leaf of tungsten, serves at the same time as an exit window for the X-rays; the object can thus be placed very close to the X-ray source, and the resolving power which is bounded above all by the diameter of this source can equal or even exceed the resolution of an optical microscope. It is also very easy to record stereoscopic images (see § 22.4).

Alternatively, we can take advantage of the fact that a truly "X-ray" microscope can be built by using the focusing properties of a system of crossed cylindrical mirrors which totally reflect the radiation. Even with soft X-rays and a reflecting surface with a high atomic number, however, there will be total reflexion only for extremely small angles of incidence, of the order of half a degree. The numerical aperture of such an objective is, therefore, very limited, and the resolving power has barely reached a micron at the present time.

16.1.4 Diffraction in the Electron Microscope

As we saw in the last chapter (§ 15.2.3), the wavelength which is associated with an electron beam accelerated to 40 kV (a normal value of the potential in electron microscopy) is of the order of 0.06 Å, or 100,000 times shorter than the mean wavelength of visible light ($0.6\,\mu$). All other things being equal, therefore, it seems that the final resolving power should be 100,000 times better (or of the order of 0.002 Å); and the useful magnification 100,000 times larger (and hence about 50,000,000).

Unfortunately, no lens is stigmatic† in electron optics, and so we cannot work with the large apertures of glass optics; aberrations will play a considerable role, for it is they which limit the aperture which can be employed. The notion of a final resolving power determined solely by diffraction is meaningless in electron microscopy, and we must now introduce a second factor into the resolving power, the spherical aberration coefficient.

16.2 THE OPTIMUM RESOLVING POWER;

DIFFRACTION AND SPHERICAL ABERRATION

16.2.1 The Geometrical Aberrations

We have already studied these aberrations (Chapter 7); they are of varying degrees of importance, and in microscopy, only the spherical aberration has any effect upon the optimum resolving power. Moreover, only the spherical aberration of the objective is relevant as the beams of electrons at the projective are of very small aperture (§ 15.4). The spherical aberration of the objective is, however, very important, especially as no way is known of correcting it (§ 7.7.2) at least so long as we keep to axially symmetric lenses (§ 7.7.6). The only way of correcting the objective would be to use a mirror as projective (§ 7.7.3), and this has never been tried; Mahl and Pendzich (1942) and Le Rutte (1952) have certainly tried—with success—to replace the

† In this context, "stigmatic" is to be understood as meaning "capable of forming a point image of a point object", and not simply "devoid of astigmatism" (Translator's note).

projective lens by a mirror, but this was only to simplify the construction and to make the instrument less cumbersome by folding the electron trajectories back upon themselves.

16.2.2 Spherical Aberration

An electron lens converges marginal rays more rapidly than rays close to the axis. The image in the Gaussian image plane which corresponds to an axial point object is spread out over a spot of radius ϱ_s which is proportional to α^3 (§ 7.2.2) thus:

$$\varrho_s = C_s \alpha^3.$$

With the aid of geometrical optics, we can determine the distribution of intensity across the emergent beam (see Fig. 194a) and the minimum beam cross-section is to be found in a plane situated at a distance $\triangle f = -\dfrac{3}{4} C_s \alpha^2$

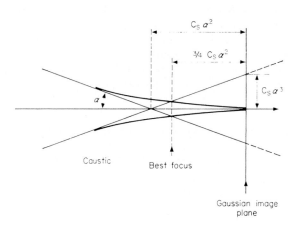

FIG. 194a. The structure of an electron beam, symmetrical about the horizontal axis.

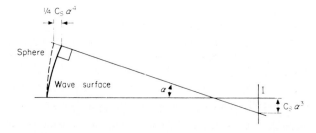

FIG. 194b. The wave surface of an astigmatic axially symmetric pencil.

from the Gaussian image plane where the radius of the beam is $\frac{1}{4}C_s\alpha^3$.
The beam is described by a caustic curve and would become infinitely intense if diffraction were ignored. The intensity falls as soon as we move away from the caustic.

The constant C_s, which has the dimensions of length, depends both upon the form of the lens and upon its convergence. It satisfies a relation of the form

$$C_s = K_s f.$$

which shows why lenses with short focal lengths are of interest: the spherical aberration will be correspondingly small as well. Numerous values of this constant K_s are quoted in §7.6, and in Chapter 17. Here, we simply mention that the best values at present are $K_s \simeq 1$ for magnetic objectives (Liebmann, 1952) and $K_s \simeq 50$ for electrostatic objectives (Bruck and Grivet, 1947). C_s cannot be reduced to zero, so that to reduce spherical aberration, we can only reduce the aperture; in practice, apertures of 0.01 radians or less are used, but this means that the diffraction phenomena, which we had hoped would be negligible at such short wavelengths, reappear.

16.2.3 The Diffraction Aberration

Diffraction, the effect and the importance of which in a perfect lens we have already discussed. is of course also present in an uncorrected lens; to a first approximation, we assume that the effect is the same as the effect in a perfect lens of which the aperture is α. The waves which are associated with the particles are diffracted by the diaphragms and spread the image of a point object out over a disc of radius ϱ_d given by

$$\varrho_d = \frac{0.61\lambda}{\sin \alpha} \simeq \frac{0.6\lambda}{\alpha},$$

for small values of α.

The diffraction aberration therefore decreases as λ which, we recall, is related to the accelerating voltage Φ_0^* by the formula

$$\lambda = \frac{h}{p} = \frac{h}{(2m_0 e\Phi_0^*)^{1/2}},$$

and for electrons,

$$\Phi_0^* = \Phi_0(1 + 0.978 \times 10^{-6}\Phi_0).$$

It is therefore advantageous to work at very high voltages.

16.2.4 Calculation of the Theoretical Resolving Power

All we have to do now is to combine the effect of diffraction with the effect of spherical aberration. As Fig. 195a shows, the spherical aberration coefficient ϱ_s and the diffraction aberration coefficient ϱ_d vary in opposite senses with the aperture of the beam; there must, therefore, be an optimum aperture, α_{opt}, at which the resultant aberration is least. A number of different methods of making an approximate calculation of this optimum condition all give comparable results only slightly different from the rigorous theory which was first established by Glaser (1943) and later by Bremmer (1950).

Von Ardenne (1940) regarded the total aberration as the quadratic mean of the two aberrations; he assumed that the defects would be combined like the luminous intensities of two beams with independent phases, so that

$$\varrho = \sqrt{\left(\frac{0\cdot6\lambda}{\alpha}\right)^2 + (C_s\alpha^3)^2},$$

which is minimized when

$$\alpha_m = C_1\sqrt[4]{\frac{\lambda}{C_s}} \quad \text{with} \quad C_1 = 0\cdot77,$$

where it takes the value

$$\varrho_{opt} = C_2\sqrt[4]{C_s\lambda^3} \quad \text{with} \quad C_2 = 0\cdot91.$$

In a note by Bruck (1947a) a comparison of the various conventions which have been suggested by the different authors is to be found—a straight-

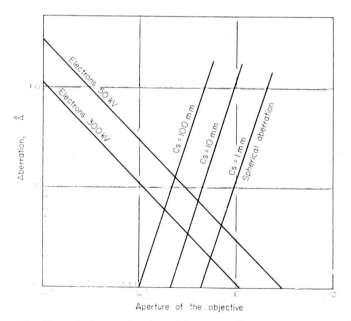

Fig. 195a. The variations of the radius of the spot produced by spherical aberration and diffraction.

Distribution of particles in the beam	C_1	C_2	References	Position of the screen
Lambert's cosine law	1·4	0·43	Conrady (1919) Haine and Mulvey (1954)	Plane of least confusion
	1·13	0·56	Glaser (1943)	Gaussian image plane
Gaussian law	0·92	0·783	Glaser (1950)	

forward linear superposition would give $C_1 = 0.47$ and $C_2 = 1.2$ (Grivet, 1948).

The full calculation of the size of the diffraction disc either in the Gaussian image plane or in the plane of least confusion of a lens with third order spherical aberration has been made by Conrady (1919), by Glaser (1943, 1950), by Bremmer (1950) and by Haine and Mulvey (1954). The Kirchhoff formula has to be used, and some expression for the angular distribution of the electrons which emerge from the object point must be assumed. We quote below some of the values which have been obtained for the optimum aperture and the optimum resolving power; an example of the calculation involved in obtaining these is given in the book by L. de Broglie (1950).

Both the empirical methods and the complete calculations, therefore, lead to values of C_1 and C_2 close to unity. It is interesting to notice that the parameter C_1 has a physical significance; if we consider the spherical (Gaussian) wave surface and the wave surface perturbed by spherical aberration (Fig. 194b), it is easy to show that the extreme disparity between these two surfaces at the edge of the diaphragm is $\dfrac{1}{4} C_s \alpha^4$, or in the optimum situation, $C_1^4 \dfrac{\lambda}{4}$.

This brings us back to the well-known rule formulated by Lord Rayleigh, which states that the spherical aberration should be corrected just up to the point at which the two wave surfaces mentioned above are a quarter of a wavelength apart. To correct further is no advantage, as it is by the diffraction that the resolution will be limited. Lord Rayleigh's calculations, therefore, give $C_1 = 1$.

The parameter which characterizes the quality of an objective is therefore $(C_s)^{1/4} \lambda^{3/4}$; with a magnetic objective, this quantity is about 5 Å (Liebmann, 1952), while it is about twice this value for an electrostatic objective. For the best magnetic and electrostatic objectives which are at present available, the best theoretical resolution which can be obtained varies between the following limits, according to the value of C_2 which is chosen:

ϱ_{opt} (magnetic): between 3 and 10 Å,

ϱ_{opt} (electrostatic): between 7 and 20 Å.

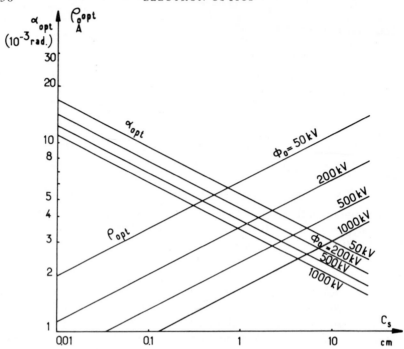

FIG. 195b. The resolving power and the optimum aperture, shown as functions of the spherical aberration constant C_s (using the constants C_1 and C_2 of Glaser, 1943).

In Fig. 195b, the curves which describe the behaviour of the resolving power and the optimum aperture as a function of the spherical aberration constant are depicted. Detailed discussions of these problems have recently been given by Liebmann (1955) and by Haine (1954).

The calculations that led to the preceding values were based upon the best objectives used in practice, with which the object was not (or hardly) immersed in the field. Recent experimental work by Riecke (1962a, b) and Ruska (1962, 1964) has, however, shown that by using a special type of condenser, it is possible to employ an objective (condenser-objective) in which the object is situated at the centre of the field; we can thus take advantage of the fact that the spherical aberration is much lower than in the standard kinds of objective. Even with high fields (and saturated pole-pieces), the focal length of this condenser-objective remains short (1·64 to 2·1 mm) and the spherical aberration coefficient C_s varies from 0·5 to 0·6 mm for electrons between 50 and 200 kV. At 1000 kV, we should still have $f = 3\cdot6$ mm and $C_s = 1\cdot1$ mm. The table below (Ruska, 1966) shows the theoretical resolving power of such an objective as a function of the accelerating voltage Φ_0 when the objective is operating at an excitation corresponding to $k^2 = 3$ (the distribution $B(z)$ is treated as a Glaser bell-shaped curve).

$\Phi_0(kV)$	50	100	200	500	1000
$f(mm)$	1·64	1·82	2·08	2·71	3·61
$C_s(mm)$	0·49	0·55	0·62	0·81	1·08
$\alpha_{opt}(10^{-2})$	1·16	1·03	0·90	0·73	0·62
$\rho_{opt}(\text{Å})$	2·94	2·29	1·77	1·23	0·98

The values of α_{opt} and ρ_{opt} have been calculated for $C_1 = 1·13$ and $C_2 = 0·56$ (Glaser, 1945).

Despite the increase of C_s with Φ_0, the immense value of raising the accelerating voltage of electron microscopes is obvious.

16.2.5 The Resolving Power for a Crystalline Lattice

The angular aperture of the beam after it has passed through the object is restricted by the objective aperture (usually to a value $\alpha_0 = \alpha_{opt}$). Electrons that are scattered through an angle $\alpha_1 > \alpha_0$ by a point in the object are halted, and this mechanism provides the main source of the contrast seen in the image; two neighbouring grains of a body, A, will be distinguishable only if the scattering of electrons by the support, B, has a different amplitude and provides enough contrast between the image of A and that of the background B (see Chapter 19).

In this case, the angular aperture of the beam illuminating the object is equal to α_0—this situation has been investigated in the preceding section: the image-forming electron beam is incoherent (there is no phase relationship between two electrons emerging from an object point).

If the object is a thin monocrystal, a new mechanism comes into play: diffraction of a proportion of the electrons at the lattice planes. A point A on a lattice plane is the source of two classes of electrons, "direct" electrons that are scattered by varying amounts and electrons "diffracted" through an angle θ_1; between the two families, there is now a certain degree of coherence. If $\theta_1 < \alpha_0$, the objective will give a normal image of the point A, formed by the cone of scattered electrons, together with a pseudo-image where the direct rays and the diffracted rays intersect. Because of the spherical aberration of the objective (the longitudinal abberation is $\triangle z = C_s \theta_1^2$), the images will not be in exactly the same plane; on defocusing slightly, we can thus bring an "image" of the lattice planes into view. If the angular aperture α_2 of the illuminating beam is considerably diminished ($\alpha_2 \ll \alpha_0$), without altering the objective aperture, the diameter of the aberration patch corresponding to the pseudo-image A' will be reduced; furthermore, the slight defocusing makes the normal image of A less distinct, which enhances the contrast, and we can thus observe an "image" of the lattice planes on the screen with a much better resolu-

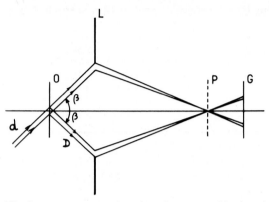

FIG. 196. The best arrangement for observing a set of lattice planes. d: direct
beam; D: diffracted beam; G: the Gaussian image plane; P: the plane on which
to focus to observe the crystalline lattice.

tion than that suggested by the formulae of the preceding section. The
resolution can be increased still further by tilting the illuminating beam
through an angle $\beta = \theta_1/2$ (Fig. 196). This "image" may be regarded as the
interference pattern formed by the direct beam and the diffracted beam
(see §19.2.6 and Dowell, 1963). The contrast and hence, the resolution,
can be yet further increased by illuminating the object with as coherent a
beam as possible; in particular, a special gun with very high brightness
using a pointed cathode is employed.

16.3 THE RESOLVING POWER IN PRACTICE;
THE INFLUENCE OF THE CHROMATIC ABERRATION

16.3.1 "Practical" Defects

The theoretical resolving power which we have just defined represents
a limit which is attainable only if the instrument is constructed with extreme
mechanical precision and is perfectly adjusted. It is an indication of the
kind of resolving power we should hope for at the final image under the
best conditions of observation—a very thin object, with very strong contrast. These optimum conditions are, however, extremely difficult to realize,
and normally, the resolving power of every actual instrument is limited
by the various *mechanical or electrical defects*, and by the difficulty of
focusing the image exactly on the screen. In everyday working conditions,
the practical resolution lies between 10 and 25 Å. The disparity between

this and the theoretical resolving power depends on a wide variety of factors, of which it is not easy to give a complete list.

The practical resolving power is a function of certain defects which we can place at first in the same category, the category which contains all defects which can always be reduced below some limit chosen beforehand by a suitable design and by taking special care. These are the defects which can be corrected; they are very numerous, and can originate in several different ways. Among the most important are the chromatic aberration, the ellipticity aberration, misalignment of the lenses, the influence of parasitic magnetic fields, and mechanical vibrations. These practical defects can be rendered negligible in the exceptional conditions which we have been calling "optimum", by ingenious construction or correction procedures and by an astute choice of the conditions of observation; in this way, the practical and the theoretical resolving powers become almost identical in the best instruments: a factor of perhaps 2 or 3 separates them.

With the best modern magnetic microscopes, working at 80 or 100 kV, a practical resolution of 4 Å has been attained (Engel *et al.*, 1962) at the image of a non-periodic object.

In the category which we have just defined, three classes can clearly be separated:

(a) Chromatic aberration, which may depend either upon the object or upon the electrical supply.

(b) Instabilities in the convergence of the objective, arising from fluctuations in the magnetizing current.

(c) Imperfect focusing (which we can regard as an error in the magnetizing current of the objective of a magnetic microscope, or as a voltage shift in an electrostatic microscope).

(d) Ellipticity aberrations (astigmatism).

(e) Residual aberrations (vibrations, parasitic fields, movement of the object heated by the beam, and other causes).

The last aberrations, (e), are negligible in good modern microscopes. We have therefore to arrange that the total aberration, (a)–(d), is as small as possible by careful construction of the lenses and their supplies.

We may consider these aberrations as the result of a number of statistically independent causes, as von Ardenne does, so that the practical resolving power is given by

$$\rho_{\text{pract}} = \sqrt{\rho_{\text{opt}}^2 + \rho_{\text{chrom}}^2 + \rho_{\text{ellipt}}^2}$$

$$= \rho_{\text{opt}} \left\{ 1 + \left(\frac{\rho_{\text{chrom}}}{\rho_{\text{opt}}} \right)^2 + \left(\frac{\rho_{\text{ellipt}}}{\rho_{\text{opt}}} \right)^2 \right\}^{1/2}.$$

Following Ruska (1966), we can gather all the aberrations (a), (b), (c) into the term ρ_{chrom}.

The form of this expression reflects the method which is used to improve the practical resolving power by making adaptations to the objective; first the optimum resolving power is made as good as possible, then, once the instrument has been built, we endeavour to check this by direct observation. To do this, the aberrations of the other kind must be reduced empirically until they no longer enlarge the "image spot" perceptibly.

16.3.2 Chromatic Aberration

This aberration has been examined in detail in § 7.4. It is a consequence of the fact that the trajectories of the electrons depend upon their energy. As the electrons which leave some point on the object do not all have the same velocity, they will not converge into a point image but onto a small spot. The radius of this chromatic aberration disc (referred back to the object plane) is given by

$$\varrho_c = K_c \cdot f \cdot \alpha \frac{\varepsilon}{\Phi},$$

where K_c is a dimensionless constant which characterizes the shape and the disposition of the electrodes of the lens, and which we can call their "form factor", α is the angular aperture of the beam, f is the focal length, and ε is the variation in the energy of the electrons about the mean value Φ. Any particular lens is characterized, therefore, by the value of the product $C_c = K_c \cdot f$ which has the dimensions of a length, and which can either be calculated by numerical integration if the form of the field is known (Glaser, 1940; Dosse, 1941; and Liebmann, 1953) or measured experimentally. We quote a few of the results:

	K_c	f (mm)	C_c (mm)	
Magnetic objective	0·75	2·4	1·8	Dosse (1941) Liebmann (1953)
Condenser-objective 100 kV	0·6	1·82	1·1	Ruska (1966) (100 kV)
Electrostatic objective	3·5	5·6	20	Boersch (1954)
	5	5	25	Bruck and Grivet (1950)

The coefficient K_c cannot be reduced to zero by a judicious choice of the field distribution within the lens. It is ε, therefore, that we must modify. We can, on the other hand, reduce the chromatic aberration considerably by working at very high voltage (Dupouy and Perrier, 1966).

16.3.3 The Causes of Velocity Dispersion

(i) *The spectrum of the initial velocities*

The electrons which leave the cathode do not all have the same velocity, and each source is characterized by a velocity spectrum which represents the "density" of electrons which have an initial energy eV; the most usual example is provided by the thermal electrons which have a Maxwellian velocity distribution which is described by

$$n(V)\,\mathrm{d}V = \left(\frac{e}{kT}\right)^2 \exp\left(-\frac{eV}{kT}\right) V\,\mathrm{d}V,$$

in which $n(V)\,\mathrm{d}V$ is the fraction of the electrons emitted with an energy between eV and $e(V + \mathrm{d}V)$, k is Boltzmann's constant, and T is the absolute temperature. The maximum of the curve, the abscissa of which represents the most probable energy, occurs at

$$V_M = \frac{kT}{e} = \frac{T}{11{,}600},$$

or about 0·25 volt for normal working conditions ($T \simeq 2700°\mathrm{K}$) with a tungsten filament. The majority of the electrons are emitted with velocities between zero and one volt, and about one half of these have a velocity below about 0·45 V, so that $\varepsilon \simeq 0·45$. Calculating numerically with these data, we find that for a magnetic objective, the resultant aberration can be reduced to 0·03 mμ, but that even when it is ten times larger (as it is in an electrostatic objective), its effect is still negligible in comparison with the other aberrations. Finally, we should point out that in principle we can reduce this aberration by using a cathode which works at a lower temperature (a matrix or sintered oxide cathode).

We can equally well produce the electrons in a gas discharge at reduced pressure. In this way, Induni (1947) and Wegmann (1953) have obtained images with a resolution of about 20 Å, which is an indirect proof that the dispersion of the initial velocities remains below ±0·5 V. Möllenstedt and Düker (1953) have recently been able to measure the velocity distribution curve of this source directly, and of a variant upon it suggested by Hailer (1938). The two curves have the same half-width, of about 2 eV, irrespective of the nature of the gas and of the material of which the cathode is composed. Figure 197, which was obtained by Möllenstedt and Düker, shows the distribution curves (A) for a discharge tube at 40 kV and (B) for the thermal electrons which are emitted by a tungsten wire at 2900°K. The spectrum for a cold cathode is two or three times more widely spread out than the curve for thermal electrons. At the larger velocities, it has a long tail but fortunately the corresponding intensities are small enough to be negligible.

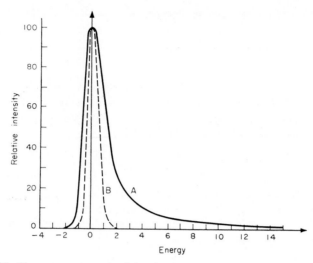

FIG. 197. The energy spectrum of the electrons emitted (Möllenstedt and Düker): (A) by a cold cathode; (B) by a tungsten wire held at 2900°K. The energies are measured in electron volts, with the most probable energy as origin.

(ii) *The spectrum which describes the interaction with the object*

Transmission through the object results in energy losses of the order of a few electron volts (up to 10 or 30 V), and we shall examine them in some detail. Since the losses are discrete, we can filter out the electrons which have been slowed down, and on this point, the reader is referred to Chapter 19. At present, we shall simply emphasize that these discrete losses represent the ultimate source of chromatic aberration in very thin objects (20 mμ). When the thickness is made greater, these losses become increasingly more important, and affect more and more of the electrons, but the width of the energy spectrum of the electrons that have been slowed down decreases very rapidly when the operating voltage is increased. Dupouy and Perrier (1966) have calculated that the chromatic aberration arising from these losses is twenty-five times smaller at 1000 kV than it is at 100 kV.

(iii) *Fluctuations of the potential and the current*

A variation in the energy of the electrons can also be a result of a *fluctuation of the accelerating potential*. The problem is very different according as the lenses are electrostatic or magnetic. In the magnetic case, a fluctuation of the lens current is effectively equivalent to a fluctuation of the voltage.

(a) *Magnetic objective*. The trajectory equation in a magnetic objective (§ 5.2.2) depends on the function $\dfrac{\Phi}{B^2(z)}$, in which Φ is the accelerating potential and $B(z)$ is the magnetic induction on the axis, and is a function of the current, i, in the coils. If i is small, the iron is not saturated and $B(z)$ varies in proportion to i. If we are trying to work under the optimum conditions, (small spherical aberration) however, we are forced to use strong lenses and hence large values of i. The iron, in this situation, is saturated and the function $B(z)$ is deformed so that when i is increased, the ordinate increases too but less at the centre of the bell-shaped curve by which $B(z)$ is represented than at the tails. The curve flattens out and its half-width increases, for the reason that the permeability of the iron falls while B increases.

In general, the convention suggested by Haine (1954) is adopted, to fix the degree of stabilization. This requires that the radius of the chromatic aberration disc which is produced by fluctuations of the high tension shall always be less than one tenth of the optimum resolving power. This condition is stringent but feasible.

If we are far from saturation, Φ/i^2 must remain constant since the focal length, for any given geometrical arrangement, depends upon this parameter (§ 9.4.3). Experiment shows that the stabilization must be to at least a fraction of one part in ten thousand, both for the high tension and for the current. This is a technical problem which can now be solved commercially, while in the laboratory the limit can even be one part in 10^5 (Haine, 1950; Bruck, 1953).

With saturation, the phenomena become more complex; Liebmann (1953) and Lenz (1953) have thoroughly explored this situation to discover whether it is in fact true, as many designers have believed, that the increase in the half-width of the field can compensate the increase in the total excitation. If this were so, we could relax the conditions on the stability of the supply to the lenses appreciably, and then operate them in these conditions at the maximum magnification. This research shows that it is possible to construct lenses with very little aberration, but not, however, without any. The anticipated advantage does indeed exist, but is not as pronounced as had been hoped. Even in this case, extreme stability is necessary, but 10^{-4} or even a little less can be tolerated. One further point is that the chromatic aberrations of the magnification and the rotation must be considered; fortunately, these latter can be completely corrected (Ito, 1955). This problem can be solved most elegantly by coupling the fluctuations of Φ and those of i electronically, when both are caused by fluctuations of the mains supply to the whole installation, in such a way that the condition

$$\frac{2\triangle B(z)}{B(z)} = \frac{\triangle\Phi}{\Phi}$$

remains satisfied as exactly as possible (Dupouy *et al.*, 1964).

We should mention that there are no current fluctuations at all when superconducting lenses running in the persistent-current mode are employed (see Chapter 9).

(b) *Electrostatic objective.* In the electrostatic version, stabilization of the potential is far less important an exercise. The differential equation of the trajectories is, in fact, homogeneous so far as the potential is concerned (§ 4.2.2) so that if all the potentials are multiplied by the same factor, the focal length does not vary. So that this condition shall be satisfied, all the potentials are obtained from a single source by a system of potentiometers. For tensions of the order of 60 kV, this precaution becomes inadequate, as there is a slight aberration which is relativistic in origin (§ 4.4) and the differential equation of the trajectories is no longer homogeneous with respect to the potentials; a stabilized high tension supply is again necessary, but stabilized to a lesser degree than in the case of magnetic lenses since only a correction term has to be eliminated: a stabilization of about one part in a thousand is adequate. This simplification is one of the main advantages of using an electrostatic system.

16.4 THE PRACTICAL RESOLVING POWER; THE INFLUENCE OF DEFECTS IN THE MECHANICAL CONSTRUCTION

16.4.1 The Various Causes of Asymmetry

So far, we have been assuming that the fields within the lenses have perfect axial symmetry. In practice, however, this ideal state can be achieved neither for electrostatic nor for magnetic lenses, and at present it is not possible to reach resolving powers of the order of 10 Å even when the lenses are very carefully constructed. Even though the most expensive machine-tools are available, it is not possible to achieve a mechanical precision sufficient to avoid an ellipticity of the field large enough to have a detectable effect. Further, in the magnetic case, there will be local variations in the permeability of the iron which have a more pronounced effect the nearer the iron is to saturation; at present, this heterogeneity cannot be eliminated by metallurgical treatment.

The magnitude of the mechanical astigmatism of a magnetic lens is also related to the size of the metal grains. An examination by Leisegang (1956) showed that for an identical treatment, a nickel steel (with 50 per cent Ni) with grains of 30 μ produces an astigmatism which is thirty times smaller than that of carbon steel with three millimetre grains.

In electrostatic objectives, the leads also produce a lack of symmetry in the field (Bertein, 1949).

A supplementary source of astigmatism appears when the instrument has been in use for some time, as the field inside the lens becomes modified by the small quantities of insulating deposit which are produced by the interaction between the electrons and the residual gas in the evacuated tube and also, quite often, by fragments of the object which break off and become fixed to the electrodes of the objective. Also, a brown substance is deposited continuously on the stop of the objective in the region which is struck by the electron beam; Ellis (1951) and Ennos (1953, 1954) have made a particular study of this contamination, as this insulating layer, which sometimes increases at 60 Å per second, conducts away the incident charge only with difficulty, and is not necessarily axially symmetrical. The parasitic electric field which is produced can result in an important astigmatism, and the influence of this defect is increased as the dimensions of the diaphragm are reduced. This contamination of the objective diaphragm seems to be less rapid when we use a mercury diffusion pump but in fact oil pumps are usually employed because of the cumbersomeness of the liquid-air supply which is required for trapping mercury vapour in the pump. We should, however, point out that although the use of a mercury diffusion pump does not eradicate all the sources of organic vapour (elastic joints, spots and residual deposits of cleaning fluids, for example, always exist within the system) it does reduce them in number.

The real solution of this contamination problem is, however, to surround the zone containing the object and aperture as completely as possible by a metal screen; the latter is vigorously cooled with the aid of a reservoir of liquid nitrogen (Heide, 1964; Ruska, 1966), although the temperature of the specimen-holder and the aperture are kept close to the ambient temperature. The same zone of the specimen can then be studied for several hours at very high magnification without any noticeable decrease in the resolving power (and without any carbon contamination appearing at the object). Reduction of the size of the electron beam is highly advisable; the quality of the vacuum in the microscope column should also be improved, by lowering the pressure but above all, by achieving a "cleaner" vacuum, free of organic molecules. The use of an ultra-high vacuum (10^{-8}–10^{-9} mm Hg) with ion pumps, in conjunction with cooled zones, seems to be inescapable in future.

16.4.2 Ellipticity Astigmatism (§ 7.5.2)

Ellipticity of the diaphragms of the lenses produces an astigmatism of the beam which we have considered in detail in § 7.5.2. The best "focus" is obtained when the circle of least confusion lies at the screen—the radius of this circle ϱ_e, is given by

$$\varrho_e = C_e \alpha,$$

in which C_e is the constant of ellipticity aberration which has been calculated by several authors (Bertein, 1947a, 1947b and 1948; Bruck and Grivet, 1947; Sturrock, 1951; and Hillier and Ramberg, 1947). We mention only that the order of magnitude of the coefficient is given by

$$C_e = \delta a,$$

in which δa is the difference between the axes of the ellipse which forms the outline of the diaphragm. Present-day methods of machining are not capable of reducing the difference between the axes to below a micron, so that if the optical aberration is to be reduced, α must be diminished; we distinguish, therefore, between two cases.

In the first of these, the value of ϱ_e which corresponds to the optimum aperture of § 16.2.4 is considerably smaller than ϱ (more than ten times, for example). This implies that

$$\varrho_e = \delta a \cdot C_1 \sqrt[4]{\frac{\lambda}{K_s f}} < 0.1 C_2 \lambda \sqrt[4]{\frac{K_s f}{\lambda}},$$

or, since $C_1 \simeq C_2 \simeq 1$,

$$C_e = \delta a < 0.1 \sqrt{\lambda K_s f}.$$

In this case, we use the optimum aperture, but this does require an objective which has been carefully machined and corrected.

In the other case, where the objective is only passably good, we must not work at the optimum aperture, as ϱ_e would then be much larger than ϱ_{opt} which would be absurd. The aperture must be reduced, but this means that the diffraction becomes all-important while the spherical aberration which behaves like α^3 has virtually no further effect. The roles have changed, for the problem now is to determine the optimum aperture and resolving power by combining the astigmatism and the diffraction which are now the two defects which are important. Using a convention similar to the one described in § 16.4.2, Bruck and Grivet (1947) found that

$$\varrho'_{opt} = 0.85\lambda \sqrt{\frac{C_e}{\lambda}}, \qquad \alpha'_{opt} = 0.71 \sqrt{\frac{\lambda}{C_e}}.$$

We must, however, verify that the spherical aberration is indeed considerably less than ϱ'_{opt} (more than ten times smaller, say). This requires that

$$\varrho_s = K_s f (0.71)^3 \left(\frac{\lambda}{C_e}\right)^{3/2} < 0.1 \times 0.85\lambda \sqrt{\frac{C_e}{\lambda}},$$

or in practice, replacing C_e by δa,

$$\delta a > 2\lambda \sqrt{K_s f}.$$

When this condition is satisfied, we should work at the aperture α'_{opt}.

If, in conclusion, the constants of the objective are such that δa lies in the transition region between the above limits, the three defects have to be combined to give the optimum operating conditions.

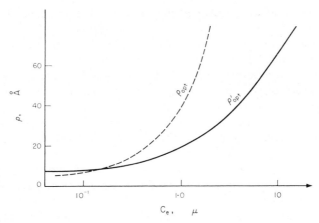

FIG. 198. The *practical* resolving power. The dotted curve represents operation with the optimum aperture calculated as in § 16.3.1; the full curve represents the case when the optimum aperture is calculated as in § 16.4.2.

To obtain the practical resolution, the various aberrations are combined in the way described in § 16.3.1. Figure 198 shows the way in which the resolution varies as a function of C_e in two cases, in one of which the aperture is α_{opt}, and in the other, α'_{opt}. An elliptical objective will be operated at α'_{opt}, but it will never be very bright since α'_{opt} is always markedly less than α_{opt}, and the difference between these has a very pronounced effect upon the brightness, which is proportional to α^2.

16.4.3 Observation of the Ellipticity Astigmatism

Measurement of the ellipticity astigmatism of an objective is extremely valuable, as it gives us a direct overall measure of the quality of its construction, and suggests the characteristics of the "stigmator" which will be necessary to correct the aberration. Various methods of detecting and measuring the astigmatism in the image have been suggested.

One of the simplest methods (Hillier and Ramberg, 1947) is based on an examination of the Fresnel fringes which are formed at the edge of an opaque object, the objective being slightly defocused in such a way as to produce dark fringes outside the shadow (the "real diagram"; see Fig. 199a).

We suppose first of all that the objective is free of astigmatism. The image of any object (specks, of smoke-black, for example) is surrounded by a diffraction fringe which is particularly clearly visible if the object is illuminated at a small angular aperture. The appearance of this fringe is markedly different according to whether the convergence of the objective is slightly higher (overfocused or positive fringe) or slightly lower (under-focused or negative fringe) than the value which corresponds to an exact focus.

The negative fringe appears as a brilliant line, attached to the edge of a dark image. The edge of this image is, therefore, extremely clear, as the contrast is far sharper than it is when the focusing is exact, the situation which in fact corresponds to minimum contrast. This explains why one tends always to underfocus the image slightly when focusing by eye. The positive fringe, on the other hand, is a dark fringe, again running around the image, and separated from it by an interval which becomes wider as the overfocusing is made more pronounced. When the image is exactly in focus, the fringes disappear, and the edge of the image seems slightly blurred; the contrast is clearly a minimum, and a trained operator can use this as a criterion.

We consider now an objective which has a certain ellipticity astigmatism; if the image is observed between the two line foci, the edges of the image which are parallel to one direction will be accompanied by a positive fringe, while the edges parallel to the perpendicular direction will have a negative fringe. Detecting a slight astigmatism, therefore, involves being able to distinguish positive fringes which are only very slightly separated from the image and which correspond to a very slight overfocusing (since the image which is being examined lies between two focal lines which are themselves very close together). The possibility of doing this is itself related to the clarity of the fringes which certainly depends upon the aperture of the illuminating beam but also on the stability of the image. It is clear, therefore, that for the astigmatism to be visible, the apparatus must be absolutely clean, and free of mechanical or electrical instabilities. When this has been achieved, however, the fringes can be made clearly visible simply by reducing the aperture of the illumination, even though the overfocusing be very small indeed, and it is easy to make them symmetrical with the aid of a correction system. Towards the end of the correction process, however, it will be necessary to work very close indeed to the true focus, as any but a small overfocusing produces a fringe far from the object and symmetrical in appearance even when the objective has a pronounced astigmatism (on this point, examine the top part of Fig. 199 b).

The correction procedure consists of adjusting the stigmator (§ 7.7.5) until the Fresnel fringes are identical in all directions. The best test object is provided by the holes in a film of collodion covered with a thin layer of some heavy metal or of carbon (Koenig, 1951) to increase the contrast of the fringes. The sensitivity of the method is limited by the problem of seeing the fringes on the screen, as the distance between the fringes and the image is of the same order of magnitude as the resolving power. According to Haine and Mulvey (1954), if a sufficiently intense illumination were used (between 2 and 5 A cm^{-2}), with a fine grain screen, and a high direct magnification of the order of 50,000 (200,000 is possible in the Siemens UM 100)

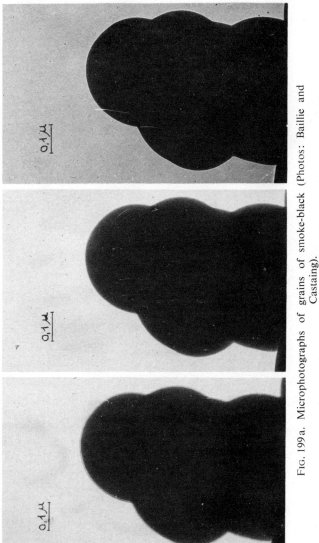

FIG. 199a. Microphotographs of grains of smoke-black (Photos: Baillie and Castaing).

The objective is free of astigmatism. *On the left*, overfocusing with a positive fringe; *in the centre*, in focus, minimum contrast; *on the right*, underfocusing, with a negative fringe.

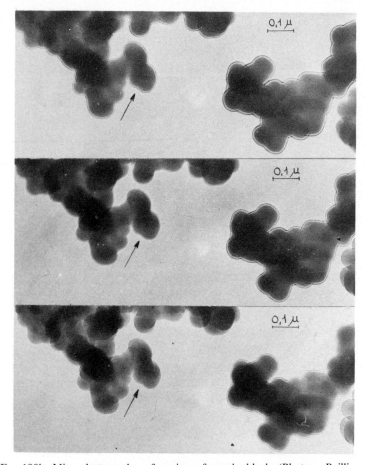

Fig. 199b. Microphotographs of grains of smoke-black (Photos: Baillie and Castaing). The grain in question is indicated with an arrow. The objective is slightly astigmatic, and the arrow points in the direction of one of the line foci; *top* over-focusing — positive fringe; *centre*, focusing onto the circle of least confusion — the fringe is partly negative and partly positive; *bottom*, underfocusing — negative fringe.

together with an additional microscope with a magnification of ten times, it would be possible to correct an astigmatism visually until the focal lines became less than $0.05\,\mu$ apart. Leisegang (1954), who followed an analogous procedure, only achieved $\triangle f = 0.16\,\mu$. A better correction can be obtained, although more laboriously, by directing the correction procedure step by step photographically.

Other methods of observing the astigmatism have been suggested, and for further information the reader is referred to Part 1 and to the original publications (Castaing, 1950a and 1950b; and Bertein and Regen-streif, 1949). Van Dorsten (1948, 1950) has shown that the astigmatism can be observed directly by magnifying it artificially; to do this, the beam is deviated to and fro by an alternating magnetic field in the vicinity of the gun. The effect is of increasing the aperture, and a very faint astigmatism becomes visible.

16.4.4 Correction Systems (Stigmators)

To correct the ellipticity astigmatism, a correcting device has to be inserted within the field of the lens; this corrector has to be free to rotate about the axis in such a way as to restore perfect axial symmetry to the field, and it was this type of system which was used by Hillier and Ramberg (1947). At first, they tried to obtain a sufficiently symmetrical system by constructing the pole-pieces with the utmost care, but they found that the field asymmetries could not be reduced below $0.5\,\mu$ (which corresponds to 10 Å aberration) as the inhomogeneities in the iron and the parasitic fields of the small cracks which remained were still uncorrected. To improve this arrangement, they then inserted eight iron screws, each inclined at $45°$ to the next, in the outer part of the space between the pole-pieces. These screws modified the field distribution, and were adjusted in such a way as to eliminate all perceptible asymmetry from the image. RCA microscopes have been fitted with a device of this type for a long time now. The correction process is long, as the operation of the microscope has to be interrupted a great many times to withdraw the pole-piece, and it may take a whole day to achieve complete correction. A considerable improvement is to use a magnetic objective with double windings (Le Poole and Le Rutte, 1949), which leaves a space free for the correction screws which can then be adjusted from outside the instrument while it is still working. Leisegang (1954) has described a mechanical system of this kind, in which two small blocks of soft iron (0.1 mm^3 in volume) are moved about in the neighbour-hood of the lower face of the objective, while remaining diametrically opposed; by adjusting their height and their azimuth, a good degree of correction can be obtained in a few minutes. This method cannot, however, be transferred very simply to the case of electrostatic lenses, as the region near the central electrode becomes highly susceptible to electrical break-down if electrodes with small radii are introduced there.

The other method of obtaining correction, which is far more flexible, is copied from an optical method. The basis of this method, which was proposed by Bertein (1948),† is the creation of a supplementary potential along some particular stretch of the beam, the effect of which is to

---

† This method may well have been thought of earlier, and independently, by Mahl, who mentioned it in 1944 in a request for a patent, but never published it.

correct the harmful effects of ellipticity in the neighbourhood of the optic axis by producing an equal asymmetry of the field of the opposite sense. We have explained the principle of this in § 7.7.5.

A very simple corrector can be constructed by producing a transverse electric field with a combination of four plates (Fig. 200) each at right angles to the next. These constitute a negligibly convergent cylindrical lens which is markedly astigmatic, in that it is convergent in the plane xOz and divergent in a plane normal to this; it is, in fact, a very weak quadrupole lens (cf. Chapter 10). As the line foci of this lens are not necessarily parallel to those which are produced by the astigmatism of the lens, provision has to be made for some kind of adjustment with which either the correcting device or the field alone can be rotated about the axis. This can be arranged either mechanically (Haine, 1954), or by rotating the field by adding a second system of electrodes, rotated through 45° with respect to the first set, which are of a new shape (Rang, 1949; see Fig. 200). With the two electrical controls both the direction and the intensity of the astigmatism can be varied. The earlier design has been simplified by earthing the negative electrodes which reduce to a single pair (Fig. 200).

In another of the possible methods of obtaining correction, a field with elliptical equipotentials with variable eccentricity is created in the neighbourhood of the optic axis by means of six electrodes in the form of a regular hexagon: a "sextupole" lens. To these is applied a potential of the form

$$V_k = r \cos\left(\theta - 2\frac{K\pi}{3}\right),$$

in which K represents the number ascribed to one of the electrodes which are numbered from 0 to 5. This produces an astigmatism which is proportional to r in magnitude, and of which one of the line foci is inclined at an angle $\frac{1}{2}\theta$ to the zero-th electrode. Systems of this kind have been built by Rang (1949) and by Castaing (1950a).

Each of these corrective systems is usually placed as close to the exit face of the objective lens as possible, as this is the place at which the diameter of the beam is smallest. The corrector can therefore be smaller, and as the

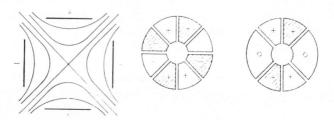

FIG. 200. Diagrams showing some stigmators.

distances both between the electrodes themselves and between the electrodes and the beam are slight, only small potentials are required, which is very convenient. Bertein (1948) calculated that for an ordinary electrostatic objective, with electrodes at a radial distance of 5 mm, potentials of the order of a hundred volts would be adequate for correcting an aberration as high as $\Delta f = 10\,\mu$. Only the form of the field in the neighbourhood of the axis matters; the electrodes can be bars either parallel or perpendicular to the axis, or planar or curved plates, and generally speaking a well-designed stigmator requires a supply of only a few hundred volts.

The optic axis of the stigmator must be aligned to coincide with that of the objective, as otherwise the consequent lack of symmetry would prevent a complete correction from being possible. To satisfy this condition, either the stigmator is provided with a mechanical adjustment or alternatively, deflecting plates with which it can be displaced are set round the beam at some point before it reaches the stigmator; and there is yet another possibility which is to supply the stigmator itself unsymmetrically, which shifts its optic axis; in any case, it is very easy to check whether the alignment has been successfully achieved since if the centring is poor the image moves when the intensity of the field in the stigmator is varied. The optimum position can be found very rapidly by temporarily applying a small alternating potential to the corrector.

16.4.5 Defects Produced by Misalignment of the Lenses

These are less serious defects than the ones discussed above, as they can be made negligible by building the system carefully. The electrostatic lens case has been studied theoretically (Bruck, 1947a; Bruck and Grivet, 1947). We differentiate between defects in the centring, and defects in the inclination. Let us suppose that all the diaphragms are perfectly round and parallel to one another. If the central plate of one of the lenses is eccentric by a distance Δe with respect to the common axis of the two outer plates, an astigmatism is produced with constant

$$C_e = C_s \left(\frac{\Delta e}{f} \right)^2.$$

If a perfectly round lens is tilted through an angle β with respect to the optic axis of the system, an astigmatism is again produced,

$$C_e = C_s \beta^2.$$

The calculation for a normal electrostatic lens (which has $\alpha = 3 \times 10^{-3}$, $C_s = 10$ cm, and $f = 0.5$ cm) shows that a 10 Å aberration corresponds to $\Delta e \simeq 10^{-3}$ cm or $\beta \simeq 2 \times 10^{-3}$. It is easy to centre the three electrodes to within a micron, by using a reference spindle which is provided with an optical microscope for sighting. Any mechanical tilt of the electrodes can be compensated by suitably inclining the beam, by adjusting the gun in fact. The beam has to be made to pass through the middle of the central electrode and the adjustment is checked by varying the potential of the objective by a few hundred volts (for a total potential of 50 kV); if the system is well adjusted, the image on the screen should not move, as the effect of a variation $\triangle V$ in the potential is a displacement given by

$$\triangle y = G \cdot K_c \frac{\triangle V}{V} \cdot \triangle e,$$

in which G is the total magnification and K_c is the chromatic aberration constant of the objective. With the aid of this relation, we find that if the aberration is to be less than 10 Å, the image must not move more than 1 or 2 mm for $\triangle V = 250$ V. In conclusion, we would mention that it is difficult to ensure directly when the instrument is being built that the inclination of the objective with respect to the axis of the projective lens or of the beam with respect to the axis of the objective is smaller than 2×10^{-3} radians.

16.4.6 Parasitic Fields

(i) *Electric fields*

The metallic case of the instrument eliminates the influence of any outside electric field, but very often, charges which originate in the electron beam itself (direct, scattered or secondary electrons) accumulate on any insulating materials which are directly in the path of the beam and produce parasitic electric fields which usually vary in an irregular way. The result is an irregular displacement of the image which depends on the intensity of the beam. These insulating materials can be fragments which have broken away from the object and fallen into the lens, or insulating deposits which have formed on the metal surfaces. Further, any metallic part which is struck by an intense beam of electrons rapidly becomes covered with a brownish layer of some insulating substance, or contamination, as we have already mentioned in § 16.4.1.

Generally speaking, a microscope cannot be kept in proper working order without a thorough cleaning at least every six months, and a weekly cleaning of the parts which are close to the beam. Wherever possible, steel wool is used for the cleaning while the more delicate parts are rubbed with a tuft of cotton-wool impregnated with alumina. The grids upon which the object is supported are cleaned in 20 per cent formic acid, or heated to red-heat (Eaves, 1952).

The contamination proves to be particularly troublesome in that it forms both on the object under observation and upon the contrast diaphragm of the objective, and this prevents the observation of any one region of the specimen from being very prolonged. Indeed, in certain unfavorable cases such as thin metal layers, a few seconds bombardment contaminates the object sufficiently badly for observation at high resolution not to be possible; focusing under these circumstances is most difficult, and various ways of reducing the contamination or even suppressing it entirely have been proposed. Heating the surface which is being bombarded to 200° is most effective (Ennos, 1953), but the object which is being examined is not always left unscathed by such treatment. Leisegang (1954) places the object in a region which is chilled, so that any organic vapours will be condensed. Sweeping the instrument with a gaseous current (which often happens involuntarily, as a result of small leaks) seems to reduce the contamination within the system considerably as well; directing a jet of air at low pressure onto the surface of the specimen reduces the contamination appreciably and can even get rid of it completely in the regions which are bombarded heavily (Castaing and Descamps, 1954). At present, it is contamination which represents the main obstacle to obtaining high resolutions conveniently. If the specimen can support this, the contamination can be remedied by bombarding it with ions during the examination itself—thin metallurgical sections (Castaing, 1954) and massive specimens observed by reflexion (Fert, 1954) are examined in this way.

Another cause of instability can originate in the electrostatic charge of the specimen itself. If an intense illumination is used, the electrons which are stopped by the object may not be able to leak away sufficiently rapidly if the object is a poor conductor, and the object charges up. Elsewhere, where the secondary emission coefficient is high, there may be areas which are positively charged. It is possible, therefore, for different zones to become charged with opposite signs, and hence to repel or attact one another. The object, as a result, becomes deformed as various parts creep slowly about and these electrostatic forces are even sufficient, quite often, to tear the thin film which supports the object. A metal layer deposited **under vacuum first of all strengthens the plastic film mechanically, but** above all it has the valuable effect of making it conductive, which gets rid of the harmful charges by conducting them away to earth, or at least reduces their influence considerably.

(ii) *Magnetic fields*

The electron beam is deflected at points along its trajectory by parasitic magnetic fields. If this were absolutely necessary, the influence of constant fields could be eliminated by adjusting the centring of the beam, but this would introduce an astigmatism which would in turn have to be corrected with a stigmator, and it is better to eliminate any appreciable field by adequate screening. Variable fields produce a lateral vibration of the beam, which blurs the image in a particular direction; these magnetic fields are usually the result of transformers, motors or leads through which intense currents are passing, which have been placed too close to the instrument. The need to choose a good single earth connection for the whole instrument has been demonstrated by Grube (1949), as the use of "multiple earths" can lead to intense currents, and hence to harmful magnetic fields.

Von Ardenne (1940) finds that if the resolution is to be 1 mμ, any residual alternating field perpendicular to the axis must be reduced to about 10^{-5} gauss, which is a difficult condition to achieve. The most sensitive parts of the beam are those at which the velocity of the electrons is reduced, and in particular, the cathode, the centres of electrostatic lenses (Mahl, 1947), and the region immediately beneath the objective. In an electrostatic microscope, the beam is screened by surrounding the whole of its trajectory with one or several concentric mumetal (or permalloy) cylinders between 1 and 5 mm thick; at the objective, the beam is much more carefully screened, and certain parts of the lens are solid mumetal. In the magnetic microscope, the same precautions are taken for the regions between the objective and the projective lenses, and between the latter and the screen; nevertheless, there are other critical zones, at the lenses themselves and above all at the objective—the metal from which the pole-pieces are made is too close to saturation to be much help, and it is the yoke which is primarily effective which explains why it is of generous dimensions. Tables of the "coefficient of reduction" which describe the effect of cylindrical screens are to be found in the books by von Ardenne (1940) and by Kaden (1950) and in the article by Charles (1947) for various geometrical arrangements; Charles also describes a method of verifying the efficiency of the screening by a measurement of the induction.

16.4.7 Indistinctness and Disturbances Produced by Mechanical Movements

It is difficult to distinguish these defects from the one described above simply by examining their influence upon the image, but their effect is often most pronounced.

A common occurrence is thermal migration at the object; every part of the object which is bombarded by the electron beam has a rather high temperature, which produces local expansion (cf. § 20.7). The thermal equilibrium (at which this displacement ceases) is only reached quickly if the object-support is a good enough conductor of heat; for this reason, and also because they are more heat-resistant, electrolytic grids are preferable to woven grids. Conversely, it is harder to reduce this migration if a transparent object-support is used, as this is subject to a number of stresses, and is very flexible. An effective solution is to work at as low an intensity as possible, and the heating of the object is certainly reduced if the illumination is reduced to a very narrow region (a few microns, say) around the point being examined. A condenser lens and suitable diaphragms seem, therefore, to be an advantage, since the brightness of the beam is increased, and hence the total intensity at the object may be diminished. The mechanical vibrations which are transmitted to the system from the ground, from the ducts of the pump (as a result of the motor of the pump or of boiling oil or mercury) or which are created by certain of the mechanical accessories of the instrument (the shutter of the camera, for example) have the particularly troublesome effect of displacing the object-support relative to the objective; in some instruments, they can also produce a slight oscillation of the lens system and hence of the optic axis.

16.5 "REVOLUTIONARY" ATTEMPTS TO IMPROVE THE RESOLVING POWER

We shall conclude this chapter with a few details of the attempts which have been made to improve the resolving power by modifying the very principle upon which the instrument operates. Theoretically, spectacular progress is to be anticipated, but in practice, the same is not true as the host of technical problems are as yet only poorly solved. The instruments which we are about to describe cannot compete with the ordinary electron microscope.

16.5.1 The Ion Microscope

(i) *The reason why ions are of interest*

The expression for the optimum resolution

$$\varrho_{\mathrm{opt}} = C_2 \, \lambda^{3/4} \, C_s^{1/4},$$

shows that we could make a great deal of progress if we were to use ions instead of electrons, as the wavelength of the waves which are associated with ions is about a hundred times shorter than the value which corresponds to electrons accelerated through the same potential. If the ion which is used is of mass M, and the electron mass is m, the gain in resolving power is described by the factor $(M/m)^{3/8}$, or 17 for H^+, 43 for Li^+ and 105 for Cs^+. The optimum aperture

$$\alpha_{\mathrm{opt}} = C_1 \left(\frac{\lambda}{C_s} \right)^{1/4},$$

falls in the ratio $1 : \left(\dfrac{m}{M} \right)^{1/8}$; for example, with Li^+ ions, the aperture at which the best resolution would be obtained would be 3·2 times smaller than the optimum aperture of the objective in the case of electrons. In so small an aperture, it is difficult to obtain a sufficiently intense ion beam, but as the potential gain in resolving power is so huge, it is still appreciable if a larger aperture than α_{opt} is used. Unfortunately, the chromatic aberrations and the mechanical aberrations are as important in the ion microscope as in the electron microscope and although modern technical advances have enabled these defects to be reduced to less than the optimum resolving power *for electrons*, the relative importance is reversed in the ion situation. At present, there is always a mechanical or chromatic aberration which is the essential factor in limiting the practical resolving power.

(ii) *Experimental arrangements*

The suggestion to use ions instead of electrons in the microscope was made originally by Knoll and Ruska as early as 1932. Koch (1936) and Walcher (1943) obtained the image of a grating with a resolution of a millimetre.

The problem was later taken up again more successfully by Boersch (1942–7) who had at his disposal an instrument which performed excellently with electrons (with a resolution of 28 Å). With ions, the photographs which he took of zinc oxide corresponded to a resolution of only about 500–800 Å, and after various attempts, Boersch abandoned his work, faced as he was with such extreme experimental difficulties; in his conclusion, he expressed the opinion that it would be better to find ways of perfecting the electron microscope than to try to use ions. The source which he was using was of the Kunsman (1926) type, modified in the ways suggested by Koch (1936) and Walcher (1943), and produced Li^+ ions of too feeble an intensity to produce a *visible image* on the screen. Boersch was obliged, therefore, to make the preliminary adjustments with electrons. The direct magnification of his photographs was between 1000 and 2000, but even with five or ten minute exposures, the pictures were very under-exposed.

A more systematic attack upon the problem was made by Magnan and Chanson (1945, 1949), who demonstrated the value of using protons, and a high tension of 300 kV. They designed a microscope which was adapted to operate in this new way, but which was extremely difficult to build and adjust. Photographs which correspond to a useful magnification of 3000 have been published (Chanson and Magnan, 1951) with an operating potential of 50 kV. More recently (Magnan and Chanson, 1955), the practical resolution has been forced down to 10 or 20 Å in visual examination, but so far no photograph at a high magnification has been published. The protons were provided by a high-frequency source of the Thonemann (1948) type, with the modifications of Moak *et al.* (1951). Certain improvements enabled Magnan and Chanson (1955) to obtain a very stable source, with a high electrical efficiency, providing 90 per cent of atomic ions, but the velocity spectrum of the ions (which is wider than a volt) produces a large chromatic aberration which limits the resolution.

A much simpler arrangement has been built by Gauzit (1951, 1953 and 1954a) who converted an ordinary commercial microscope by replacing the electron source by a source of alkali ions, the latter due to Couchet (1951, 1954). At 60 kV, the practical resolving power which he was able to obtain was 80 Å, effectively half as good as the value obtained with electrons. The photograph in Fig. 3 was taken under these conditions.

The experimental difficulties are very numerous; photographic plates and fluorescent products which give good results with electrons are no longer so good with ions. Plates with a high concentration of AgBr, which had been intended for nuclear physics, can be employed. The vacuum must be excellent, as otherwise the ions may be neutralized by collision with the residual gas.

16.5.2 The Shadow Microscope

In the shadow microscope, the image of a thin object is obtained by conical projection from an electron source of very small dimensions. The source may be real, as it is in the cold emission microscope (§ 13.6) but in the shadow microscope in the proper sense (Boersch, 1939, 1940) the source is virtual and consists of the highly reduced image of a real source. The aberrations of the system of reducing lenses have no effect on the fineness of the image, as the clarity of the latter depends only upon the Gaussian diameter of the image-source (and not upon its real diameter, which may well be made considerably larger by the aberrations of the reducing system) and upon the Fresnel diffraction at the specimen. The aberrations of the optical system by which the image-source is formed only have the effect of producing a distortion in the image of the specimen which increases indefinitely when we try to reduce the importance of the Fresnel diffraction by increasing the nominal magnification (cf. § 7.6.2 and Castaing, 1950a, 1950b). The result is not a limitation on the resolving power, but upon the possibility of using a theoretically infinite resolving power, and this limitation can only be overcome if we have some means of reconstructing the real image from the actual image obtained. The diffraction microscopy due to Gabor (1949) is based on just such a procedure.

16.5.3 Gabor's Microscope

In the method which Gabor suggested, the object is illuminated by a virtual electron source extremely small in (Gaussian) diameter, that is, by a highly coherent radiation. A photographic plate is placed beyond the object, just as in the ordinary shadow microscope, and records the diffraction pattern which is formed by the interference between the electron waves which arrive directly from the source and the waves which are scattered by the object. This pattern (which Gabor calls the "hologram") is transformed into a transparent positive, and reprojected with the aid of a light source, S', *which has the same aberrations as the original electron source, S.* By sighting an eyepiece through the hologram onto the plane originally occupied by the object, a highly magnified "reconstructed" image is seen, in theory free of aberrations. The electron wave which is incident on the plate contains all possible information about the structure of the object, and with a knowledge of the form of this wave, it is, in principle, possible to reconstruct an image, the resolving power of which is the same as that of an electron lens with all the geometrical aberrations corrected, in which the illuminating beam has the same semi-aperture as the projecting beam. Unfortunately, the photographic plate only records the amplitude distribution in the wave, irrespective of the phase relations—an analogous limitation is present in the Bragg X-ray microscope. The ensuing loss of information is revealed physically in a most singular fashion by the simultaneous appearance of a second reconstructed image placed symmetrically behind the source. This "false image" becomes more troublesome the nearer it lies to the real image; this, most inconveniently, is the case when the projection source is placed very close to the object in order to illuminate only a very small area of the latter, and to reduce the exposure to an acceptable value. For further information about this brilliant method, the reader is referred to Dyson (1950), Haine and Mulvey (1952), Haine and Dyson (1950) and Gabor (1951).

16.6 NEW TENDENCIES IN ELECTRON MICROSCOPY

We have just reviewed the extremely numerous factors which broaden the spot which we call the "image" of a point. The principal part of this chapter has been given to defects which in theory can be corrected, but which are less easy to eliminate in practice. This explains the asymptotic form of the curve which illustrates the progress which has been made in improving resolving power. Technical advances allow us to approach the limit which is fixed by the optimum resolving power, but every fresh step is harder than the last. We might well imagine that the resolution would be improved by an increase of the operating potential. We have already seen that the theoretical resolution limit is lowered when the associated wavelength λ is reduced (§ 16.2). The theoretical chromatic aberration too is diminished when Φ_0 is increased, since the energy spectrum of the electrons emitted by the filament and the discrete amounts of energy which the electrons lose in passing through the object (see § 19.1) remain constant in absolute value. Furthermore, when the object is thick, the velocity dispersion introduced by the passage of the electrons through matter decreases very rapidly as Φ_0 increases. When Φ_0 exceeds 1 MV, however, it becomes ever more difficult to keep the fluctuations of the accelerating voltage within the range 10^{-5}–10^{-6}, as it is not easy to obtain stabilities comparable with those of standard microscope generators (which operate in the range $50 < \Phi_0 < 100$ kV). At present, therefore, the aggravated experimental problems have prevented the resolving powers attained in the various attempts to use very high tensions in microscopy (Müller and Ruska, 1941; von Ardenne, 1941; Zworykin, Hillier et al., 1941; van Dorsten, Le Poole et al., 1947; Tadano, Sakaki et al., 1956; Tadano et al., 1958; Kobayashi et al., 1958; Dupouy and Perrier, 1960, 1964) from reaching the level of those obtained with ordinary electron microscopes. The improvements that have been achieved during the years 1962–6, and which are the fruit of unremitting labour, have been such that the practical resolution has come considerably closer to the theoretical limit and we can reasonably hope that the same will sooner or later be true of very high voltage microscopes. In addition, the new perspectives that have been opened up before the experimentalists by the possibility of studying thick metallurgical or biological specimens together with this real hope that a practical resolution of about an Ångström will be reached, explain the spectacular increase in the number of very high voltage microscopes (500 kV microscopes: Shimadzu et al., 1966, Tadano et al., 1966; 750 kV: Smith et al., 1966; 1 MV: Watanabe et al., 1966, Nishigaki et al., 1966). The absolute record is still held by Dupouy and his colleagues (Dupouy and Perrier, 1960, 1960a, 1964a, 1964b; Dupouy et al., 1960b, 1961) with $\Phi_0 = 1\cdot2$ to $1\cdot5$ MV, but even more ambitious designs are being studied (3 to 5 MV, Gilroy et al., 1966). To go beyond $1\cdot5$ MV while keeping the focal lengths of the lenses short and their aberration coefficients small, new lines of approach will be necessary, however: superconducting lenses with very high fields and extreme temporal stability (see §17.4.5) or quadrupole lens systems equivalent to round lenses. The familiar types of high voltage generators and electrostatic accelerating columns will certainly give way to linear accelerators and, at the other end of the instrument, new methods of recording the image that are satisfactory at very high energies will have to be found.

A further point is that microscopes are being operated more and more with a "clean" vacuum, in the vicinity of the object at least, so as to avoid contamination of the specimen by organic vapours and molecular fragments of carbon compounds, broken by the electron beam; if deposits of these are allowed to form, image contrast rapidly deteriorates and hence the resolution becomes poorer.

The results which are being attained allow us to hope nevertheless that resolving powers of the order of an Ångström unit will be reached in a few years, resolving powers which are, that is to say, of the same order of magnitude as the radius of an atom. The conditions for the microscopic examination of an atom have been studied by Gabor (1948) and Louis de Broglie (1950), who showed that as the beam electrons set the atoms in vibration, they will disturb the image. In the light of the recent experimental results described above which show that crystal lattices can be observed, we may conclude that these predictions were too pessimistic. Since the resolving power (for periodic specimens) has been reduced below 10 Å, a considerable amount of work has been devoted to this problem, and especially to the determination of the contrast in the image plane. If an atom is to be "seen", the contrast between the image point and the continuous background must be at least 5 per cent. Heidenreich and Hamming (1965) have shown that the classical amplitude contrast arising from the inelastic scattering of electrons outside the aperture would be of the order of 1 per cent for isolated atoms. Phase contrast, caused by interference between the diffracted waves and the direct wave near the image plane, could on the other hand be higher. An analytic investigation of this problem must, however, take the spherical aberration of the objective into account, for this deforms the wave-surfaces, as well as possible defocusing at the objective. A first calculation of the interference in the vicinity of the image plane was performed by Glaser (1949); this was reconsidered by Heidenreich and Hamming (1965) and later by Eisenhandler and Siegel (1966a).

It is not unreasonable to think that it will be possible to observe atoms in the near future, when objectives with little spherical aberration become available, and we operate at very high voltage (since the contrast is proportional to the relativistic increase in electron mass) and very low temperature (to reduce the thermal motion of the atoms in the crystal lattice to the utmost). The phase contrast will be enhanced by placing an aperture resembling a "Fresnel zone plate" at the exit plane of the lenses (Hoppe, 1963; Lenz, 1963; Hanszen et al., 1964, 1965; and Eisenhandler and Siegel, 1966b).

CHAPTER 17

THE OPTICAL ELEMENTS OF A MICROSCOPE

17.1 THE OBJECT ILLUMINATION

17.1.1 The Structure and the Function of an Electron Gun

Formally, the gun resembles the electron gun of an oscillograph, but the properties are rather different (§ 8.3). The source of electrons is in general a filament of tungsten or tungsten–molybdenum, about 0·1 mm in diameter, bent into the form of a V (Dlugosz, 1950), and raised to a temperature between 2600 and 2800°K by Joule heating† (which entails a current of a few ampères). The grid, which surrounds the filament, is an electrode in which a circular hole between 1 and 3 mm in diameter has been bored and which is centred on the point of the filament. The accelerating potential is applied between the filament and the anode—the latter is normally earthed. A small potential difference (one-hundredth of the total potential drop, for example, or about 500 V) is applied between the filament and the grid. The diameter of the opening in the anode is also a few millimetres, but its size is not critically important.

Microscope guns are of two types. The first type is wholly analogous to the gun of an oscillograph, and provides a divergent beam of aperture α_0 (which is of the order of a few thousandths of a radian) which appears to have originated in a virtual source of extremely small diameter ($2 r_0 \simeq 10$ to 40μ, for example). This beam could be used directly to illuminate the object, but in fact a condenser—usually magnetic—is always added, so that the way in which the object is illuminated can be modified.

To avoid using a condenser, Bruck and Bricka (1948) and Steigerwald (1949) have modified the form of the grid so that the beam converges slightly as it leaves the gun, and forms a real crossover a distance L from the accelerating anode. The angular apertures are of the same order of magnitude as those quoted for the preceding case if L is of the order of 10 or 15 cm; the density of the current which passes through the object

---

† A thorium oxide-coated tungsten filament is sometimes used—this increases the emissive power considerably.

is then high, and may reach one or two A cm⁻². In Fig. 201a, Bruck and Bricka's gun is shown schematically, and in Fig. 201b that of Steigerwald, **which is called** *Fernfokus*. **It is not possible to obtain a crossover as** small as the virtual source of a gun provided with a condenser lens with guns of this type; the diameter $2\,r_0$ is reduced only with difficulty below

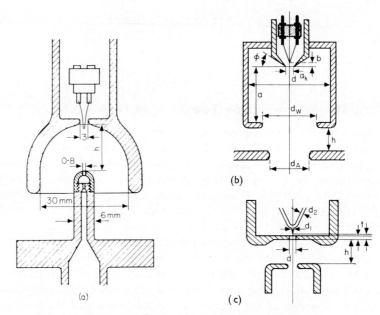

FIG. 201. Two types of gun which work without a condenser.
(a) Gun with spherical electrodes (Bruck and Bricka, 1948); $h = 17$ mm.
(b) *Fernfokus* (Steigerwald, 1949). The essential figures for a possible version (Braucks, 1958) are as follows:
$a \simeq 45$ mm, $d_w = 50$ mm, $d_A = 20$ mm,
$d \simeq 60$ mm, $b = 0$, tan $\varphi = 0.55$,
$h = 10$ mm, $a_k = 1$ mm ($a_k =$ diameter of the bore in the conical inset of the Wehnelt).
c) Gun employing a point cathode (Sakaki and Möllenstedt); the various parameters have the following values: $d_1 = d_2 = 0.2$ mm, $h = 8$ mm, $d = 1$ mm, $l = 0.7$ mm, $t = 0.5$ mm, $V_g \simeq 550$ V (or $d = 2$ mm, $l = 1.4$ mm, $t = 1$ mm, $V_g = 200$ V).

0.3 to 0.5 mm, for $L \simeq 10$ to 15 cm. It is, however, possible to retain only the central part of the beam, in which the current density is greatest, by placing a very small diaphragm (20 to 50 μ in diameter) beneath the anode.

The fundamental quality of a microscope gun is the electron brightness

$$B = \frac{I}{\pi \alpha_0^2 \cdot \pi r_0^2},$$ which is measured in A cm⁻² sterad⁻¹. This expression

defines a mean brightness. Placing a stop close to the anode is a very useful artifice in electron probe devices, as a source of reduced area is obtained, with the same brightness as the central zone (see Chapter 21).

The maximum brightness, B_m, can be roughly estimated with the aid of Langmuir's formula:

$$B_m = \frac{\varrho_c \, \Phi}{\pi \, k \, T} \simeq 3\cdot 7 \times 10^3 \, \frac{\varrho_c \, \Phi}{T} \, ,$$

in which ϱ_c is the maximum current density at the centre of the cathode measured in A cm^{-2}, Φ_0 is the accelerating potentials in volts, k is Boltzmann's constant $\left(\dfrac{1}{11,600} \, \text{eV} \, {}^\circ\text{K}^{-1} \right)$ and T is the absolute temperature of the cathode ($^\circ$K).

For a tungsten cathode held at temperature $T = 2500^\circ$K, with $\Phi = 50$ kV, we find that the brilliance is $2\cdot 2 \times 10^5$ A cm^{-2} sterad^{-1}; the order of magnitude accuracy of this result has been confirmed by experiment (Haine, 1947, 1950, 1957; Haine and Einstein, 1952; Boersch, 1954). If T is increased to 2800°K, a brightness of the order of 3×10^5 A cm^{-2} sterad^{-1} can be obtained but the emission is then restricted by space charge and the lifetime of the filament is also considerably shortened. Haine, Einstein and Borcherds (1958) have measured the brightness at this temperature, and found values $B_m \simeq 1\cdot 6 \times 10^5$ A cm^{-2} sterad^{-1} in a divergent gun with a condenser.

The whole illuminating system forms the "head" of the microscope—it is articulated in various ways; with the knobs provided, the filament can be moved laterally and in height with respect to the grid, and the beam can be "aligned", by moving the whole system bodily transversally, or by rotating it around the object (for the case of the gun without a condenser).

17.1.2 Electron Sources

(i) The heated cathode

A tungsten filament is currently used. The electrical power which has to be dissipated to obtain a suitable emission is of little importance, as in microscopy it is the lifetime of the filament which is all-important; a value of 100 hours is readily acceptable, and von Ardenne (1940) finds that the temperature dependence of lifetime is given by the values in the following table:

T ($^\circ$K)	2300	2420	2525	2640	2750	2825	3000
Lifetime (hours)	287000	30700	4360	770	150	53·5	5·5

Haine *et al.* (1958) give the following empirical formula for the lifetime t:

$$t = \frac{50}{\varrho_c} \text{ hours}$$

(ϱ_c, measured in A cm^{-2}, is the maximum current density at the cathode).

Again, it is of no particular interest to increase the specific intensity incident upon the object beyond the values which are ordinarily obtained with a tungsten wire, as the power which is dissipated as heat in any object which is not exceptionally thin is already close to the maximum which can be tolerated. For this reason, the improvements which are being made at present consist in giving the gun a more complicated structure, so that only the minimum region of the object which lies within the field of the microscope is illuminated (an area of some $20\,\mu$ by $20\,\mu$).

It is only with this kind of gun, and in exceptional circumstances (a very short exposure, or a very thin object, for example) that it can be advantageous to increase the specific emissivity of the source.

In general, the tungsten filament which is employed is bent into the form of a hair-pin (see Fig. 201), and it is the point of the filament, situated on the axis, which effectively constitutes the electron source. To obtain beams of very small aperture, and higher brightness, Hibi (1954) and later Sakaki and Möllenstedt (1956) and Swift (1960) have studied cathodes in the form of a very fine point, of which the useful radius of curvature is a few microns—this point is formed at the tip of a wire by electrolytic polishing. The wire is heated from behind the point by leading in a lateral current. This wire may either be of tungsten or of platinum, coated with barium and strontium oxide, which require only very slight heating. The aperture of the resulting beam is of the order of 1 to 3×10^{-5} radians, and this type of source is particularly well adapted for electron interference experiments, as the beam is highly coherent. A gun using a point cathode is shown in Fig. 201 c.

(ii) *Induni's cold cathode*

In the first electron microscope which Knoll and Ruska (1932) constructed, the illuminating electron beam originated in a gas discharge, at low pressure (about 10^{-2} mm Hg), between an anode and a cold cathode; the discharge, however, was unstable and difficult to control, and variations of potential of more than one per cent were found between the terminals, which produced a variation in the energy of the electrons of the same order. For this reason, a hot cathode was used instead in subsequent instruments, and it was generally admitted that a cold cathode is not suitable for a microscope (von Ardenne, 1940; Gabor, 1945; and Zworykin *et al.*, 1949).

Induni (1947, 1955), however, showed that this conclusion must be revised, by designing a source which resembled that of Knoll and Ruska in appearance but which he could control easily and with which he could obtain images with a resolution of about twenty Ångströms. Figure 202 indicates the structure of this type of source—a potential of 40 kV is applied to the central rod (which is made of aluminium or 18/8 stainless steel); the anode,

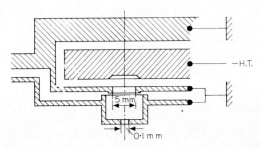

FIG. 202. Induni's cold cathode: cross-section of the electrodes.

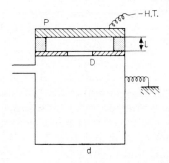

FIG. 203. Diagram illustrating the principle behind the cold cathode; the separation of the ionization zone from the acceleration zone is shown.

which is connected to earth, is a concentric cylinder about 5 mm away, with a large opening cut in it. A very fine diaphragm limits the flow of gas. The lifetime of this gun is virtually unlimited, and the geometrical features and the specific intensity of the beam which it provides are satisfactory and comparable with those which are obtained with a hot cathode. The electrons which are provided by this type of gun are virtually monoenergetic; the reason for this can be understood by examining Fig. 203 which illustrates the principle of the gun. The chamber contains air at a low pressure; any ion in this chamber (which is virtually equipotential) is slowly drained off towards the diaphragm D by the weak field which overlaps through the aperture in this diaphragm. It then flies towards the high tension plate P, which in turn emits secondary electrons with a very small initial energy. These electrons cannot create ions before they reach the chamber, as the distance l between P and the diaphragm D is less than their mean free path.

Some of them do produce ions inside the chamber, and thus maintain the discharge, but these ions can only subsequently produce electrons at the plate P, and neither in the chamber (where their energy is too small) nor in the accelerating region between P and D (where the trajectory is too short). The electrons all originate in P and will therefore be mono-energetic when they leave the diaphragm d; this is not the situation in an ordinary discharge tube, where the electrons may be produced anywhere within the field and emerge (through a Lenard window, for example) with widely varying velocities. The basic advantage of the Induni cathode is that the ionization region and the acceleration region are separate.

17.1.3 The Electrical Supply Circuitry—Autobias

Our aim is to achieve stable behaviour at the gun, in the sense that we strive to obtain an electron current I_e of constant intensity. Variations in I_e would alter the brightness, affect the specimen heating (and hence give rise to transverse migrations of the object) and above all, create fluctuations in the high tension itself since high tension sources have a relatively high impedance. The variations in I_e are essentially due to fluctuations of the filament temperature T, as the emissivity varies exponentially with T. Present-day high tension sources are stable enough (one part in 10^5 or 10^6) for their fluctuations to have virtually no effect upon the emission.

Figure 204 shows two ways in which the potential of the gun can be supplied. Every effort must be made to make the supply as stable as possible. In the arrangement (a), the filament is heated with the aid of a high frequency transformer (100 kc/s) which is easily made as both the primary and the secondary windings have only a few turns; any thermal fluctuation at the filament is eliminated by using a high frequency. The high tension is applied through a potential divider in such a way that the point of the

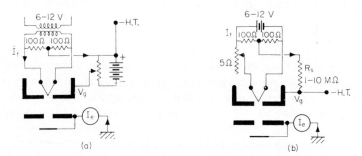

FIG. 204. The two possible ways of supplying an electron gun.
(a) The Wehnelt, or grid, is polarized with batteries. (b) The Wehnelt is polarized automatically by the potential drop created by the gun current as it passes through the polarization resistance R_s.

filament is held at a constant potential. The grid is supplied with the aid of a potentiometer connected to batteries.

The system (b) which is biased automatically is, however, more usual. The filament may still be heated at a high frequency, but often (as illustrated in Fig. 204 b), it is fed by an accumulator, while the current is varied with a well-insulated rheostat. The bias is produced by the potential difference across a high resistance (between 0.5 MΩ and a few dozen MΩ). A condenser is often added to eliminate a.c. ripple. The negative bias V_g of the grid is proportional to the current, I_e. When I_e increases, the bias also increases and hence I_e tends to diminish: we therefore have a self-regulating mechanism for stabilizing the current.

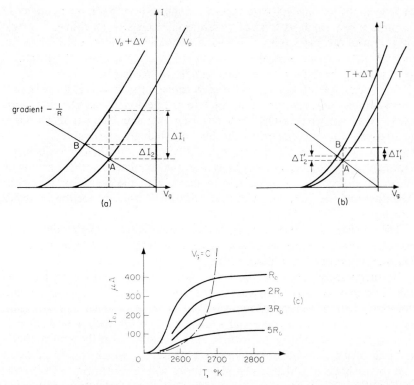

FIG. 205. The difference between the behaviour of the gun supplied in these two ways (see Fig. 204).
(a) The anode potential is increased by an amount ΔV. The yield of the gun with independant polarization is increased by ΔI_1, whereas that of the automatically polarized gun is increased only by ΔI_2. (b) The filament temperature increases by an amount ΔT. The increase in the output of the independently polarized gun is considerably larger than that of the automatically polarized gun. (c) The influence of the resistance R_s on the characteristic $I_e = f(T)$ of a triode gun; the appearance of a kind of saturation.

Figure 205a shows I_e as a function of the bias, V_g. When V_g is constant (batteries), a variation $\triangle V$ of the potential of the high tension supply results in the variation $\triangle I_1$, shown in Fig. 205a; when, on the contrary, the system is autobiased, the working-point moves from A to B, and the change $\triangle I_2$ is far smaller. In Fig. 205b, two curves representing the emission at slightly different temperatures T and $T + \triangle T$ are plotted; the current is stabilized by an analogous mechanism, using autobias. In this latter case, the stabilization is better to the extent that the two curves are close together. If, however, we examine the curves which describe the variation of I_e, as a function of the temperature, T, for a given voltage V (Fig. 205c) it is apparent that if V_g = constant, the curve is of the exponential form which we should expect, whereas if the system is autobiased, the variation in I_e tends to become stabilized, even though at first it is more acute as V_g is initially very small. A kind of "saturation" appears, which has its origin in the autobias mechanism itself; the saturation is more noticeable the larger is R_s or, what is equivalent, the closer V_g lies to the cut-off potential V_{gc} of the gun. The total current falls rapidly as R_s increases.

For each type of gun, therefore, we must bear two factors in mind: (i) we should select a working point in the region where the saturation curve is as flat as possible, after ensuring that the filament temperature is relatively low in order that the lifetime shall be reasonable; (ii) at the same time, we should, by modifying one of the dimensions, for example (the filament-grid distance, or the diameter of the opening in the grid), make this working point coincide with the region of maximum brightness of the gun, which is a function of V_g (see the section on triode guns).

The focusing of a microscope gun therefore involves a meticulous experimental study. An example of such a study is to be found in an article by Haine, Einstein and Borcherds (1958).

Further improvement on the above arrangements is still possible, with the result that we can obtain a saturated region in which I_e remains virtually constant when the heating current varies by as much as 30 per cent. A biasing system composed of a potential source V_g in series with a resistance R_s is used; this system may be formed very straightforwardly with a triode, the grid of which is held at a positive potential, v, which can be adjusted, and the cathode of which is placed in series with a resistance r (Dolby and Swift, 1960). The valve then behaves like a voltage generator, with e.m.f. $V = -\mu v$ and internal resistance $R_s = (\mu + 1) r$; μ is the amplification factor.

The current I_e can be described by a law of the form

$$I_e = K \left(\frac{V_{gc} - V}{V_{gc}} \right)^n \qquad n \simeq 2,$$

in which V_g is the bias, V_{gc} the cut-off potential of the gun, and K a coefficient which depends only upon the emission of the cathode (which is a function of T). The accelerating potential V and the geometry of the gun only affect V_{gc}, since for all guns, $n \simeq$ constant.

If we define a stabilization index, ε, for fluctuations of T, in the following way:

$$\varepsilon = \frac{dI/I}{dK/K},$$

it is clear that $\varepsilon = 1$ when $V_g =$ constant (battery-operated polarization) which corresponds to a complete absence of regulation. When V_g tends towards V_{gc}, ε decreases, which clearly displays the fact that the stabilization is much better in an autobiased gun the nearer it is to cut-off. For $V_g = 0.8 V_{gc}$ for example, we should find that $\varepsilon \sim 0.1$, a tenfold gain with respect to fixed polarization guns.

With the final arrangement, we can always hope to gain yet another factor of ten when $V_g = 0.8 V_{gc}$.

17.1.4 The Role of the Condenser

The condenser is a magnetic lens, with which we can adjust simultaneously the angular aperture of the beam which falls on a point of the object, and the current density which passes through the object.

Consider the situation shown in Fig. 206, in which the crossover, which, as we explained in § 17.1.1 plays the part of the electron source, illuminates the object directly. The pencil of electrons which passes through a point on the object has an angular aperture which is given by

$$\alpha_0 = \frac{r}{p_1},$$

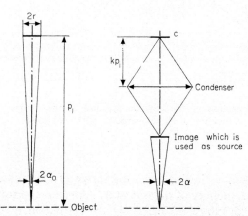

FIG. 206. In the absence of a condenser lens, the object is illuminated at a fixed angle $2\alpha_0$. The condenser produces an image of the crossover which acts as the source and with which the angular aperture of the illumination can be varied.

in which r is the radius of the crossover. We now insert a lens of convergence C at a distance kp_1 from the crossover. This lens will produce an image of the crossover which will now play the part of the source of the illumination. A simple calculation then shows that

$$\alpha = \alpha_0 \frac{1}{k(1-k)p_1 C - 1}.$$

It must not be forgotten, however, that the condenser contains a stop of radius R, so that the aperture of the beam can certainly not exceed

$$\alpha_m = \frac{R}{(1-k)p_1}.$$

Finally, therefore, we shall obtain the curve shown in Fig. 207, which displays the dependence of α upon the convergence of the condenser. There are, therefore, two possibilities: (1) We can obtain a large aperture, accompanied by a high electron density at the object, which we shall use

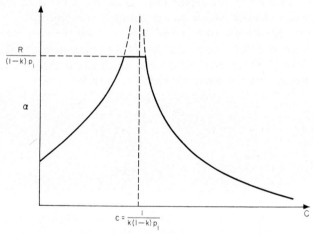

FIG. 207. The variation of the angular aperture of the illuminating beam as a function of the convergence of the condenser.

for focusing visually; in this case, the image of the crossover is formed in the immediate vicinity of the object. (2) A small aperture, on the other hand, has the advantage of increasing the resolution of the instrument, but compels us either to go over immediately to photographic recording because of the low intensity or to take special precautions and accept the resulting fatigue to the observer. The latter has to remain for at least a quarter of an hour in darkness, before attempting the adjustment, while his eyes become adjusted into a state of maximum sensitivity; the optimum physiological condition characterizes the properties of what is known as "scotopic" sight.

17.2 GENERAL CHARACTERISTICS OF
MICROSCOPE LENSES

It may perhaps surprise the reader to meet a chapter on electron lenses at this point in the volume, but recently, important work has been performed on these lenses and it is useful to have a rapid summary of the results as the lenses are the vital organs of the microscope; the construction of the objective represents the summit of perfection of experimental technique. All the lenses with high magnification which are used in microscopes possess certain characteristics as a result of which they form a well-defined class within the plethora of lenses which are used in television tubes, image converters, and mass spectrographs. It is, therefore, a fit point at which to bring together the new results.

17.3 ELECTROSTATIC LENSES

17.3.1 Significant Geometrical Parameters

The electrostatic einzel lens, with its three electrodes, has so far always been used in the symmetrical form. To avoid intense fields (the effect of a point), the rims of the holes must be rounded, and the lens is defined geometrically by a rather large number of parameters (Fig. 208). These are:
the interelectrode spacing $\triangle = z_2 - z_1$;
the thickness of the central electrode $e = 2z_1$;
the diameter of the diaphragm of the central electrode $D = 2R_1$;
the diameter of the diaphragms of the outer electrodes $d = 2R_2$;
the radius of curvature ϱ of the edges of the electrodes.
The parameter $\triangle$ cannot be less than a certain limit, as it represents the distance between two metal parts at different potentials, and the minimum is determined by the breakdown condition.

Experiment shows that the influence of the thickness of the outer electrodes on the distribution of potentials is negligible. The curve shown in Fig. 209 (Regenstreif, 1951) shows that the axial potential directly below an external electrode is virtually Φ_2. This result shows that even a marked lack of symmetry between the entry aperture and the exit aperture only modifies the focal length very slightly.

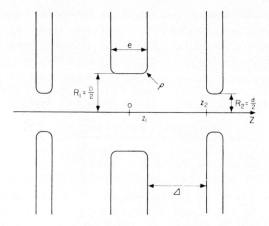

FIG. 208. The geometrical parameters of an electrostatic microscope lens.

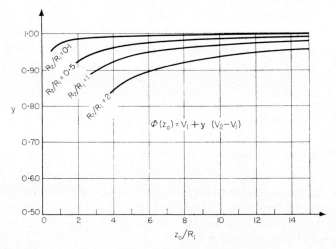

FIG. 209. The value of the axial potential directly below the outer electrode (the abscissa denoted by z_2 in the text is here rechristened z_0). The potentials of the central and outer electrodes are V_1 and V_2 respectively. It can be seen that y is always close to unity (Regenstreif, 1951).

So far, the influence of ϱ has not been studied. Certain workers have chosen ϱ to be equal to $e/2$ (Bruck and Romani, 1944) while others select fixed values for ϱ/L (Lippert and Pohlit, 1952). We write

$$L = 2\Delta + e.$$

Finally, the diameter of the outer electrodes only plays a minor role, at least provided d is small, as the electrons are moving at high velocities when they are level with these electrodes.

We conclude, therefore, that the parameters which are optically important are $\triangle$, e and D. Regenstreif's results have been given in the first part.
We shall complete this description of lenses with a résumé of the results
of a series of precise measurements which have been made recently by
Lippert and Pohlit (1952, 1953 and 1954) which have yielded valuable data.

17.3.2 The Results Obtained by Lippert and Pohlit

These results relate to lenses of which the central electrode is at cathode
potential. Figure 210 gives values of L/f as a function of the parameters
D/L and e/L.† The lenses which are defined by $\partial f/\partial v = 0$, where v is the
initial velocity of the electrons ("apparent" achromatism), and which are

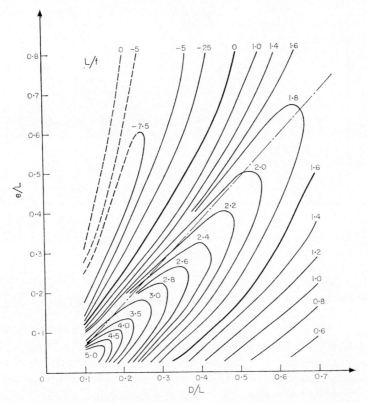

Fig. 210. Curves showing the convergence of a series of einzel lenses (Lippert
and Pohlit, 1952).

† These curves are slightly different from the ones which Lippert and Pohlit (1952)
have published, as they are reproduced from unpublished papers which have been kindly
put at our disposal by these authors. A few points which represent results obtained
in our own laboratory have been added.

free of distortion lie on the dotted line near $e/L = D/L$. For $e/L < D/L$, the lens has cushion distortion and $\partial f/\partial v$ is positive. Between the achromatic lenses and the curve $L/f = 0$ (at which the lens is afocal) lies a region in which the lenses suffer from barrel distortion, for which $\partial f/\partial v < 0$. Beyond

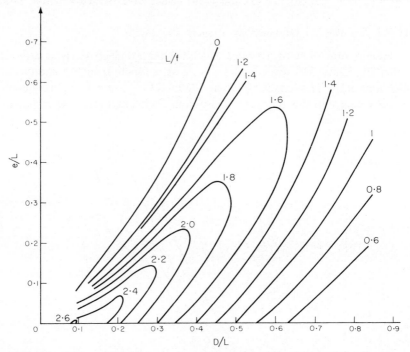

Fig. 211. Curves giving the convergence of a series of einzel lenses (after Regenstreif).

is the transgaussian region (where the focal length is negative), at the beginning of which the distortion is again cushion, but $\partial f/\partial v$ is still negative.

Figure 211 shows the same set of curves, as calculated theoretically by Regenstreif. The correspondence between theory and practice is satisfactory, as we can see, making allowance for the imperfection of the theoretical model adopted which had to be simplified before the calculations could be concluded.

Figure 212a represents the behaviour of the abscissa of the focus in the same coordinate system. Any of the lenses which correspond to $p/L > 0.5$ can be used as objective. The first curve, $p/L = 0.5$ corresponds effectively to the parallel curve $L/f = 1.6$ of the preceding diagram; the focal length of such an objective is in practice determined only by L, therefore—the interelectrode spacing fixes the lower limit of the focal length; the focal

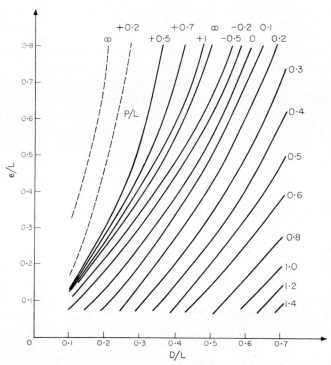

Fig. 212a. Curves giving the abscissae of the foci of a series of einzel lenses
(Lippert and Pohlit, 1952).

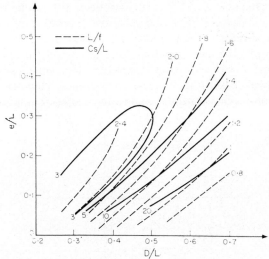

Fig. 212b. Curves giving the spherical aberration and the convergence of a
series of electrostatic einzel lenses, plotted from the results obtained by Lippert
and Pohlit (1953) and Bruck and Romani (1944).

length of an "optimum objective", working in the unipotential fashion, is therefore about 60 per cent of the total thickness of the lens.

Figure 212b gives the values of the reduced spherical aberration coefficient C_s/L as a function of the parameters e and D. This set of curves is based upon the results of Lippert and Pohlit, together with those of Bruck and Romani (1944); only the region which corresponds to conditions in the neighbourhood of an objective for which $L/f \geqq 1·6$ is represented. The spherical aberration constant has a minimum in a region which, on the "convergence diagram" lies between distortion-free lenses and objectives. It can be seen that the spherical aberration constant for objectives is reduced according to the extent to which the focus lies within the electrode.

17.3.3 Einzel Lenses with Unsymmetrical Electrodes

At present, lenses of this kind have rarely been studied, still less used in practice, despite the fact that in an experimental investigation of the spherical aberration constant of various einzel lenses, Seeliger (1948) found that one very unsymmetrical lens, with a central electrode, had an extremely small spherical aberration constant ($K_s \simeq 3$).

To determine experimentally the optical elements of these lenses is a particularly delicate task. Heise's method (1949) which is suitable for symmetrical lenses has been extended by Everitt and Hanssen (1956) to the case of lenses with any kind of asymmetry. Hanssen (1958) has also shown that the optical properties of any unsymmetrical lens are well represented by those of a doublet of two thin lenses, the more convergent of which lies on the side in which the field $E_z = -\varphi'(z)$ is stronger. It has long been known in ordinary optics, however, that the spherical aberration of an objective is least when the surface of the lens with the greatest curvature—where the convergence changes the most rapidly—faces the object. In the same way, an unsymmetrical einzel lens should produce less spherical aberration than a symmetrical lens with the same convergence. Hanssen (1958) has been able to confirm this hypothesis by making such a comparison. Septier

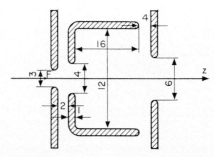

FIG. 213. Asymmetrical lens with small spherical aberration for an object placed in the neighbourhood of F ($C_s \simeq 2·5$); all quantities in mm; $f_{min} \simeq 10$ mm.

(1960) has recently determined the optical properties of a number of electrostatic lenses used with very wide aperture incident beams, and gives values
of K_s for several unsymmetrical lenses together with an analysis of the
influence of the various geometrical parameters on K_s. It is in fact possible
to obtain values of K_s of the order of 2·5, whereas even the best symmetrical
lenses only give

$$12·5 < K_{s(min)} < 20.$$

These high quality lenses (Fig. 213) do, however, have the disadvantage
of being relatively weak ($f_{min} \simeq 8$ to 10 mm) and can thus only be used as
low magnification objectives either in the composite immersion objectives
described in §8.3.2, or finally in electron probes where we need the focus
to be outside the electrodes (see §21.5).

17.4 MAGNETIC LENSES

17.4.1 Construction of the Electromagnet

(i) *The form of the magnetic circuit*

A magnetic lens is an axially symmetric electromagnet or permanent
magnet, through the pole-pieces of which is bored a channel along which
the beam of electrons passes.

The field in the gap of a magnetic lens must be intense, and localized into
a short axial distance; a field which overlaps the edges of the gap too far
is not only useless but harmful, as it produces centring problems. There is
one factor, however, which limits the extent to which the field can be
localized and intensified, which is the saturation of the pole-pieces; this,
therefore, it is convenient to reduce to the maximum if not to eliminate

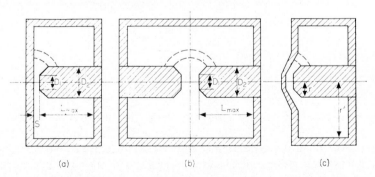

Fig. 214. The possible forms of the core of a microscope lens. (The central channel
is not shown.)

altogether by a judicious choice of lens design. The designer is likewise guided in his choice of the shape of the pole-pieces by considerations of a practical order: for example, the objective has to be designed in such a way that the object can be easily displaced radially.

The great majority of magnetic lenses lies somewhere between two extreme arrangements, the asymmetrical version shown in Fig. 214a, and the symmetrical version shown in Fig. 214b. For an objective, the principal advantage of the asymmetrical arrangement is that the focus can be completely detached and the mechanism for displacing the object can, in consequence, be simplified considerably. More extreme arrangements have been used in certain exceptional cases (Fig. 214c); one of these is the electron probe microanalyser, where the beams of X-rays have to leave the object at an appreciable emergent angle, but this is a very special case, in which we are forced to use the lens in unfavourable conditions (Castaing, 1954; Mulvey, 1958).

(ii) *The distribution of magnetic field; the optimum form for the pole-pieces*

We now consider the symmetrical arrangement. The lines of force are distributed as if the symmetry plane P of the gap had been replaced by a plane of some infinitely permeable substance. A symmetrical lens, therefore, can be regarded as two asymmetrical lenses in juxtaposition and it was this

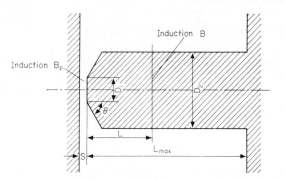

Fig. 215. The significant parameters in a magnetic lens.

property which led Mulvey (1953) to study the distribution of magnetic flux in a single asymmetric lens; he began with a lens with unbored pole-pieces, away from the saturation conditions, and performed the experiments on the lens itself, using a fluxmeter. The meaning of the various parameters is shown in Fig. 215, and to begin with we shall suppose that $\theta = 0$, so that $D_1 = D_2$.

A short coil is moved along the core, starting from the gap, towards the base of the core which is a distance L_{max} from the gap. The field distribution clearly shows that the flux which passes through this coil increases steadily with the distance between the coil and the gap; the induction B in the core at a distance L from the gap increases from the value B_p (which is obtained in the gap) for $L = 0$ to a considerably larger value which is obtained for large values of L. The graph of Fig. 216a, which Mulvey obtained, shows the way in which B/B_p varies with L/D_1; this variation is more rapid when the separation S between the pole-pieces is wider, and consequently, the field within the gap itself is less localized.

This is a particularly disadvantageous form to give to the pole-pieces— if the lens excitation is increased, the value of the induction rapidly reaches the saturation induction B_s of the mild steel which is used at the junction between the pole-piece and the end-face. The useful induction B_p in the gap is always less than B_s; if we require B_p to be equal to B_s (16 to 20 kG for ordinary mild steel; 22 kG for Armco iron; 25 kG for special cobalt steel), the form of the pole-pieces must be chosen in such a way that the induction decreases steadily within the iron as we move away from the gap, despite a magnetic flux which, on the contrary, increases steadily since the lines of force of the overlap field are to be added to those of the useful field. The solution is to construct pole-pieces in the form of a truncated cone ($\theta \neq 0$); the cross-section then increases with L, and for any given value of S/D_1, we can find a value of θ such that B/B_p is a decreasing function of L (see Fig. 216b). For too wide an angle, however ($\theta \geq 70°$), the total flux becomes too large and the dimensions of the magnetic circuit too great, to no purpose; $\theta = 50°$ is a suitable value.

There is no point in flaring the core indefinitely; as soon as the induction reaches a low enough value, the diameter of the pole can remain constant (Mulvey, 1953), equal to D_2. The induction begins to increase again with L, and D_2 must simply be chosen sufficiently large for B to reach its maximum at L_{max}, and to remain of the order of $2 B_s/3$, for example. D_2 must be kept as small as possible to avoid giving the windings too large a mean radius, as too much Joule heat would be dissipated in supplying the lens normally, and to prevent the whole assembly from becoming prohibitively cumbersome. In Fig. 216c, curves describing the variation of the induction with the distance L are shown, in the two cases $S/D_1 = 0.2$ and $S/D_1 = 0.3$. The full curves correspond to a core of total length $L_{max} = 3 D_1$, and the dotted curves to $L_{max} = 6 D_1$. From Mulvey's study, we can extract the optimum characteristics of the magnetic circuit: $\theta = 50°$, $S/D_1 \simeq 0.2$ and $D_2/D_1 \simeq 3$. The presence of an axial hole of diameter D has no appreciable effect upon these conclusions—the total flux is practically the same after the channel has been bored. If D is increased, the induction within the pole-pieces increases; to diminish this effect, we have only to choose a value of the diameter D_1 such that $D_1 \geq 3 D$.

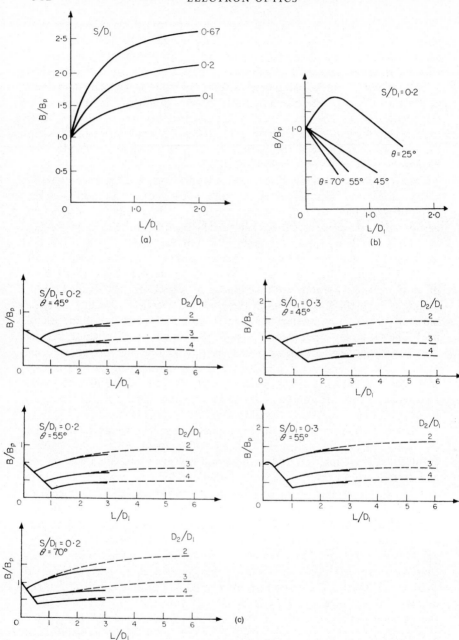

FIG. 216. The value of the magnetic induction on the axis of the magnetic core.
(a) For a core of constant cross-section ($\theta = 0$). (b) For a core which ends in
a truncated cone (Mulvey, 1953). (c) The value of the magnetic induction on the
axis of a magnetic core for various values of the parameters (Mulvey, 1953).

A very complete study has been made by Durandeau (1955, 1956), Durandeau and Fert (1957), and Dugas, Durandeau and Fert (1961), which allows us to build lenses with the minumum weight in which saturation appears at the position of the gap so that the induction is highest in the useful zone. The form of the gap is shown in Fig. 217a, and the shape of

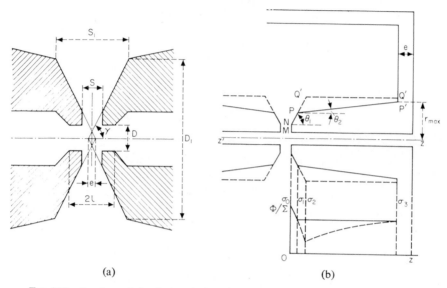

(a)   (b)

FIG. 217a. Section of the faces of the poles. The optimum values of the parameters are: $e \simeq s/3$; $(\pi/2 - \gamma) \simeq 25$ to $30°$; $S_1 \geq N I/500$; $2 l \geq S + \sqrt{(S^2 + D^2)}$ (Durandeau and Fert, 1957).

FIG. 217b. The magnetic circuit of a high powered lens and the variation of the mean density, Φ/Σ of the induced flux in the polar core. The broken line corresponds to the pole-faces in the form of truncated cones with cylindrical cores (Mulvey, 1953); the full line corresponds to the case where both pole-faces and cores are in the form of truncated cones (Durandeau, 1956; Dugas, Durandeau and Fert, 1961).

the pole-pieces in Fig. 217b. The truncated conical section, of angle γ is followed by a second truncated cone of angle $\theta_2 \simeq 12°$, rather than by a cylinder of diameter D_2; the magnetic flux then remains constant over the whole length of the second cone (Fig. 217b). The thickness of the cheek is such that $e \simeq \dfrac{r_{max}}{2}$, so that at the junction the cross-section is effectively constant.

For small excitations, the field distribution along the axis is the same whatever the form of the poles in any lens in which the ratio S/D remains unaltered (S now represents the total width of the gap). Replacing the actual distribution of induction by the distribution which would be produced by a solenoid, we arrive at the following formulae for the maximum induction B_m on the axis (of a symmetrical lens) and the width, $2a$, of the curve representing the distribution $B(z)$ at the point at which it has fallen to half the maximum height (Fig. 217c):

$$B_m = \mu_0 \frac{N I}{\sqrt{S^2 + 0 \cdot 45 D^2}} = B_1 \frac{S}{L},$$

in which

$$L = \sqrt{S^2 + 0 \cdot 45 D^2},$$

and

$$2a = 0 \cdot 96 L.$$

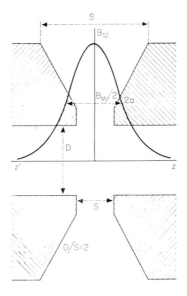

FIG. 217c. The field distribution on the axis in the region of the gap (symmetrical lens).

These formulae are valid for low excitations ($B_1 < 12{,}500$ gauss for ordinary mild steel; B_1 represents the field in the gap S when $D = 0$). B_m is always smaller than B_1. The length L which we have just defined plays an extremely important role in these lenses, and as we shall see later, it allows us to argue in reduced coordinates.

Durandeau has pointed out that the central hole of diameter D must be cylindrical for a certain distance $2l$, as otherwise $B(z)$ is deformed; it is convenient, therefore, to increase the diameter, both to facilitate cleaning and to avoid the possibility of electrons being reflected from the walls—in general, we choose $2l \geq S + \sqrt{S^2 + D^2}$.

(iii) *The influence of saturation*

According to Durandeau, we can establish a threshold of the excitation NI, in ordinary mild steel, beyond which the first saturation symptoms will appear in the vicinity of the gap. If S is expressed in millimetres, we find that

$$NI = 1000 S$$

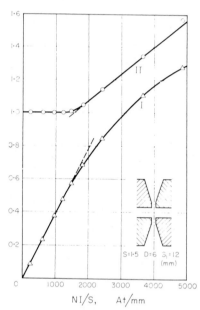

Fig. 217d. The variations of the maximum induction B_m and of the width a as functions of the excitation, expressed in NI/S as unit; the pole-pieces are of soft iron and the lens is symmetrical. Curve I: B_m (in teslas, or webers m^{-2}). Curve II: a/a_0 (Dugas, Durandeau and Fert, 1961).

is virtually true. For $NI > 1000 S$, the width $2a$ of the distribution $B(z)$ increases slightly, and the maximum induction, B_m, increases more and more slowly when NI increases (Fig. 217d). The lens properties are then well represented by a fictitious increase of S, D and a; for these new values S', D' and a' of S, D and a, we have

$$\frac{S' - S}{S} = \frac{D' - D}{D} = \frac{a' - a}{a} \simeq 0 \cdot 14 \frac{NI - 1000 S}{1000 S}.$$

When the induction is very large, the difference can reach 10 per cent. In every case, we can write $\dfrac{a B_m}{n I} = $ constant.

Further, if the magnetic circuit is poorly chosen, a parasitic induction B appears in the central channel near the zones where saturation occurs. This induction is extremely harmful, as it behaves like a weak lens placed before or behind the main lens, and the optical centring of the lens becomes very difficult.

If we are to be confident that the saturation will set in at the gap, the distance between the bases of the two truncated cones which form the first part of the poles (see Fig. 217a) must be greater than S_1 (mm), given by:

$$S_1 = 500\,N\,I.$$

This condition determines the initial diameter D_1 of the core of the pole. When the excitation is small ($N\,I < 1000\,S$), the circuit can be considerably simplified.

(iv) *The power which is necessary for excitation*

The minimum internal radius r of the winding is dictated by the diameter D_2 of the core. Let r' be the external radius of the coil of length h, s the cross-sectional area of the copper wire which is used, and N the number of turns. The total cross-section $S = h(r' - r)$ of the winding will be most efficiently filled by contiguous turns of carefully wound copper wire coated with the minimum thickness of enamel which will provide the necessary insulation. We have $N\,s = k\,S$, in which k is a factor smaller than unity, and equal to at most 0·7, say, to fix our ideas.

If $N\,I$ is the excitation, and ϱ the resistivity of the copper, the resistance of the whole winding is

$$R = N\pi\varrho\,\frac{(r + r')}{s},$$

and the power expended in the winding by virtue of its excitation is

$$W = R\,I^2 = \frac{R}{N^2}\,(N\,I)^2 = \frac{\pi\varrho(r' + r)}{k\,S}\,(N\,I)^2 = \frac{\pi\varrho(r + r')}{k\,h(r' - r)}\,(N\,I)^2.$$

The power which is required for any given excitation is therefore independent of the cross-sectional area of the wire, and hence of the driving potential also. Accumulators (6 V) are therefore equivalent to the modern regulators (250 V). There is an advantage in reducing the power, not so much to ease the load on the supply as to reduce the heating of the lens; this can be done by increasing h, which means increasing the size of the lens generally and produces an undesirable elongation of the core, but above all, by reducing r as far as possible; it is necessary, therefore, to prevent D_2 from being increased unnecessarily. To give an idea of the magnitudes involved, consider a copper winding with internal radius $r = 2$ cm, external radius

$r' = 6$ cm, and length $h = 10$ cm; for $k = 0.7$, we find

$$W = 1.5 \left(\frac{NI}{1000} \right)^2,$$

or about 6 watts for 2000 A turns, and 24 watts for 4000 A turns (a very high excitation). In the hypothetical case—clearly most unfavourable—in which the lens is thermally insulated, the heating is about $4°$ hour^{-1} for 2000 A turns and $16°$ hour^{-1} for 4000 A turns (the total mass of the lens is assumed to be 15 kg and its mean thermal capacity 0.1). In reality, the thermal contact between the windings, the yoke and the mounting provides an efficient and adequate cooling.

17.4.2 The Optical Properties of the Lenses

(i) *The normal elements*

In the projective lens or "projector" which is used to magnify the image formed by the lenses which precede it, it is the *normal* focal length f_1 which interests us, and above all, the very strong convergence. The spherical and chromatic aberration coefficients are of little interest, as the aperture of the beam is very small, but we must try to work in conditions of minimum distortion.

A detailed study of the variation of f_1 with the excitation for a number of different geometries has been carried out by Liebmann and Grad (1951) and Liebmann (1952). The excitation can be characterized by the quantity

$$k = \frac{NI}{\sqrt{V^*}},$$

in which V^* represents the potential V, after correction for relativistic effects, at which the microscope is operating:

$$V^* = V(1 + 0.98 \times 10^{-6} V).$$

The coefficient of r in the trajectory equation in a rotating coordinate system:

$$r'' + \frac{e B^2}{8 m_0 V^*} r = 0,$$

is thus proportional to $(NI/\sqrt{V^*})^2$.

Further, the image rotation θ is also proportional to k. The focal length f_1 at first decreases as k increases, passes through a minimum f_{1m} for $(NI) = (NI)_0$, and subsequently increases again. The abscissa, z_1, of the image focus, measured with respect to the centre of the lens (which is assumed to be symmetrical) decreases steadily, passes through zero and becomes negative for very high values of k.

Liebmann, and later Durandeau, have pointed out that the curves giving f_1 as a function of k for different geometries can, to all intents and purposes, be reduced to a single curve which describes the variation of f_1/f_{1m} as a function of the ratio of the excitation actually applied to the excitation

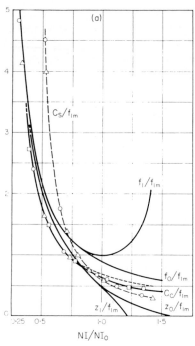

FIG. 218a. The optical elements of a magnetic lens, shown as a function of the excitation; the reduced curves relate to symmetrical lenses. f_1: asymptotic focal length; f_0: immersion focal length; z_1: abscissa of the asymptotic focus; z_0: abscissa of the immersion focus; C_c: chromatic aberration constant; C_s: spherical aberration constant; f_{1m}: minimum asymptotic focal length, for $(NI) = (NI)_0$
(Durandeau and Fert, 1957).

$(NI)_0$ (Fig. 218a). The same is true of the curve z_1/f_{1m}, to an approximation which is in practice adequate, within the useful ranges of D/S (from 0·5 to 5) and of $(N I)/(N I)_0$ (0·25 to 1·6); the differences remain below 4 or 5 per cent. If, further, we plot the curves giving $(N I)_0/\sqrt{V^*}$ and f_{1m}/L as functions of D/S, we see (Fig. 218b) that for $0·5 < D/S < 5$,

$$(N I)_0 \simeq 13·5 \sqrt{V^*},$$

and

$$f_{1m} = 0·5L = 0·5 \sqrt{S^2 + 0·45 D^2}.$$

From Fig. 218b we see that L is a better choice of reference length than $(S + D)$, the quantity selected by Liebmann (1955). Given $(N I)_0$ and f_{1m}, we can obtain the optical properties of any symmetrical lens from the universal curves. For asymmetrical lenses, L has simply to be replaced by

$$L_A = \sqrt{S^2 + 0\cdot45 \left(\frac{D_1 + D_2}{2} \right)^2},$$

in which D_1 and D_2 are the diameters of the channels bored through the two pole-pieces (Durandeau, Fert and Tardieu, 1958). Here we can again calculate f_{1m}, and obtain a single curve for f_1/f_{1m}, which coincides with the symmetrical lens curve. The curve which gives the abscissa of the focus is slightly separated from the curve which corresponds to $D_1 = D_2$. If $D_1/D_2 < 1$, the new curve lies below the curve for a symmetrical lens; if $D_1/D_2 > 1$, the converse is true. The distance between the curves always remains slight, however.

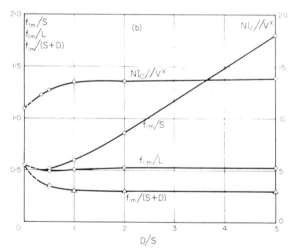

FIG. 218b. The minimum focal length f_{1m} and the corresponding excitation as functions of D/S (Durandeau and Fert, 1957).

The possibility of obtaining extremely short focal lengths has been studied by Liebmann together with van Ments and Le Poole (1947). Were it not for the fact that the iron becomes saturated, the minimum focal length could be made arbitrarily short simply by reducing the width S of the gap. The field in the gap is limited by saturation to about 26 kG, however, with the best steel known (Permendur or AFK 2 produced by the French Imphy company). In this way, Liebmann (1952) calculated that for $D = 0.6$ mm, the following absolute minima of f_1 are reached: $f_1 = 0.63$ mm for $S/D = 2$ and $V = 40$ kV; $f_1 = 0.93$ mm for $S/D = 3$ and $V = 90$ kV; and $f_1 = 1.43$ mm for $S/D = 4$ and $V = 210$ kV—these calculated results agree well with the measurements made by Ruska (1944). The value f_{1m} is reached for relatively wide gaps.

If the distribution $B(z)$ is represented not by a bell-shaped curve, which is the appearance which it usually has, but by a rectangular model of height B_0 and length L_1, the cardinal elements can be calculated easily; we have simply to write $B = $ constant in the differential equation above. The trajectory then has the form:

$$r = r_0 \cos \omega L_1 \quad \text{with} \quad \omega = \frac{B_0}{\sqrt{V^*}} \sqrt{\frac{e}{8m_0}},$$

for a ray incident parallel to the axis. The value of B_0 is defined by

$$B_1 S = B_0 L.$$

We then find:

$$\frac{f_1}{L_1} = \frac{1}{\omega L \sin \omega L_1} \quad \text{and} \quad \frac{z_1}{L_1} = \frac{1}{2} + \frac{1}{\omega L \tan \omega L_1},$$

or in the reduced coordinates

$$\frac{f_1}{f_{1m}} \simeq \frac{\sin 2.03}{\dfrac{(NI)}{(NI)_0} \sin \left[2.03 \dfrac{(NI)}{(NI)_0}\right]},$$

when the excitation is weak $[(NI) < 1000 S$ or $(NI)/(NI)_0 < 0.774]$.

(ii) *The immersion elements*

It is most important to know the immersion focal length f_0 and the position of the immersion foci when an electron microscope objective is being designed, since the object is placed at the immersion object focus.

The immersion focal length f_0, expressed as a function of $k = NI/\sqrt{V^*}$, does not pass through a minimum; instead, it decreases steadily as the excitation is increased, for the values of k used in practice at least. If, moreover, it were to reach a minimum for a certain value of the excitation, with pronounced saturation present, the objective would not be achromatic. Only f_0 would be stationary—the magnification of the objective would be achromatic, but since the abscissa of the immersion object focus continues to decrease when k is increased, the object would not be corrected for axial chromatic aberration; the practical resolution is thus limited (see § 16.3). The only effect of increasing NI or diminishing V is to increase the curvature of every point on the trajectory, and to bring the point at which the latter first intersects the axis steadily closer to the object.

Just as for f_1, universal curves for f_0 and the abscissa z_0 of the immersion focus can be constructed; these quantities need simply to be divided by the length f_{1m}. These curves are shown in Fig. 218a. When the lenses are asymmetrical, the curve f_0/f_{1m} remains very close to the corresponding symmetrical lens curve for $0.5 < D_1/D_2 < 2$; the same is true of the curve z_0/f_{1m}, always provided that $L = \sqrt{S^2 + 0.45\,D^2}$ be replaced by

$$L = \sqrt{S^2 + 0.45\left(\frac{D_1 + D_2}{2}\right)^2}$$ in the formula which gives f_{1m}.

(iii) The aberration coefficients

A knowledge of the spherical aberration coefficients of lenses which are to be used as objectives is most important, as the beams which pass through them are of considerably wider aperture than the beams which pass through projective lenses. The spherical aberration coefficient, C_s, and the chromatic aberration coefficient, C_c, are thus determined for an object lying at the immersion focus F_0 with image at infinity, which is the condition in which the lens is normally employed.

For slight excitations, F_0, moreover, coincides with the asymptotic focus F_1 (according to Durandeau, this is true when $(N\,I)/(N\,I)_0 < 0.774$).

For any given lens, irrespective of the excitation it may have, the coefficients C_s and C_c decrease when $N\,I$ increases. In Fig. 218a, the curves which show the variations of C_s/f_{1m} and C_c/f_{1m} as a function of $(N\,I)/(N\,I)_0$ are plotted. These are virtually universal curves, and here again, the difference between these curves and the curves obtained for asymmetrical lenses is very slight. The coefficient C_c is always smaller than f_0, and for very high excitations, $C_c/f_m \simeq 1$. C_s, on the other hand, is much larger than f_0 for weak lenses, while for strong lenses, $C_s/f_{1m} \lesssim 0.5$.

A detailed study of the problem of determining the objective lens with the smallest aberration coefficients, C_c and C_s, has been undertaken by Liebmann (1951, 1952). If we consider a weak lens in which D is varied for a given value of S/D, the axial induction B_m and the working potential V^* being held constant, we find that at first C_s decreases rapidly as D is increased, then passes through a minimum, and subsequently increases slowly. Liebmann uses the parameter

$$k^2 = 0.022\,\frac{R^2\,B_m^2}{V^*},$$

in which $R = D/2$, and whatever may have been the value of B_m initially selected, the minimum corresponds to $k^2 \simeq 5$. We thus obtain an optimum value of R, which is given by

$$R_0 \simeq 15\,\frac{\sqrt{V^*}}{B_m}\ \text{cm},$$

and so
$$(C_s)_{min} = 0.267\,R_0 = 4.05\,\frac{\sqrt{V^*}}{B_m}\ \text{cm}.$$

For other values of S/D, slightly different values of $(C_s)_{min}$ are obtained. The search for the absolute minimum of $(C_s)_{min}$ is bounded by saturation phenomena, which are in turn linked to the value of the induction B_p which prevails between the pole-pieces off the axis, and which differs the more widely from B_m as D is made larger. Liebmann (1951) finds that this absolute minimum corresponds to the zone $1 < S/D < 2$; with a carefully constructed objective, we can hope to obtain $C_s \simeq 1$ mm. An analogous study, with the object of reducing C_s, reveals that again an optimum lens exists which resembles closely the minimum spherical aberration lens.

The defect which it is essential to eliminate from a projective lens is the distortion. In another article by Liebmann (1952), the ways of determining the parameters which correspond to a projective with minimum distortion are to be found. The distortion of the magnification always vanishes for a particular value of $(N\,I)/(N\,I)_0$. For every value of D/S, Liebmann recommends that the lens should always be operated near $(NI) = (NI)_0$.

(iv) *The resolving power of a magnetic objective*

Using the results which had been obtained by Liebmann and Grad (1951), Mulvey (1953) plotted the curves which represent $C_s^{1/4}\lambda^{3/4}$—for various values of the ratio S/D and of the field in the gap—as a function of the diameter D of the opening in the pole-pieces. To each value of S/D and of the field, there corresponds a minimum of the parameter which describes the resolution (the optimum resolution); this minimum is reached for an excitation of about 6000 A turns when $V = 50$ kV, and about 12,000 A turns when $V = 500$ kV.

Liebmann (1955) plotted analogous curves to show the behaviour of $C_s^{1/4}\lambda^{3/4}$ as a function of S for various values of the field B_p in the gap and the ratio S/D. The curves vary very little with S/D provided this ratio remains between 0.5 and 2 (which is usually the situation). These curves have very flat minima, and by sacrificing a completely negligible degree of resolution, a great many ampère-turns can be saved. For example, $(C_s^{1/4}\lambda^{3/4})_{opt} = 5$ Å for $B_p = 24,000$ gauss (2.4 T), $V = 50$ kV and $S = 0.36$ cm; without noticeably reducing the resolution, however, S can be made 0.2 cm, which reduces the excitation from 6700 A turns to 3700 A turns.

The chromatic aberration always remains close to the focal length (see Liebmann, 1952).

17.4.3 Some Practical Examples

From the article by Durandeau Fert and (1957), we have extracted the numerical data and diagrams necessary to construct a magnetic microscope with high resolution.

All the lenses have the same magnetic circuit, but the extremities of the pole-pieces are of different shapes.

For $V = 75$ kV, which corresponds to $V^* = 80$ kV, and values of D/S between 0·5 and 2, we have

$$(N\,I)_0 \simeq 13\cdot5\, | \overline{V^*} = 3900 \text{ A turns}.$$

(i) The objective

For an immersed object lying in the plane of the upper pole-piece (see Fig. 219a), we should use $S = 3$ mm, $D = 4$ mm; f_{1m} is then approximately 2 mm, but $z_0 = 1\cdot5$ mm, or $z_0/f_{1m} = 0\cdot75$. The curves of Figs. 218a and 218b

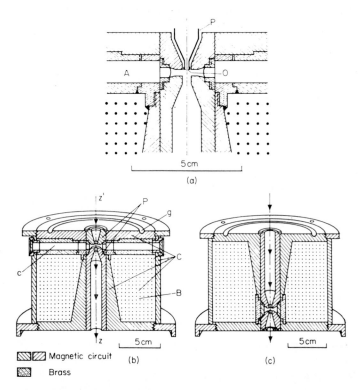

Magnetic circuit

Brass

FIG. 219a. The pole-pieces of an objective (with an immersed object). O: object; P: object-support; A: access to the gap (diaphragm); $S = 3$ mm; $D = 4$ mm; $f_0 = 2\cdot1$ mm (Durandeau and Fert, 1957).

FIG. 219b. A magnetic objective. ($V = 50$–75 kV); $z'z$: axis of the lens; P: pole-pieces; C: body of the lens; B: windings; c: channel to give access to the gap; g: socket for the joint (to prevent leaks) (Durandeau and Fert, 1957).

FIG. 219c. Section of a projective lens $S = D = 1\cdot7$ mm; $f_{1m} = 1\cdot20$ mm for $N\,I = 3900$ A.t. and $V = 75$ kV (Durandeau and Fert, 1957).

give the following data:

$$\frac{(N\,I)}{(NI)_0} = 0\cdot 85 \quad \text{or} \quad N\,I = 3300 \text{ A turns} \quad \text{for} \quad V = 75 \text{ kV},$$

$$\frac{f_0}{f_{1m}} = 1\cdot 04 \quad \text{or} \quad f_0 = 2\cdot 1 \text{ mm},$$

$$\frac{C_c}{f_{1m}} = 0\cdot 8 \quad \text{or} \quad C_c = 1\cdot 6 \text{ mm},$$

$$\frac{C_s}{f_{1m}} = 0\cdot 9 \quad \text{or} \quad C_s = 1\cdot 8 \text{ mm}.$$

Since the specimen-holder must be convenient to move, the arrangement shown in Fig. 219b is adopted.

(ii) *The projective lens*

The angular aperture of the beam is small, but the emergent beam may well be inclined at a large angle to the optic axis, and hence the form shown in Fig. 219c is to be recommended.

We can select

$$S = D = 1\cdot 7 \text{ mm, with which } f_{1m} = 1\cdot 02 \text{ mm}.$$

For $(NI) = (NI)_0 = 3900$ A turns $= 2300\,S$, saturation has set in—the actual value of f_{1m} will then be

$$(f_{1m})_{\text{actual}} = f_{1m}\left[1 + 0\cdot 14\,\frac{(N\,I) - 1000\,S}{1000\,S}\right] = 1\cdot 18 f_{1m} = 1\cdot 2 \text{ mm}.$$

The curves shown in Fig. 218a cannot be used to calculate C_s and C_c, as these have been calculated for an object lying at the immersion focus F_0.

17.4.4 The Single-field Condenser-objective

The possibility of obtaining very short focal lengths and low aberration coefficients C_s and C_c by situating the object at the centre of lens (where $B_z = B_{\text{max}}$) was pointed out by Glaser as early as 1941, but has not been exploited until recently, by Riecke (1962a, b) and Ruska (1962, 1964), because of the practical difficulties of using such a lens. The first half of the field acts as a strongly convergent condenser (Fig. 219d) so that if the diameter, the angular aperture and the intensity of the illuminating beam are to be adjusted with any precision, the entry diaphragm must be centred perfectly and two high quality condenser lenses must be placed in front of the objective.

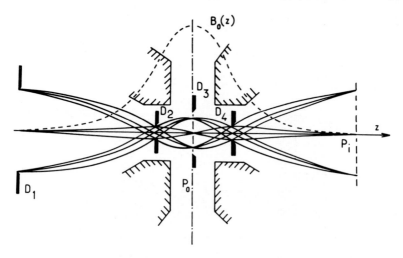

FIG. 219d. The trajectories in the condenser-objective. D_1, D_2, D_3 and D_4 represent the condenser aperture, the condenser-objective aperture, the field diaphragm in the object plane P_0 and the objective aperture.

The excitation of the objective is such that

$$k^2 = \frac{eB_{\max}^2 a^2}{8m_0\Phi_0^*} = 3,$$

where a is the half-width of the curve $B(z)$. For $\Phi_0 > 100\,\text{kV}$, the lens is highly saturated but the curve $B(z)$ only broadens very slowly. We have

$$a\,(\text{mm}) = 1\cdot27 + 0\cdot057\,I\,(\text{kA}) \simeq f\,(\text{mm}).$$

In consequence, C_s and C_c do not vary much as Φ_0 is increased (Ruska, 1966) and C_s remains smaller than 1 mm even when $\Phi_0 = 200\,\text{kV}$. At 100 kV, this lens is better than standard types of objective at 400 kV. Its chromatic aberration is given by $C_c \simeq C_s/2$.

A further advantage is that the object is placed half-way between the poles; in the practical design, therefore, the specimen-holder and the apertures can be inserted straightforwardly.

17.4.5 Lenses with Permanent Magnets

No external current source is required by these lenses, and they are therefore very stable. It is, however, difficult to vary their optical characteristics over a continuous range.

For these lenses, we have

$$J = \int\limits_{-\infty}^{\infty} B_0(z)\, \mathrm{d}z = 0,$$

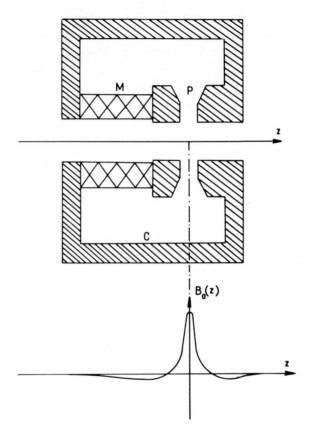

FIG. 219e. Permanent magnet lens with a single gap. *Above*: schematic cross-section (P = pole-pieces, M = magnet); *below*: the distribution $B_0(z)$.

so that the image rotation is always zero. In a lens with a single gap, which cannot be adjusted (Fig. 219e), the form of the curve $B_0(z)$ is more complicated than that of the analogous electromagnet, since the above relation ($J = 0$) must be satisfied; the focal length is always greater. The fringe field is suppressed by using two gaps in tandem (Fig. 219f), forming a double lens. To alter the convergence, a soft iron magnetic shunt must be moved in the vicinity of the permanent magnets, which will diminish the magnetic flux in the axial zone.

The magnets used for this purpose are usually made of a special steel containing aluminium, nickel and cobalt of the "Alnico V" type (Al 8 per cent; Ni 13·5 per cent; Cu 3 per cent and Fe 51 per cent).

The induction B in the gaps of the lens illustrated in Fig. 219f may be calculated approximately as follows.

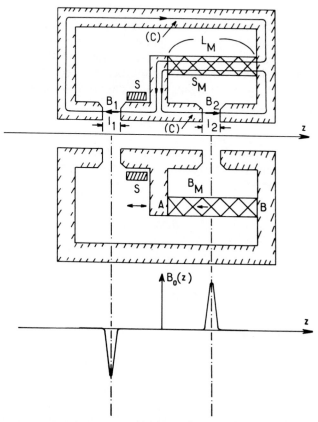

FIG. 219f. Two-gap lens, and the corresponding curve $B_0(z)$. S is a mobile shunt of soft iron, which enables us to modify $B_0(z)$ slightly.

The line integral $\int H dl$ of the magnetic *field* along a curve (C) passing through the gap, length l, and magnetized material, length L_M, is given by

$$\int_{A^{(C)}}^{B} H dl = H_M L_M.$$

If we assume that the permeability of the soft steel is infinite, we have

$$H_2 l_2 = H_M L_M = H_1 l_1.$$

(H_2 and H_1 denote the fields in the gaps G_2 and G_1 respectively, and l_2 and l_1 are the lengths of these gaps.) Hence

$$B_2 = \mu_0 \frac{H_M L_M}{l_2}; \quad B_1 = \mu_1 \frac{H_M L_M}{l_1}.$$

The flux of magnetic induction is conserved over the whole magnet. If Ψ_1 denotes the flux in G_1 and Ψ_2 the flux in G_2, we can define equivalent surfaces S_1 and S_2 for each gap, such that $\Psi_1 = B_1 S_1$ and $\Psi_2 = B_2 S_2$. If we assume that during its construction the magnet has been subjected to a constant induction B_M at every point, the flux passing through it is

$$\Psi_M = B_M S_M.$$

The induction B_M is related to the magnetic field H_M by the characteristic $B = f(H)$ of the magnetic material, which is provided by the manufacturer.
We have therefore

$$\Psi_M = \Psi_1 + \Psi_2,$$

$$B_M = \frac{B_1 S_1 + B_2 S_2}{S_M}.$$

We know that

$$H_M = \frac{l_1 B_1}{\mu_0 L_M} = \frac{l_2 B_2}{\mu_0 L_M},$$

and finally, we obtain

$$\mu_0 (S_M L_M)(B_M H_M) = B_1^2 S_1 l_1 + B_2^2 S_2 l_2$$

or

$$\mu_0 V_M (B_M H_M) = B_1^2 V_1 + B_2^2 V_2.$$

(V_M, V_1 and V_2 denote the volumes of the magnet and of the two gaps, respectively.) In the symmetrical situation ($B_1 = B_2 = B$, $V_1 = V_2 = V$), we have $l_1 = l_2 = s$; we obtain the simple relations

$$B = \frac{\mu_0 L_M H_M}{s}, \tag{1}$$

$$B^2 = \frac{\mu_0}{2} \frac{V_M}{V} (B_M H_M). \tag{2}$$

If B is fixed in a gap of given geometry, formula (2) shows that we must use a magnetic material having as high a value of the product $(B_M H_M)$ as possible, and work at the point on the characteristic $B = f(H)$ at which this product is greatest; H_M is then determined and formula (1) tells us what length of magnet to use.

For Alnico V, for example, we have $(B_M H_M)_{max} = 3 \cdot 5 \times 10^4$ (m.k.s.a.) $= 3 \cdot 5 \times 10^6$ gauss-oersted. The maximum field on the axis, B_0, is obtained in terms of B_1 and the dimensions of the gap, s and D, from the same relation as we use for electromagnets:

$$B_0 \simeq \frac{B_1}{s \sqrt{1 + 0 \cdot 45 \, (D/s)^2}}.$$

To alter the convergence of such a lens, we move a magnetic shunt of soft steel near the gap by some mechanical means; this deflects part of the flux in the principal gap and thus causes a small variation in B_1 or B_2.

17.4.6 Superconducting Magnetic Lenses

Since the development of new superconducting materials (see §9.4) with which high current densities and very high field stability can be attained, lens designs for microscopes and diffraction devices have appeared which take advantage of their properties. These lenses are of several kinds:

(a) lenses without iron;
(b) coils with external magnetic shield;
(c) classical lenses with ferromagnetic pole-pieces and superconducting windings.

Lenses of the first type, suggested by Laberrigue and Levinson (1964 and 1966) and Fernández-Morán (1965) have the advantage that their properties are easily deduced, since $B(z)$ can be calculated and then matched to a Glaser bell-shaped curve, but the field falls off very slowly. It is therefore advantageous to enclose the superconducting coil in a magnetic shield of soft steel or a special cobalt steel. With very high fields, this shield will saturate locally near the axis, but the curves $B(z)$ will still be shorter.

The magnetizing coil can be made in various ways:

(a) Niobium–zirconium wire will allow us to obtain $nI = 30,000$ A in 1 cm² cross-section for fields of 20 to 30 kG (Fernández-Morán, 1965, 1966a, b; Ozasa et al., 1966).

(b) With niobium–tin ribbon, appreciably higher fields can be attained (50 to 60 kG); a ribbon 30 to 40 μ thick and 6·4 mm wide can operate with a steady current of 200 A at 30 kG, with 110 A at 60 kG and with 75 A at 90 kG. The behaviour of such a coil can be calculated by using the $I(B)$ characteristic and the formula given in §9.4.

(c) Niobium–tin rings may be used; these are obtained by vacuum evaporation of the compound Nb_3Sn onto thin layers of platinum. Each circular layer supports several concentric rings and the beam passes through a hole at the axis; these layers are then piled together to form a coil. In this case, the current cannot be obtained from an outside source. The assembled discs are now placed in a field B_{ext}, greater than the critical field, produced by an auxiliary solenoid and B_{ext} is then gradually reduced; when the material becomes normal again ($B = B_c$), high induced currents are produced in the discs so that the flux passing through them remains constant. When $B_{ext} = 0$, the lens is excited to its maximum field strength. With such a lens, Siegel et al. (1966) and Kitamura et al. (1966) have obtained $B_{max} = 22·3$ kG, $a = 2·6$ mm, $f = 1·6$ mm, $C_s = 0·59$ mm. The technique used for constructing the rings is described in articles by Hanak (1964) and Petzinger and Hanak (1964).

In all these lenses, in which the field is essentially created by conductors, departures from rotational symmetry can be caused by imperfect positioning of the wires, ribbons or rings (either during construction, or after cooling which may bring non-uniform forces into play at the windings). The lenses must therefore be provided with facilities for correcting astigmatism.

To obtain fields that are highly stable in time, it is preferable to use the persistent-current or short-circuit mode; this situation automatically arises with Nb_3Sn rings. With coils, we have only to provide a short-circuit across the input terminals, made of superconducting wire which can be held at a higher temperature than its critical temperature by heating. The coil is supplied in the normal way from an external source and when i has been satisfactorily adjusted, the short-circuit is allowed to cool (and becomes superconducting), and the external source is removed. In theory, the current trapped in the coil will continue unaltered indefinitely. In reality, however, the resistances at the contacts will not be exactly zero and the current will decay very slowly, but the consequent decrease in B is not detectable even after several hours (Fernández-Morán, 1966a, b) at an image with a direct magnification of 20,000.

The exact fine adjustment of the focal length of the lens can be achieved by means of a small auxiliary coil supplied from the outside, which is placed close to the principal lens (Ozasa et al., 1966). Another solution is to use a ring transformer with a ferromagnetic core, the secondary of which (having only a few turns) is connected in series with the winding of the principal lens. By altering the primary which is connected to an external source, the flux in the secondary can be varied as slowly as we may wish, and hence so too can the lens current (Fernández-Morán, 1966a, b).

CHAPTER 18

THE ANCILLARY PARTS OF AN
ELECTRON MICROSCOPE

AN ELECTRON microscope is equipped with various appendages, which are devices of the kind which are found in every electron optical instrument (power supplies and vacuum systems for example). In the transmission microscope, however, these devices are exceptional in that they must be of extremely high quality. The provision of the potential for a magnetic microscope barely differs from the corresponding arrangements in a diffraction camera or an electron image-converter, save in that the degree of stability in the former must be exceptionally high.

18.1 THE ELECTRICAL SUPPLY

18.1.1 The High Tension Generator

The current which is taken from the supply is very small, at most a few hundred microampères; the potential, however, has both to be highly stabilized and variable, if possible continuously; the maximum value to which we are restricted usually lies between 50 and 100 kV, and we shall cite a few details of some of the various solutions which are employed to-day.

(i) *Rectification of the output of a transformer*

An X-ray tube generator can be used and the ensemble will consist of a transformer (50 c/s high tension) and diodes with the aid of which either half-wave rectification or full-wave rectification can be obtained. The stability is achieved by holding the frequency of the alternating potential at 50 c/s and placing two saturated iron regulators connected in series, at the input of the transformer; the output is smoothed with large condensers (about one microfarad). This instrument is robust and simple, but the output which it provides is insufficiently stabilized for a modern microscope, although we should point out that an expert of world-wide renown was still using a system which had its origin in this type of instrument in 1950

(Ruska, 1950). The stabilization can be improved (Bauer, 1943; van Dorsten, 1948) by using an electronic feed-back system. This combination has now, however, given way to systems of the type described in (iii) below.

(ii) *A potential multiplier*

With a device which multiplies the potential, the potential at the output of the transformer can be reduced and the inverse potentials which the valves have to stand-off are smaller (the diode rectifiers used in television will then be suitable, for example, or alternatively, a chain of silicon or selenium rectifiers can be used). As a particular case, we might mention the potential multiplier of Vastel (1951) which is used in the electrostatic CSF electron microscopes. The maximum high tension is 80 kV, stabilized at best to one part in a thousand, with an output of 100 μA. A 600 c/s oscillator, the output of which is stabilized and variable, provides an alternating voltage with a high enough frequency for a transformer and a filter of small dimensions to be used.

This oscillator can be replaced by a low frequency valve or transistor oscillator, the H.T. of which is easier to stabilize. When the frequency is of the order of a few kilocycles per second, miniature transformers and selenium rectifiers can be used. The final high tension is then stabilized in the same way as it is for the higher frequency systems described below. Stabilities of the order of one part in 10^5 can be obtained for about half an hour when the high tension is 100 kV. The stability is measured in the following way: the high tension V_0 is fed into a chain of very large resistances ($R = 1000$ MΩ for example) which are arranged in such a way that there will be neither continuous discharges nor sparking between the terminals of the individual resistances. A small resistance, r, is placed in series with R; r for example might be equal to $R/10,000$. Between the terminals of r we find a potential $\triangle V_0 \simeq 10^{-4} V_0$, and this we connect in opposition to a steady potential $\triangle V_1$ provided by a cascade of highly stable batteries. The fluctuations of the resulting potential,

$$\delta V = \triangle V_0 - \triangle V_1,$$

represent those of V_0 apart from a constant factor. The difference δV is traced directly on a pen-recorder, which has been calibrated beforehand.

(iii) *High frequency generators*

Devices of this type were perfected in the U.S.A. for supplying television tubes. Later, they were used in conjunction with the electron microscope (Vance, 1941) and nowadays a great many different types of instrument are equipped with them. Generators of this type have many advantages; in particular, the use of a high frequency allows us to use very small condensers in the filter circuits to obtain the same ripple. In addition, the speed of adjustment, which is related to the frequency, is very rapid, and finally, the magnetic screening is very straightforward.

Diode rectifiers like those used in television have to be employed, as semi-conductor rectifiers have too large a parasitic capacity. These valves can be heated with a high frequency potential, however, by using a special oscillator and a special type of transformer with a straight ferrite core, the

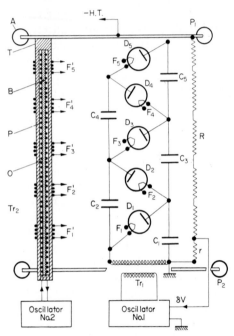

FIG. 220a. Diagram showing a high tension supply of the Cockcroft–Walton type (potential multiplier), giving 50–100 kV in air. $C_1, C_2, \ldots, C_5$: condensers; $D_1, D_2, \ldots, D_5$: diodes; Tr_1: supply transformer; R: resistance chain; r: resistance used for adjusting the principal oscillator by feed-back; Tr_2: transformer for heating the diodes; P: primary (connected to earth), supplied by oscillator number 2; B: ferrite cylinder; O: insulating oil; T: perspex envelope; $F_1', F_2', \ldots, F_5'$: secondaries, at progressively higher potentials, connected to the filaments $F_1, F_2, \ldots, F_5$ of the diodes; P_1, P_2: metal plates with anti-corona rings A.

ends of which are left open. The primary is earthed and is wound directly onto a ferrite cylinder; it is enclosed within a perspex column with walls one centimetre thick. The secondaries, which are held at progressively higher potentials, are wound straight onto the perspex tube and connected to the corresponding valves. The insulation between the various stages of the multiplication and earth is provided by the perspex tube (see Fig. 220a).

Again, the high tension column can be built in a very compact form, and immersed in an insulating tank filled with a special oil; the high tension cable is then led away from the upper part, and the supply is fed in at the base of the tank. In this way, "H.T. units" giving 100 kV can be constructed within a cylinder 20 cm in diameter and 35 cm high.

The whole column assembly can, alternatively, be left exposed to the atmosphere, provided we take precautions against continuous discharges by building the column between two metal plates, P_1 and P_2; the lower plate is earthed, and the upper one is connected to the high tension; the potentials of the several parts of the assembly, which increase from bottom to top, effectively correspond to those of the equipotential surfaces within the parallel plate condenser, $P_1 P_2$. A diagram of this kind of generator is to be found in an article by Vastel (1951). An oscillatory high tension is produced in a high-Q pilot-oscillator, and then passes through a stage of power-amplification which also assists the regulation. This potential is then increased by an air-cored HF transformer, rectified, and multiplied by a four-stage Cockcroft–Walton (or Greinacher) multiplier. The stabilization is effected by means of feedback, using a potential divider followed by a differential two-stage direct current amplifier which acts on the grid potential of the power-amplifier. If the H.T. V_1 increases by an amount $+\triangle V_1$, the high frequency potential with which the column is supplied varies in such a way as to reduce $\triangle V_1$ to zero. The high tension which is thus obtained is very stable and can be varied between 20 and 60 kV (simply by altering one of the potentials of the power amplifier); in this way, in a small space, we can obtain a potential which can be continuously varied between 20 and 60 kV, which has a ripple of less than two parts in 10^5, and which has a long-term stability of the same order for a ± 5 per cent variation of the mains. When the current which is taken is varied between zero and $100 \mu A$, the voltage varies by one part in 10^4; above $500 \mu A$, however, the potential falls suddenly to a negligible value. This is a most useful phenomenon, especially in the case of the electrostatic microscope, as it automatically prevents electrical break down from destroying the surfaces of the electrodes. This source of potential is, furthermore, hardly dangerous at all; an accidental contact only produces the discharge of one of the condensers of the cascade, of a comparatively small capacity. Moreover, a safety resistance of 5 or 10 MΩ reduces this discharge current to a practically harmless value.

The Metrovick microscope is provided with an analogous device (Haine, Page and Garfitt, 1950); the H.T. ranges between 25 and 100 kV, and is stabilized to one part in 25,000. The very simplified arrangement which is used with the Zeiss A.E.G. gives ± 50 kV with a stability of one in 5000. The reader is referred to a detailed fundamental examination by Mautner and Schade (1947) for the sophisticated theory of the high frequency generator of high tension and to an article by Baldinger (1959) devoted to H.T. generators of the "cascade" type.

(iv) *Electrostatic machines*

A few attempts have been made (Grivet, 1948) to use Felici's electro-static machine which has a metallic belt and no regulator. A system of this kind is attractive by virtue of its simplicity and its reduced dimensions (it is cylindrical, 20 × 50 cm), but so far is in its infancy so far as microscopy is concerned; at present, the H.T. which it provides is sufficiently stable to be used for visual scrutiny and straightforward photography, but in-adequate if the best resolution is to be attained. Recently, more encouraging results have been obtained by Felici (1957) using a machine with an in-sulating belt, provided with an electronic regulator.

H.T. supplies of this type are at present in use, which have a range of 20 to 100 kV and can be adjusted to within one part in 10^5, and which yield a few hundred microampères; they take up very little space, and are easy to handle.

(v) *Very high tensions*

Sources of a steady H.T. between 1 and 4 MV have been produced for many years now by specialist firms, but the stability has not been adequate for the very delicate experiments of electron optics. The exacting demands which nuclear physics has recently made have, however, spurred designers on to study ways of regulating and stabilizing these very high tensions with precision, with the result that it is now possible to use them in electron microscopy.

These very high tension sources are of two types:

"Cascade" models, which resemble those already described, which may be either open to the air (for tensions up to 2 MV; they are then very cumber-some) or enclosed within pressurized chambers (10 or 20 kg of nitrogen or freon) which allows the dimensions to be considerably reduced.

Electrostatic machines, which may use a flexible belt, of the "van de Graaff" type; such machines are capable of giving a very high tension of the order of 5 MV, and for $V > 500$ kV, they are built within pressurized chambers. Alternatively, they may use a cylindrical insulating conveyor placed within a pressurized chamber (these are the SAMES machines, studied by Felici). We can place several machines of this type in series, and obtain potentials from one to two MV with an output of several mA. The potential can be stabilized to a few parts in 10^5.

A 300 kV microscope which has been described recently (Tadano *et al.*, 1958) employs a van de Graaff generator, while the very high tension microscope at the Institut d'Optique Électronique at Toulouse (Dupouy *et al.*, 1960b, 1961) is supplied by a 1·5 MV cascade generator, similar to the one described by Bas (1960) and by Adler *et al.* (1960). The system has ten stages, and is supplied with a 400 c/s alternating potential provided by an oscillator; there are three columns instead of the two shown in Fig. 220a, and these are supplied at the mid-point by a transformer T_1, as this considerably reduces both the ripple of the high tension and the regulation effects. Two separate circuits are employed in the stabilization. The potential produced, the oscillator is "coarsely" adjusted, to within about one part in 10^3, by the first; the transformer T_1 is then supplied with a feedback signal which is obtained by extracting the fluctuations of the high tension through a potentiometer, then amplifying these fluctuations and changing their sign (just as in the other systems described above). The stability is about one part in 10^5, for a few minutes.

The preceding systems will certainly have to be abandoned if voltages of 2 MV and higher are to be obtained; cascade generators have too much energy stored in the chains of condensers, and the accelerating tube can be seriously damaged by a breakdown. In van de Graaff generators, which must be built compactly (with accelerating columns and generator in the same pressurized chamber), the motion of the belt causes troublesome vibrations at the high voltage terminal. We might then consider using electrons from a linear accelerator (Gilroy *et al.*, 1966); there is not longer a high voltage electrode, no electrostatic column and hence no danger of destructive discharges. At the present time, however, the energy spectrum of the electrons emerging from an accelerator is too wide, and a magnetic monochromator would be needed to reduce the dispersion to a reasonable figure (10 to 20 volts).

18.1.2 Slow Variation of the Potential

Feedback devices are excellent for maintaining stability for a comparatively brief interval (a few seconds, for example, which is longer than the time of a normal exposure) but all these devices drift slowly over about ten minutes. This drift is a most inconvenient phenomenon when long photographic exposures are necessary (in electron diffraction, in the velocity analyser, and in the ion microscope), but also in the case of the transmission microscope it is a hindrance, as about 30 seconds are necessary for focusing and exposure, and in any case, the H.T. has to be connected up several minutes beforehand if there are to be no modifications between the focusing and the taking of the photograph. With an electrostatic instrument, this nuisance is reduced.

A simple solution of this problem is to leave the H.T. generator working continuously. A better solution is due to Haine (1950) who suggested that the generator should accelerate an auxiliary electron beam which would pass through a magnetic deflector with 180° deviation (cf. Chapter 22).

Any fluctuations in the high tension act to displace the beam at the exit of the deflector. The beam falls on two identical electrodes which receive equal electron currents for a given H.T. value. If the voltage fluctuates, one of the electrodes receives more charge than the other. If the two electrodes are connected to the two inputs of a differential amplifier, we finally obtain an error potential which is used to regulate the H.T. generator through a feedback loop.

18.1.3 Current Regulation in Magnetic Lenses

As we have seen in Chapter 16, the magnetizing current of a lens must be highly stabilized; the degree of stabilization will depend on the following factors:

The purpose of the lens: the objective, for example, must be protected more carefully against fluctuations than the condenser, the stability of which is less important. Regulation of the objective is facilitated by the fact that the current is constant—the current variations which are needed for focusing are small enough not to complicate the structure of the regulator. In projective lenses, on the other hand, wide and carefully controlled variations of the potential have to be available to cover the vast range of magnifications and illuminations.

The structure of the objective: the conditions become less stringent when the slope of the curve of focal length as a function of current is reduced by saturation. In the most unfavourable case, however, this current has to be about twice as stable in relative value as the H.T. voltage; in a good commercial microscope of this kind, the lens current is controlled to about one part in 40,000, and the H.T. to one in 20,000.

The solutions which have been adopted have been based upon two properties. The first of these is simply the well-known stability of batteries of accumulators. Up to 1950, therefore, Ruska supplied the lenses of his large research microscopes with a battery of accumulators of 64 volts and a capacity of 100 A hr. The maximum total current was about 2·5 amperes for the whole assembly of lenses. The magnetizing coils are then wound with coarse wire, so that the current density is low (1 A mm^{-2}). The heating is therefore very small, and as the yoke is rapidly cooled, the temperature of the coils remains constant after a short running-in period (between $^1/_4$ and $^1/_2$ hr). After this time, the coil resistances are constant, as are the resistances of the manganin control rheostats—the e.m.f. of the accumulators which discharge slowly ($^1/_{40}$ of the capacity) is constant enough for the problem to be regarded as solved. The practical success of the solution depends upon the care and attention with which the apparatus is handled. Technically, the success of the method is due to the use of a thick wire with little insulation, but it is also helped by breaking up the winding into smaller parts which are separated by sheets of copper, the sole purpose of which is to conduct the heat away towards the yoke.

Nowadays, however, it is considered inconvenient commercially to use a large battery with all the careful attention which it requires, and we replace it with the aid of electronic rectifiers and regulators.

The power required is rather high, however (between 50 and 250 watts according to the number of lenses and the bulk which they can be permitted) and electronic valve regulators can only provide this at high voltages (100–800 V). A commercial regulator of this kind might employ, for example, a valve rectifier giving 750 V after filtering, the current for each lens being stabilized independently by a separate electronic regulator operating in the way shown diagrammatically in Fig. 220 b. The current is then small, of the order of 0·1 A. The windings are of fine wire, and conduct the heat away inefficiently, as the volume which the insulation occupies is appreciable; the heating is slightly worse, and the resistance varies over a long period corresponding to the slow thermal drift, so that stabilization of the current (instead of the voltage) becomes imperative. As a result of recent progress in transistor technology, however, very powerful L.T. supplies (up to 10 kW, 200 V) have been developed which are as stable as generators using valves (see Sauzade, 1960, for example). The coils can be wound with coarse wire with very little insulation when this type of supply is used, and the thermal conductivity is thus better. This conductivity can be still further increased by impregnating the coils with an epoxy resin (araldite) thickened with silica.

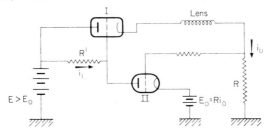

Fig. 220 b. Diagram of a current regulator. If an increase in E produces an increase in i_0, the potential at the grid of the triode II which had previously been zero becomes positive. i_1 increases and hence the potential at the grid of the triode I is lowered. The current i_0 thus tends to diminish.

Good current regulation is assured by using circuitry similar to that of the potential controls; the potential has simply to be stabilized across the terminals of a constant resistance R made of a special alloy, which is placed in series with the load (the windings of the lens, in this case). The current in the objective must be regulated with particular care; examples of this kind of regulator are to be found in the literature (Elmore and Sands, 1950; Dupouy and Perrier, 1960a for H. T. systems with valves. Sauzade, 1960 for transistorized systems). The variations of current with which the necessary adjustments are made are obtained either by altering the feedback circuit or by placing an attenuator with a constant input impedance between the regulator and the lens.

18.2 HIGH TENSION EFFECTS

18.2.1 Breakdown between the Electrodes under a Steady High Tension

In principle, the leads to the various electrodes carry only very small currents, and it is only the cathode lead which takes a steady current of a few μA to compensate for the charge led away by the beam. No other lead has to carry steady current. Nevertheless, sudden increases in the load on the high tension are often observed when a microscope is in action. This is a sign that a "breakdown" between the electrodes has allowed a discharge to pass directly from one electrode at high tension to the other which is earthed.

Even the slightest of discharge phenomena can be rendered visible by placing a fluorescent screen in the neighbourhood of the region in which the breakdown takes place, as such a screen can be sensitive to current densities of the order of 10^{-11} A cm^{-2}, but the appearance of discharges which are examined in this way gives no information about the way in which they were formed. On the observation screen, where they are particularly easy to detect, they usually have the form of a circular glow, sometimes intense, but of short duration. The discharges which are produced at the projective lens have a diameter of one or two centimetres, at the objective of about three to five centimetres, while the discharges which occur at the gun fill the whole screen. It is interesting to notice that this glimmering radiance often varies in intensity from one point to the next. In other cases, however, after the instrument has been in operation for some time, the appearance is different—a steady glow of light appears on the screen, which can be of a very feeble intensity but which gradually becomes more intense as the high tension is increased until the screen is blindingly bright. Another method of detection and study is to examine the appearance of the electrodes themselves, either in an experimental electrostatic lens or more simply in a bell-jar within which two plane or spherical surfaces are placed, one of which is earthed while the other is connected to the high tension (Anderson, 1935; Goltz, 1940; Bertein, 1946; and Arnal, 1955). The average current can be measured with a galvanometer, the shape of the impulses can be examined with an oscillograph and the number of X-rays emitted can be counted with a scintillation counter.

In normal conditions (electrodes which have not been outgassed placed in a dynamic vacuum), four classes of discharge can be distinguished.

(i) Below a certain value of the H.T., *no discharge* occurs and the vacuum is a perfect insulator.

(ii) Suddenly, between 30 and 60 kV, with a spacing of a few millimetres and a source with a rather high internal resistance (40 MΩ), very short *microdischarges* appear (10^{-4} sec), several per second, statistically distributed in time; the mean current lies somewhere between 10^{-8} and 10^{-6} A. When the high tension is increased, the rhythm of the microdischarges speeds up (to 10 or 20 per sec), and the mean current increases to several μA. The interelectrode region is the source of a diffuse glow, while at the same time certain metallic parts or certain insulators are struck by charged particles and give a bluish luminescence.

(iii) If we continue to increase the potential, a different kind of discharge appears, which *sparkles*, and has something of the appearance of ordinary sparks in the air. This type of discharge has the form of a sudden passing glimmer, highly localized, which appears on a metallic or insulating surface. It is exactly as if some kind of self-focusing mechanism were concentrating the discharge into a small beam. Local heating is produced over a microscopic zone (Bertein and Grivet, 1951) and the metal evaporates, which produces an increase in the intensity as a result of thermionic emission and ionization of the vapour. Whereas the microdischarge leaves no microscopic trace on the electrodes, the sparks leave a mark, each of which extends over a region several hundred microns in diameter, and the surface is completely covered with small points each of which corresponds to a point at which the metal has been evaporated.

(iv) If the tension is increased still further, the sparks become more and more frequent, and a new kind of behaviour, *arcs*, sets in; the current is seen to increase further, after which it becomes stable towards several tens of μA. Permanent bright points appear on the electrodes.

When an insulator is placed between the electrodes in a region in which microdischarges appear, a different mode of behaviour appears, which is characterized, in its early stages at least, by a discharge which is not very intense, but which is continuous, rather than the microdischarges described in the preceding paragraph. A feebly and diffusely illuminated region is observed at the surface of the insulator, between two and five millimetres in diameter, which seems to shift towards the anode, together with small bright permanent curves (which are visible by virtue of a blue fluorescence which appears at the metallic or insulating surface) which are usually very clearly defined; the concavity of these curves faces towards the cathode, their appearance changes with the high tension. For even higher tensions, *sparks* appear as well.

Despite any amount of research, these phenomena are still poorly understood. For this reason, we shall restrict ourselves to describing some of the features which are best understood among these mechanisms. They play a central role in electron microscopy as they disturb the focusing by lowering the potentials at the electrodes. Further, the charged particles and the flashes of light which are produced fog the photographic plate.

18.2.2 The Parameters which Influence the Breakdown Effects in the High Tension

(i) *The role of the vacuum*

The dynamic vacuum within a microscope is maintained between 10^{-4} and 10^{-5} mm of mercury. We might suppose that the best high tension conditions are obtained with the best vacuum, and blame the pump if electrical breakdown occurs too frequently.

In reality, the vacuum plays an unexpected role; the pressure within the apparatus is in general considerably lower than the value at which the results depend upon the quality of the vacuum. It has been demonstrated by van Atta and van de Graaff (1933), Anderson (1935a), Trump and van de Graaff (1947) and by Arnal (1956) that as soon as the interelectrode region ceases to be subject to gaseous ionization, the breakdown mechanism becomes independent of the quality of the vacuum. This condition is fulfilled as soon as the mean free path of the electrons and the ions which are moving through the residual gas within the apparatus is longer than the interelectrode spacing (10^{-3} mm Hg). Reducing the pressure does not improve the high tension situation in so far as the pressure of the residual gas can slowly affect the concentration of the molecular layers of gas adsorbed at the electrode surfaces. A sudden increase of the pressure may even halt the discharges momentarily, but after a few dozen seconds, they recommence; reducing the pressure has the opposite effect (Bertein, 1946; Arnal, 1956). It even seems to be true that the threshold at which the discharges appear is higher when the pressure of the residual gas is relatively high, about 10^{-4} mm Hg (according to Clifford and Fortescue, 1952, the threshold is changed by a factor which can reach two).

(ii) *The role played by the condition of the surface; formation*

A priori, the state of the surface ought to play an important role if it is cold emission which sets off the discharges, as was believed for a long time. The least irregularity ought to have a harmful effect, as it would produce an increase of the electric field and hence an intense emission of electrons as a result of the point; Anderson (1935), and Arnal (1956) have shown that it has none. Extremely careful polishing is in fact useless and often even harmful, as foreign bodies become embedded in the superficial layer with which the metal of the electrode is coated. An auxiliary bombardment of the electrodes by an electron beam produces only a very slight change of the breakdown threshold. When the electrodes are bombarded with ions on the other hand (Bruck, 1953), breakdown occurs for very small values of the electric field, of the order of 1 kV cm^{-1}.

The state of polishing of the surface has been studied by Hadden (1951) and by Arnal (1955) who used, in particular, electrolytic polishing. The conclusion to which they came is that unpolished surfaces are "neither better nor worse" than surfaces polished electrolytically. It seems that the breakdown thresholds are almost identical, irrespective of the quality of the polishing; whether the surface of any given metal is roughly turned, attacked by an acid, or polished electrolytically, it still has virtually the same characteristics in practice. In electron microscopy, however, the mechanical tolerances (due to ellipticity aberration, for example) require certain surfaces to be polished.

One parameter is, on the other hand, of cardinal importance; this is the *superficial state*, the nature and the quantity of the various substances within the superficial film of the electrodes, either adsorbed or in the form of inclusions. We know that in effect the surfaces of metals are never physically clean, as various substances (grease, water, oxygen and nitrogen for example) which come from outside can become attached to them in the course of the preliminary cleaning, just as under vacuum conditions within the apparatus, gas occluded within the body of the metal can diffuse to the surface; as breakdown phenomena consist of exchanges between one surface and another, the state of these latter plays an important role. We can appreciate this by watching the electrode surface during the pre-breakdown phase with the aid of an optical projection microscope with a high magnification (Razin *et al.*, 1960). Under the action of an intense electric field (10^5 or 10^6 V cm^{-1}), local points of microscopic dimensions appear (a few microns across) composed either of oil molecules, or of small fragments of the metal itself. These points grow, and then break off, thus engendering a microdischarge. After a certain formation time has elapsed, the process becomes more and more rare. Various treatments are applied, to "clean" the surfaces—bombardment with electrons or ions, careful chemical cleaning, or vacuum outgassing. Good results are obtained by producing a discharge in hydrogen at a pressure of 1 mm, with a current density of 0·25 A cm^{-2} for three minutes. The everyday method, however, is simply "formation of the electrodes", which consists in increasing the H.T. very slowly (after reversing the polarity of the electrodes if necessary); small discharges are produced, and then disappear, only reappearing at a higher value of the potential. In this way, electrostatic lenses (with a spacing of 4 mm) can be made to operate at more than 60 kV, whereas the usual figure is only 45 kV.

(iii) *The role of the nature of the metal and of the adsorbed gases*

The various attempts which have been made to determine the influence of these parameters have produced contradictory results. The superficial state of the metal and the quantity of occluded gas seem to be more important than the nature of the electrodes. Goltz (1940) has compared the thresholds at which the microdischarges appear in various metals (Fig. 221). He studied the variations in the threshold which result from a first break-down, and he recommended the use of chrome–nickel steel with chemically cleaned surfaces. Arnal (1955) related Goltz' observations, and above all the intensity of the microdischarges, to the absorption capacity of the metal in hydrogen. The development of the preliminary discharges would be at least partly due to an exchange of positive ions, H^+, and negative ions, H^-, with secondary emission coefficients A^+, A^- such that $A^+ A^- > 1$. Ionov (1960) too has shown how important is the ionic component of the current. The ions are produced by the adsorbed gas, and are thermally emitted by the surface which is heated locally by the arrival of a heavy particle or an electron which sets off the breakdown. So long as $A^+ A^-$ remains greater than unity, the intensity of the discharge increases, but if the particle exchange remains highly localized at some point on the electrode, the gas adsorbed at the surface of the metal may be temporarily impoverished and $A^+ A^-$ fall below unity; the discharge is halted. The adsorbed layer may then be formed afresh by diffusion of gas from the interior of the metal, and the discharge can recommence after being triggered off anew. For

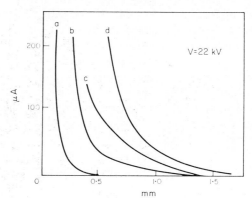

Fig. 221. The threshold at which the micro-discharges appear: (a) polished steel; (b) polished aluminium; (c) annealed steel; (d) steel covered with contamination.

pressures (p) of the order of 10^{-4} mm Hg, the high tension conditions would then be improved by a reduction in the number of ions present, which could be the result of a neutralization phenomenon, the ions coming into contact with the gas molecules between the electrodes. When $p > 10^{-4}$ mm Hg, there would on the contrary be ionization within the volume of this gas, an increase in the current, and hence the high tension conditions in the system would be worsened.

The importance of heavy particles in the initiation and maintenance of the preliminary discharges has been displayed by Pivovar and Gordienko (1958) also; these authors entirely eliminate the electron component of the current with the aid of an intense magnetic field parallel to the surface of the electrodes in question—the threshold of the preliminary discharges is practically the same, irrespective of whether or not the magnetic field is present.

By using electrodes made of different metals, it has been possible to demonstrate that it is above all the material from which the anode is formed which determines the characteristics of the microdischarges; although the cathode is probably important as a source of particles once the phenomenon has begun, its nature has very little influence upon the value of the threshold and the frequency of discharges. This is one of the more certain results of these delicate explorations, as it has been confirmed by various authors (Goltz, 1940; Heard et al., 1952; Clifford and Fortescue, 1952; and Arnal, 1955).

(iv) *The roles of the potential and of the interelectrode spacing*

If the discharges depended only upon the electric field (field emission), we should have a simple relation between the spacing and the breakdown potential, and the phenomenon would depend essentially upon V/d. This simple hypothesis is not at all valid, however, and when the interelectrode spacing is increased, the voltage threshold does not increase in proportion; the gradient of the mean breakdown potential decreases constantly as the distance increases. It is not the field at the electrodes which is relevant, but the *total voltage* which is applied across them.

For a given fixed interelectrode distance, d, we can increase the potential and determine the thresholds at which the various phenomena appear. These thresholds are rather poorly defined, as the experimental conditions are only reproducible with difficulty; in particular, the state of the surface is never the same between one experiment and another. To attach an order of magnitude to the various phenomena, however, we quote a few of the results. The threshold of the microdischarges, V_s, is tolerably well defined. Anderson (1935) suggested that for copper, V_s is related to d by a bilinear formula, of the form

$$V_s = \frac{144d}{0.97 + d}, \qquad 0.2 < d < 4 \text{ cm}.$$

The experiments of Arnal (1955) led to a law of the form

$$V_s = k \, d^{1/3},$$

in which k is of the order of 3 kV cm$^{-1/3}$ for copper. Ionov (1960) sets out with the hypothesis that ions are locally thermally emitted at the points of impact of particles which act as triggers, and offers the following explanation. The local temperature is a function of the power received, and hence of the overall potential between the electrodes, V. The ion emission coefficients, A^+ and A^-, are, however, dependent upon the local electric field E, which lowers the extraction potential (Zandberg and Ionov, 1959).

Both for sparks and for arcs, the thresholds are poorly defined. Cranberg (1952) and Heard *et al.* (1952) proposed a law of the form

$$V_s = k\, d^{1/2},$$

for the sparks, in which k is of the order of 40 for stainless steel, and 30 for copper. For the arcs, Anderson (1935) suggested that the empirical law

$$V_s = \frac{590\,d}{0\cdot77 + d}$$

is obeyed.

(v) *The role of the internal resistance of the source of high tension*

The importance of this parameter was first demonstrated by Arnal (1955). In general, the electrodes are supplied through a resistance R (the internal resistance of the source, itself, together with a safety resistance) and with respect to earth, they have a parasitic capacity C (of a few dozen picofarads in the case of an electron microscope). The discharges fall into three classes, according to the value of R. For small values (tens of kΩ), the microdischarges last between 10^{-2} and 10^{-4} sec, and have the form of a fairly constant current of a few dozen microamps—they are repeated every 10 or 20 sec.

For medium values of R (up to about 20 MΩ) a current peak is produced by the discharging of C, and is followed by a steady period of about the same duration as in the preceding situation but less intense.

Finally, for large values of the resistance ($\geqq 20$ MΩ), the signals become ever shorter (down to 10^{-4} sec). The current which flows during the breakdown reduces the potential between the electrodes (by about 1000 V) with the result that the discharge ceases by itself; the frequency, on the other hand, increases (to many dozens per second). For the electrodes to be well preserved, it is this final mode of breakdown which is the most advantageous.

18.2.3 Breakdown with a Pulsed Potential

The microdischarge process only reaches its full intensity after an interval of about 10^{-4} sec. If the potential is applied in short pulses which last less than 10^{-5} sec, therefore, the normal current will not have the time to build up completely. Herreng (1943) has shown, in the course of some experiments with X-ray tubes excited in pulses, that the threshold for sparks becomes 95 kV when the tension is applied in 1 μsec pulses whereas it had been only 56 kV for a steady high tension.

This advantage which is to be had with a pulsed potential can be put to good use in an electrostatic microscope. The first practical attempts were made by Spivak and Doubinima (1953) in the emission microscope

(2 kV pulses between electrodes 0·08 mm apart). These authors found an improvement in the maintenance of the potential of five to seven times, but only a small magnification was obtained.

More recently, Gardez (1959) has built a very high tension emission microscope; an immersion objective which could not support a potential higher than 40 kV in steady working conditions will support 95 kV if the potential is applied in pulses. (The pulses last about one microsecond, and have a frequency of 200 per second.)

18.2.4 Breakdown along the Insulators

Frequent discharges occur along the insulating supports of certain of the electrodes. Gleichauf (1951, 1951a) has shown that this is not due to the same phenomena as the breakdown between metallic electrodes. The threshold at which microdischarges appear is considerably lowered when an insulator is placed in the interelectrode region—the reduction can reach one half of the threshold which is observed in the absence of the support. The law which governs the appearance of the microdischarges, $V_s = k \sqrt{d}$, is approximately applicable to the breakdowns which occur along the insulators; the constant k depends upon the nature of the insulator—it is of the order of 20 kV mm$^{-1/2}$ for polystyrene and 30 to 40 for perspex and porcelain (Borovik and Batrakov, 1958). The best results seem to be obtained with unglazed zirconium porcelain (Trimmer and Pearlman, 1951) and with araldite considerably thickened with silica (Gribi et al., 1959). The rougher the surface, the better maintained is the potential; this result can easily be verified with porcelain, for example, moulded in such a way as to produce ribs normal to the lines of force of the electric field (and thus to elongate the line of the breakdowns).

In reality, however, the preliminary discharges in the lens always appear much sooner than the formula above would lead us to suspect. The insulators are subjected to a continual bombardment of charged particles which have their origin either in charge exchanges between the electrodes (even though these currents be well below the preliminary discharge threshold, and produce no detectable phenomena between the electrodes) or alternatively in various kinds of contamination (oil or grease molecules which leave carbon deposits after being broken down by the charged particles).

If the charges which are collected on the insulator are too numerous at certain points, very intense local electric fields will be created; these may either penetrate through the insulator, if this latter contains a conductive rod, as in the case of potential leads, or give rise to bright plumes by local ionization of the residual gas, which corresponds to an intense breakdown current and may lead to breakdown in the vacuum.

If the charge density is low, the charges manage to flow towards the metal (as the surface conductivity of the insulator is never exactly zero), following the electric field lines tangential to the insulator. The transition zone between the metal and the insulator will thus play a very important role; the lines of electric field become progressively detached from the insulator, cross the intervening space and terminate normally at the metal. Further, the surface at which metal and insulator are in contact is never perfectly plane, and little crevices remain where there is no electrical contact. The charges accumulate in this transition zone, and very often flow out into the vacuum where they may ionize the residual gas and also produce emission of secondary particles from the metal. The magnitude of this phenomenon is related both to the number of charges present and to their energy when they strike the metal (and hence to the operating potential); it is essentially the origin of the discharges which limit the H.T. which can

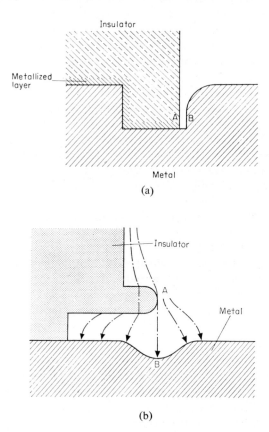

FIG. 222. Two possible ways of joining an insulator to an electrode in such a way as to reduce the risk of breakdown:
(a) by metallizing the base of the insulator the electrical contact is improved.
(b) by lengthening the path of the charges which flow between the insulator and the metal (Gribi, 1959). The broken line represents lines of electric field strength.

be applied to lenses which have been carefully designed and do not break down between the electrodes.

A first improvement is to ensure that the contact zone between metal and insulator is clearly defined; the end of the insulating support is metallized either by vacuum deposition or by a chemical method, and a proportion of the charges can thus flow steadily across. Alternatively, we can move the metal away from the end of the (non-metallized) insulator, to prevent the field lines which direct the motion of the charges from detaching themselves, as far as possible; Fig. 222a represents this type of solution.

Gribi *et al.* (1959) have suggested a further improvement which is shown schematically in Fig. 222b, and this has made it possible to increase the operating potential of a lens of normal dimensions from 45 to 50 kV, for which it had initially been designed, to 70 kV. The dotted line represents the electric field line tangential to the flange of the insulator; the shape of the electrode is selected in such a way that this line terminates normally at the bottom of a rounded groove hollowed out in the metal. At A, the electric field is lowered, and the local vacuum is better. The threshold at which microdischarges produced by ionization of the gas appear is considerably higher.

18.3 THE VACUUM

18.3.1 Pumps

The majority of microscopes are provided with a primary rotary pump with vanes, together with one or more secondary diffusion pumps. As mercury diffusion pumps require a liquid air trap, it is preferable to use an oil diffusion pump or a rotary molecular pump, and although the latter has the inconvenient property of letting oil into the body of the microscope, many instruments are still equipped with them. Further, it seems that although the pumping rate may nominally be higher, the practical efficiency of oil diffusion pumps is inferior to that of mercury pumps equipped with a good liquid air trap, as the characteristic pumping rate of the latter is higher for certain vapours which condense at the trap; when films are being used, the photographic emulsion and its base provide an abundant source of water vapour and organic vapour.

When the microscope is in use, the vacuum has frequently to be let down and then re-established; it is necessary, therefore, to have a fast and easily controlled pump system. Figure 223 shows an arrangement which is commonly used; it consists of a primary pump I and a diffusion pump II. A system of taps, valves and points at which air can be admitted allows the vacuum to be established or atmospheric pressure to be restored to any part of the arrangement at will.

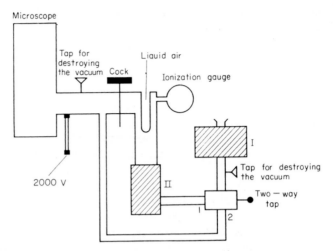

FIG. 223. The vacuum circuit of a microscope.

In the microscopes in regular use today, every effort is made to reduce the pumping time; the instruments are equipped with small "vacuum locks" which allow us to change the specimen or replace the photographic plates (or film) rapidly, without destroying the vacuum elsewhere. The object can be changed in a few seconds in this way (instead of in the one or two minutes which are required in a well-designed instrument without locks) without switching off the filament of the gun or cutting the high tension. Technical details are to be found in von Ardenne (1940), van Dorsten (1948) and von Borries (1949). To obtain the very highest performance and to avoid contaminating the object with fragments of molecules of carbon compounds that the beam has produced, by destroying organic molecules from the pump oils, grease and the joints themselves, it is advisable to use a "clean", very high vacuum, at least in the vicinity of the electron gun, the objective and the specimen holder.

Ultra-high vacuum techniques are being more and more widely used—apparatus made wholly of stainless steel, metal joints, quartz or aluminium insulators, for example—with pumping units consisting of high-speed ion pumps (§13.3.5); without baking, the limiting vacuum of 10^{-8} mm Hg is reached after a few hours, but after heating the whole evacuated region to 150 or 200°C, a vacuum of 10^{-9}–10^{-10} mm Hg can be attained. Work on the nature of the residual gases and the importance of their various partial pressures has shown that they consist principally of nitrogen and carbon monoxide (Hartman and Hartman, 1966).

The interior of the microscope may be divided into two distinct regions, to prevent vapours extracted from the photographic emulsion from rising towards the object: the zone around the object is a "clean" high vacuum while in the camera, a good vacuum (10^{-5} mm Hg) is adequate. The two regions can be separated in practice by placing a metal plate beyond the lens; a small hole is drilled in the plate around the axis to allow the beam to pass. A simpler solution that is used in many microscopes is to "clean" the region around the object by cooling the adjacent surfaces to the temperature of liquid nitrogen ($T = 77°$K) and connecting this region to a supplementary ion pump with a capacity large enough to maintain a pressure of 10^{-9}–10^{-10} mm Hg (Grigson et al., 1966, Valdrè et al., 1966).

18.3.2 Measurement of the Vacuum

In normal use, it is not necessary to measure the pressure precisely all the time, but for the detection of accidental leaks, it proves to be most useful to do so.

The crude vacuum can be estimated from the appearance of a glass discharge tube to which an alternating potential of several kV is applied. This gives an indication of the quality of vacuum between 1 and 0·01 mm Hg. At lower pressures, the discharge disappears. The colour of the discharge which depends upon the nature of the gas is useful as a guide to leaks—the doubtful part of the instrument is smeared with alcohol, which passes into the instrument if the part is defective, and produces a striped whitish glow. A Pirani gauge can also be used.

The high vacuum is measured with the aid of a Penning gauge, which gives an output of about 100 μA for a pressure of 10^{-5} mm Hg, or an ionization gauge in which the ion current is effectively proportional to the pressure, one microampère corresponding to about 10^{-6} mm Hg. Special ionization gauges (ultra-high vacuum gauges) have been developed for measuring pressures lower than this (10^{-7}–10^{-11} mm Hg). With ion pumps, the value of the discharge current in the pump gives the pressure at the pump directly. To detect a small leak which is apparent with this secondary gauge is a matter of logic and patience; the apparatus must be isolated section by section, so that the leak can be localized.

A quicker and more refined method of detecting leaks is now available, as a number of commercial firms have produced leak-detectors using helium. A very fine jet of helium gas is squirted onto the point of the wall to be tested; if the wall is porous, a certain amount of the gas will penetrate through into the evacuated chamber, which is connected to that of the detector. The latter is a simplified but very sensitive mass-spectrometer, adjusted to detect the helium line. Whenever any helium is present in the evacuated chamber, an electric signal is produced, the amplitude of which is effectively proportional to the size of the leak.

A great deal of useful information about the detection of leaks, measurement of the rate of leaking, and the use of one particular helium detector is to be found in an article by Choumoff and Laplume (1957).

18.4 THE GENERAL APPEARANCE OF A MICROSCOPE

We shall conclude with a description of three French microscopes, one of which is electrostatic, the second magnetic and the third is the very high tension microscope recently constructed by G. Dupouy and his collaborators (1960, 1961).

18.4.1 The Compagnie Générale de T.S.F. "M IX"

This is a microscope with three electrostatic lenses, which appeared in 1950–1. The appearance of the exterior is shown in Fig. 224, while Fig. 225 is a cross-section of the optical array. The microscope itself is 75 cm high, but the case and the various appendages increase this to 140 cm.

The maximum magnification is of the order of 50,000. In everyday use, the resolving power is 40 Å at a useful magnification of 25,000 which is adequate for many of the uses to which the microscope is put; if a stigmator is used, a resolution of 20 Å can be attained.

By adjusting the convergence of the first projector (the focal length of which can be varied between 2·5 and 50 mm), the magnification can be reduced to $G = 2000$, as we have already explained (§ 15.4.2). The instrument can be used as a diffraction camera, by retracting the contrast diaphragm which is situated above the objective. The illumination is produced by a gun similar to the one designed by Bruck and Bricka (§ 8.3.3). The focal length of the objective is 6 mm, and that of the second projective lens, 2·5 mm.

The object is supported either by a small capsule between 2 and 4 mm in diameter, in the centre of which is an opening between 0·5 and 0·1 mm in diameter, or by a very fine mesh grid with a similar outer diameter.

FIG. 224. The CSF microscope, M IX. All the electrical controls are collected together into the right hand side of the instrument, and all the controls for the vacuum into the left hand side; the various different taps shown in Fig. 223 are here all controlled by a single cock. The controls for the specimen-holder are to be found on the table. In the centre, reading from top to bottom, we see the gun (off which the cover has been lifted), the window through which the object is inserted, the screen with which the intermediate image can be seen (the image produced by the objective alone), the screen for examining the final image and finally the camera.

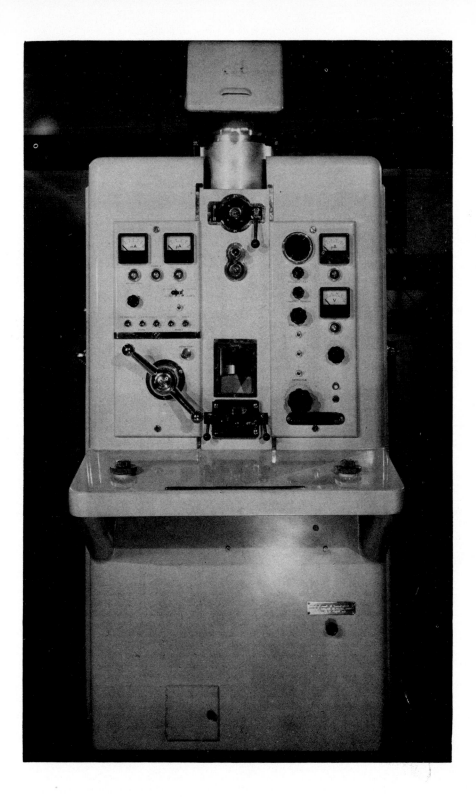

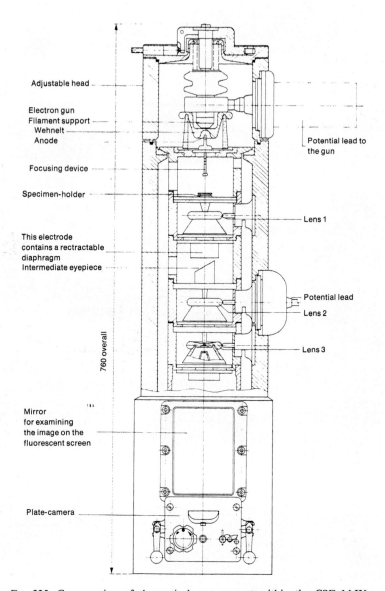

Adjustable head

Electron gun
Filament support
Wehnelt
Anode

Potential lead to
the gun

Focusing device

Specimen-holder

Lens 1

This electrode
contains a rectractable
diaphragm
Intermediate eyepiece

Potential lead
Lens 2

Lens 3

760 overall

Mirror
for examining
the image on the
fluorescent screen

Plate-camera

FIG. 225. Cross-section of the optical arrangement within the CSF M IX.

These grids or capsules are attached to a specimen-holder which can be moved in two mutually perpendicular directions; the movement is effected with the aid of mechanical demultiplication. The structure of the specimen-holder is illustrated in Fig. 226. Above all, we try to ensure that the movement is smooth, and that the device is insensitive to vibrations,

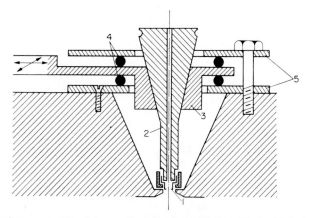

FIG. 226. Cross-section of the specimen-holder which is used in the CSF microscope. (1) Capsule for supporting the collodion film (cf. Chapter 9). (2) Detachable brass block. (3) Support which can be displaced in a plane. (4) Toroidal joints tightened by flanges (5).

as these affect the resolution very adversely. One solution is to make the holder of the specimen-holder almost an integral part of the objective with the aid of two toroidal joints gripped by two flat discs attached to the upper part of the objective. With other solutions, the object migrates slowly as a result of hysteresis of the springs and thermal expansion of unsymmetrical parts which are heated by the impact of the beam, and this is most inconvenient for long exposures.

18.4.2 The O.P.L. Microscope (Optique de Précision de Levallois)

This microscope, which is provided with a condenser and three magnetic lenses, is a recent development (Fert and Selme, 1957), and is at present in regular production by the French O.P.L. company. It represents the industrial equivalent of Fert's optical bench (1954), and its originality lies in the construction of the column. This latter consists of a tower of standardized elements, which gives a great practical flexibility (see Fig. 227a). Each element (lens, condenser, ...) is set within a cylindrical section bounded by two plane faces; the upper face contains the groove of an O-ring vacuum seal, and the lower face a centring flange. These faces are ground with great care to ensure that the column, once assembled, be rigid, and that

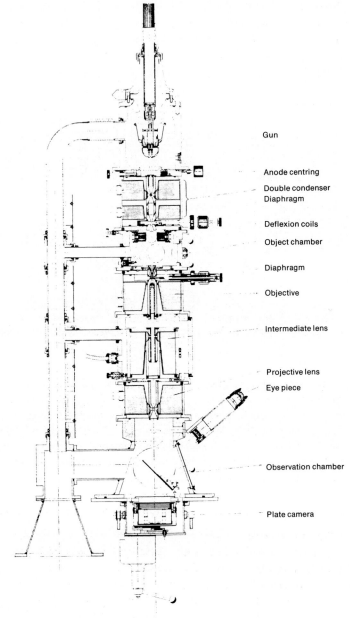

Gun

Anode centring

Double condenser
Diaphragm

Deflexion coils

Object chamber

Diaphragm

Objective

Intermediate lens

Projective lens

Eye piece

Observation chamber

Plate camera

FIG. 227a. Cross-section of the O.P.L. microscope, built in separate units which
can then be stacked up.

the optical elements be aligned. It is hence a simple matter to replace one lens by another, better adapted to the problem in hand. From the electron source to the screen, every essential point can be reached and cleaned without dismantling the column. The instrument can be modified rapidly: by adding an angled element, for example, the gun can be tilted for reflexion microscopy. Figure 227a shows a section of an early version of the instrument, and Fig. 227b a photograph of a recent model.

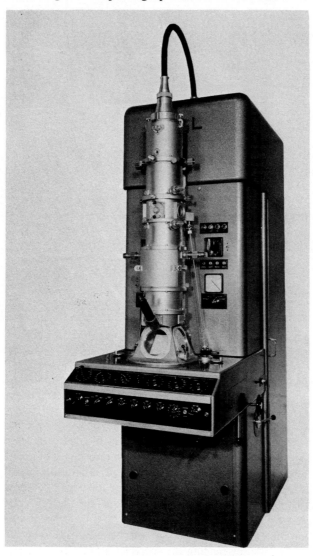

Fig. 227b. Photograph of a recent O.P.L. microscope.

The magnification can be varied continuously between 2000 and 110,000 at 75 kV and between 2000 and 75,000 at 100 kV. The gun can be supplied with a H.T. of 50, 75 or 100 kV, stabilized to about one part in 10^5, and gives about ten or twenty μA. The condenser and the objective are provided with different sets of diaphragms; a device to correct the astigmatism is placed behind the objective. The lens current is regulated to within about one part in 10^5.

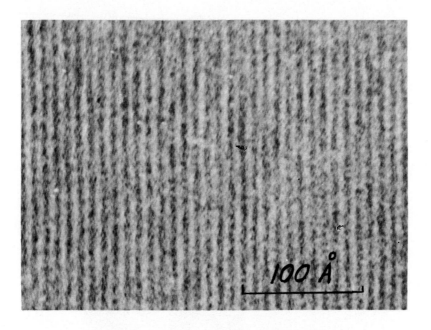

FIG. 227c. The surface of a copper phthalocyanine crystal (photographed with an O.P.L. microscope). Periodicity: 9·9 Å.

The theoretical resolving power is 5 Å, and in its present form, the instrument does indeed give a resolution which in practice is less than 10 Å (see Fig. 227c, which shows the surface of a monocrystal of copper phthalocyanine, the periodicity of which is 9·9 Å). Special attention has been given to the transitions from microscopy to diffraction and to microdiffraction, which entail the introduction of a special stop into the object plane of the intermediate lens.

Diffraction diagrams can equally well be obtained with a very high resolution (5×10^{-6}) by placing the specimen-holder beneath the projective lens; for further information, the reader should consult Chapter 21.

18.4.3 A Very High Tension Magnetic Microscope

We shall now give a brief description of the microscope which has been built by G. Dupouy and his collaborators at the Institut d'Optique Électronique in Toulouse, of which the maximum operating potential is planned to be 1·5 MV. When a very high tension is to be used, numerous modifications are necessary, not to the general optical arrangement, but to the technological features and to the overall bulk of the instrument.

(i) *The electron source*

The gun is of the ordinary kind, and works at a potential of a few tens of kilovolts in the usual way; it is placed at the top of an accelerating column, and is held at the prevailing H.T. (Φ). Just as in electrostatic

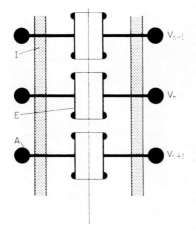

Fig. 227d. Section of an accelerating tube for very high energy electrons. *I*: insulator (porcelain or araldite); *E*: electrode; *A*: anti-corona ring.

accelerators, the column is a tower of ten porcelain rings, separated by metal discs which carry the accelerating electrodes. The latter are in the form of tubes with rounded ends; these tubes are separated by a gap which forms a two-cylinder accelerating lens (see Fig. 227d). The electrodes are held at potentials which are regularly spaced between ($-\Phi$) and earth ($\Phi = 0$), by means of a resistance chain across the H.T. source. The ratio $\dfrac{V_{n+1}}{V_n}$ varies, therefore, between the top and the bottom, since

$$\triangle V = \text{constant} \simeq -\Phi/10,$$

and the lenses are progressively less convergent.

The complete tube behaves like a convergent lens, the focal length of which is determined by the design; it is wise to insert a condenser between the gun and the tube, as we can then adjust the size and aperture of the beam at the base of the column where a diaphragm of small diameter acts as entry pupil for the optical system. The accelerating column is about twenty feet high. The radii of curvature of all the connexions and of the outside surfaces of the metal discs in the column are large enough to prevent continuous breakdown by the corona effect. The high tension installation and the accelerating column are shown in Fig. 227e.

Fig. 227e. H.T. generator of the Cockcroft–Walton type (on the right) and accelerating column (on the left) of the Very High Tension microscope at the Institut d'Optique Électronique laboratory of the C.N.R.S. at Toulouse (Dupouy and Perrier). Above the column are the electron gun and its various adjustment mechanisms, enclosed in a polished metal cover with a large radius of curvature. The whole system is installed inside a spherical Faraday cage.

(ii) *The optical system*

The optical system consists of a tower of separate elements each with its own clearly defined function, just as in the O.P.L. microscope (see Fig. 227f). There are two condensers C_1 and C_2, an objective O, and intermediate

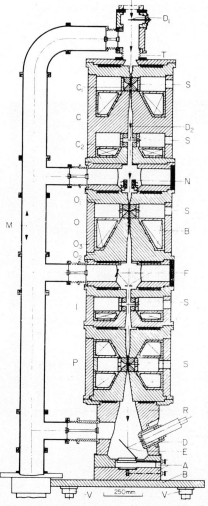

FIG. 227f. Cross-section of the optical system of the V.H.T. microscope. D_1: entry diaphragm; T: flexible membrane; C: double condenser (C_1 and C_2); O: objective; N: deflexion coils for aligning the beam; O_1, O_2 and O_3: the magnetic circuit; B: coil; I: intermediate lens; P: projective; R: observation chamber; D: observation window (with lead glass), with an eyepiece or a television camera; F, E: fluorescent screens; A: plate-camera; B: pedestal with levelling screws; S: polar units, which can be moved and centred – the objective polar unit contains the specimen-holder, a stigmator, and various different diaphragms.

sections with windows of lead glass and openings for pumping. The whole assembly is connected to the accelerating column by means of a flexible membrane T, and rests on an observation chamber R, which is itself placed on a large plate B supported on levelling-screws. The overall height is about eight feet, and, not counting the lead screens which provide protection against X-radiation, the whole assembly weighs about four tons; indeed each lens weighs several hundred kilograms.

The lenses are fixed. To align the various elements, therefore, they are provided with a metal "pole-unit", S, which contains the pole-pieces and the various accessories necessary for the role of each lens (specimen-holder,

FIG. 227g. View of the V.H.T. microscope of the Institut d'Optique Électronique in Toulouse (Dupouy and Perrier). It lies immediately beneath the accelerating column of Fig. 227f. Two lead shields (for protection against X-rays) can be seen at the base and the top. The microscope is about 2·6 m high.

diaphragms, stigmator, for example); this pole-unit can be displaced laterally by micrometer screws. The vacuum seal between the pole-unit and the rest of the lens is guaranteed by toric joints.

The objective weighs 700 kg and its focal length, f, is 5·5 mm when the current in the windings (28,500 turns) is 0·6 A and 1 MeV electrons are used. It may be used either with an ordinary specimen-holder, or with the special specimen-holder which is described in § 20.3.4; the latter may perhaps enable us to observe living tissue which is surrounded by air at atmospheric pressure.

A major problem is the protection of the operator against X-rays; this has been solved by placing lead screens around the "hot zones" of the microscope (and in particular, at the various levels at which a part of the beam is intercepted by diaphragms: at the entry, the condenser C_2, the specimen-holder, and the observation chamber); in addition, heavy concrete walls are erected (between the accelerating column and the microscope, and around the room in which the microscope is situated). The various screens and the image which is finally obtained are observed with television cameras. A photograph of the instrument is shown in Fig. 227g; the lead sheaths about the entry diaphragm and the observation chamber can be seen.

18.5 THE OPTICAL ADJUSTMENT OF THE INSTRUMENT

To obtain good images, the possession of a good microscope is insufficient—it has also to be used correctly. In the majority of high quality instruments, the column of lenses can be centred, and hence the mechanical imperfections remedied. The experimenter must be capable of aligning his microscope correctly in the minimum time possible, as this is one of the most basic adjustments to be made to the instrument.

18.5.1 Adjustment of the Gun

In the majority of modern microscopes with a heated cathode, it is easy to adjust the automatically biased gun. The point of the filament has to be centred with respect to the opening of the Wehnelt (or grid), and its height has to be adjusted, and these are operations which can often be performed visually, after withdrawing the gun from the instrument (the RCA microscope). All that is then necessary is to adjust the inclination of the gun to the axis approximately, in order to compensate possible defects of the preliminary centring, and to obtain a true and symmetrical image (on an intermediate screen for example) of the point of the filament for a convenient value of the excitation of the lenses. If the position of the filament can be adjusted while the lens is in use, the grid and the anode remaining centred by virtue of the construction, it will be necessary to adjust the height of the filament to obtain the correct value of the saturation current and then to adjust the centring until the best possible image of the crossover is obtained on an intermediate screen placed at the same level as the object (the C.S.F. microscope).

18.5.2 Alignment of the Microscope

The purpose of this alignment is, above all, to reduce the displacements of the image which result from variations of the high tension. This is a vital adjustment in the magnetic microscope, as it has become apparent that this "voltage centring" in the magnetic microscope in much more important than the "current centring" which was once foreseen, and with which it does not always coincide (the reason for this lies in the parasitic magnetic fields). As a centring criterion, therefore, we select the point at which the image is least sensitive to variations in the high tension. In an electrostatic microscope, these variations only produce second order changes in the total magnification if all the potentials are obtained from the same potential drop; if, however, the instrument is focused by modifying the polarization of the objective with batteries, these variations can become more trouble-some—the instrument must be centred in such a way that the image in-creases away from its centre, and hence the smallest possible perturbation of the whole ensemble occurs when the polarization of the objective is slightly increased.

In a magnetic instrument, a variation of the high tension produces first order effects; these can be either an increase or a decrease of the size of the image, or above all, a rotation about some point which it is convenient to bring to the centre of the screen if the perturbation of the whole image is to be as small as possible.

We shall describe the alignment procedure in two typical cases; first of all, the case of an electrostatic microscope without a condenser, with two lenses only, and after, that of a magnetic instrument with two lenses and a condenser.

(i) *The electrostatic microscope with two lenses, without a condenser*

We suppose that the objective and the projective lenses have been centred geometrically onto a common axis during construction. The centring of the instrument consists, therefore, solely in bringing the crossover to this axis.

First, the beam is directed onto the region of the object which is to be exa-mined, which it illuminates brilliantly. With the aid of the focusing potentio-meter, the convergence of the objective is then slightly modified. The image which is observed on the screen is then the image of the projection of the object which is produced by the source of illumination on a plane of variable height. Figure 228 shows that the result is that if S is not on the axis, the image point I which had previously been at the centre of the fluorescent screen is displaced. All that need be done is to adjust S until the sole effect of varying the focus is an expansion or contraction of the image about the centre of the screen. This is a particularly easy operation if the gun can be turned about a point on the axis of the apparatus in the immediate vicinity of the object; if this is not the case, the beam has to be brought onto the object by a lateral displacement of the gun after each alteration of the inclination.

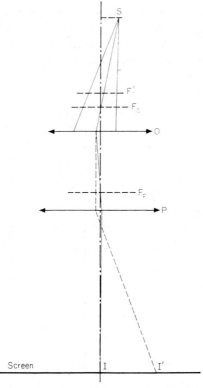

FIG. 228. The alignment of an electrostatic two-lens microscope. The crossover S does not lie on the axis, and a modification to the convergence of the objective displaces I to I'. The plane F_0 is conjugate to the plane F_p of the intermediate image when the objective has its normal convergence. When the convergence is less, it moves to F_0'.

(ii) *Alignment with respect to the potential—the case of a magnetic microscope with a condenser*

Here, it is a question of aligning the optical centre of the condenser and the crossover on the axis which joins the optical centres of the objective and the projective lenses. The simplest procedure is illustrated in Fig. 229a, where the alignment criterion which has been selected is that the image shall be as stable as possible in the presence of variations of the high tension.

(a) A preliminary approximate adjustment of the position of the gun brings the image of S into the centre of the fluorescent screen, while the condenser is adjusted to give the maximum brightness (the image of S on the object). The image is vividly brilliant, but the instrument is far from being centred.

(b) The condenser is adjusted to give a very faint illumination (which is then nearly parallel). The point I of the image which occupies the centre of the screen is then singled out; the H.T. is reduced (or cut if it is fixed), so that the supply capacity discharges. The convergence of the lenses increases; the final image is that of a projection of the object onto a plane which becomes closer and closer to the objective. Figure 229b shows that if the illumination is not centred, the point I moves away from the centre of the

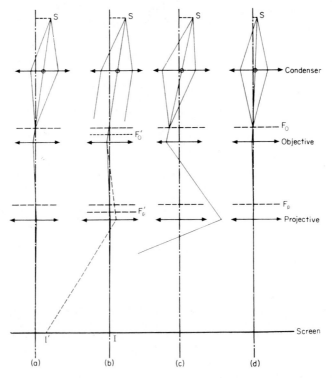

Fɪɢ. 229. The alignment of a magnetic microscope with two lenses and a condenser. When the accelerating potential is normal, the intermediate image plane is F_p, the plane conjugate to F_0 with respect to the objective. When the accelerating potential falls to a lower value, the intermediate image plane moves to F_p' in order to remain conjugate to the screen with respect to the projective. The point conjugate to F_p' with respect to the objective is F_0'.

screen (to a point I'), and we take note of the direction in which it moves. A magnification and a rotation of the image around I is seen at the same time, but with these we shall not be concerned; for the sake of clarity, therefore, the rotations have been suppressed in the figure.

(c) The high tension is restored to its normal value, and the excitation of the condenser to the value which corresponds to maximum brightness. The image of the source reappears at the centre of the screen. We then shift the *combination of condenser and gun* in such a way that the spot of light is moved a considerable distance from the centre of the screen *in the same direction as the point I had moved previously* (Fig. 229 c).

(d) The bright spot is brought back to the centre of the screen *by moving the gun alone*. The source S and the optical centre of the condenser are now both closer to the projective-objective axis (Fig. 229 d).

(e) The cycle of operations just described is repeated until the point I ceases to move when the high tension begins to fall, and a spot of the maximum brightness is obtained at the centre of the screen by a simple adjustment of the condenser. The complete centring procedure, when performed well, only takes one or two minutes, even in the case of an instrument initially totally eccentric.

18.5.3 Focusing

This is a delicate operation as the depth of focus of the instrument is so large. The best method is based upon direct observation of the Fresnel fringes at the edge of a detail of an object with adequate contrast. The search for the point at which the positive fringe disappears is facilitated by the fact that the contrast is reduced at the exact focus; a focusing microscope is also helpful.

This method is, however, only practicable for rather experienced workers, and the designers have tried to simplify the approximate focusing which is required in everyday use in various ways. With a "wobbler" (Le Poole, 1947) the illuminating beam is made to oscillate, which is equivalent to artificially widening the aperture. If the focusing is imperfect, the image at the screen vibrates, and to obtain a well-focused image, this vibration has simply to be reduced to the minimum (the Philips microscope). Oscillation of the illumination can be replaced by a multiple illumination produced by three sources (one axial, and two lateral). A poorly focused microscope gives a triple image (the Trüb and Täuber microscope). Both of these procedures lead to an image slightly out of focus, as a result of the spherical aberration of the objective.

CHAPTER 19

THE ACTION OF THE OBJECT IN A MICROSCOPE ON THE ELECTRON BEAM; CONTRAST IN THE IMAGE; VELOCITY ANALYSERS

19.1 THE TWO EFFECTS TO WHICH ELECTRONS MAY BE SUBJECTED ON PASSING THROUGH A THIN FILM

19.1.1 Definition of the Mass Thickness μ_s

Since the Lenard window was discovered in 1895, a great deal of work has gone into clarifying the behaviour of fast electrons (1000 eV to 200 keV) as they pass through a thin film of matter. A clear and valuable summary of these fundamental studies is to be found in a very thorough review by Bothe (1933) from which we extract only two essential points.

The first of these is the definition of the fundamental parameter upon which the effects of traversing the solid depend. This is the mass thickness, μ_s; in fact the actual thickness "d" through which the electron passes plays a role no different from that of the density μ, and the action of the layer depends upon the product

$$\mu_s = \mu d,$$

in which μ is the normal specific mass (per unit volume) or density. In addition, the effects which are observed can be classified into three categories according to the value of μ_s, or if the experiments concern a single material only (aluminium was a common choice), according to the value of the thickness d.

19.1.2 The Three Types of Scattering. Definition of a "Thin Film"

We can best understand the character of each of the scattering processes by examining the principle on which the theory rests. The scattering, retardation or deflexion which we observe is only the resultant effect of the interactions, in which each electron has been involved, with the nuclei and electrons of the film close to which the electron trajectories happen to have passed. Matter is virtually empty, since the diameters of electrons and nuclei are only of the order of a hundred thousandth of an Ångström unit (a millionth of a millimicron), whilst their mutual separations are a few Ångströms from one nucleus to another and a few tenths of an Ångström from nucleus to electron. It is because the space within the material is predominantly empty that a projectile has a chance of passing through a continuous layer about a hundred Ångströms thick (about 30 atomic layers)—the thinnest which we know how to make—without being either retarded or deflected. As the thickness is increased, the chances of passing near enough to another particle for the beam electron to interact increase, and it is then that the effects of retardation and deflexion appear. The same is true if the atomic number of the material is increased, as the number of planetary electrons increases in proportion, and in addition, the extent of the zone over which the nuclei have an effect increases with their charge. This elementary theory accounts well for the importance of μ_s, and explains the equivalence of μ and d. Further, it is adequate for us to separate out the three categories.

Multiple scattering

In this case, the layer is thick enough for every incident electron to be subjected to a statistically "large" number of interactions. The mean scattering angle, as well as the mean loss of velocity, is considerable. The theory is statistical in nature, as only the statistical trends can be observed with any ease. The most probable energy with which the electrons emerge, V_m, is given as a function of the incident energy, V_0 by Whiddington's law (1914):

$$V_0^2 - V_m^2 = K\mu_s,$$

in which K is a constant which is characteristic of the substance; for aluminium, K is of the order of 10^{12} $(V\ cm)^2\ g^{-1}$.

Plural scattering

Here, the interactions are too infrequent for statistical averages to be meaningful, but too numerous for a simple composition of the diverse effects to be possible. This is an intermediate case, in which the theory is of little help.

Single scattering

The layer is thin enough, and made of a sufficiently light material, for each electron to have only one chance at most of interacting with one of the particles in the barrier through which it passes.

In this chapter, we shall only be concerned with single scattering, which occurs when the substance is light, as we can see from Fig. 230, where the curve which describes the maximum thickness for which the scattering

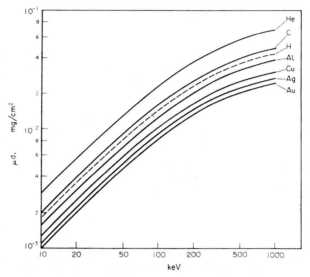

FIG. 230. The maximum thickness of a thin film for which the laws of single scattering are valid.

remains single is depicted. This maximum thickness for which the scattering is single corresponds to a value of μ_s of the order of 10^{-5} g cm^{-2}, or to a thickness which ranges between 100 and 1000 Å according to the nature of the specimen; for the heavy metals, which are used in shadow techniques, the thickness has to be reduced to about 10 Å.

19.1.3 Elastic and Inelastic Collisions

The simplicity of single scattering, the feature with distinguishes it from plural scattering, is more clearly apparent if we take into account the fact that the elementary interactions which are the origin of the scattering, and which are known as collisions since they occupy a very short time in comparison with the period of the planetary motion of the electrons around the nucleus, fall into two categories.

Elastic collisions

These are essentially collisions with the nuclei, as the impact laws of classical mechanics are valid in this case, and give a simple result; if there is to be an appreciable exchange of energy between two particles in a collision, their masses must be comparable. Here, on the contrary, the disparity is very large, for M/m is of the order of 50,000 for aluminium and 22,000 for carbon, and the collision simply deviates the electron without affecting its velocity.

Inelastic collisions

These are predominant when one of the electrons of the beam encounters one of the electrons in the specimen, as the two participants in the interaction are of the same mass, which is the case most favourable to a mutual exchange of energy—this always entails a loss for the incident electron which was much the faster. To verify this, we need only point out that whereas the energy quanta which are associated with the Bohr orbits (which appear in the experiments performed by Frank and Hertz on absorption in gases) are only a few volts in magnitude, the energy of the incident beam may reach some tens of thousands of volts.

The difference between elastic and inelastic collisions is clearly apparent in the general formula which embraces both cases and expresses the relative value $\delta V/V$ of the energy loss δV which corresponds to a given angle of deflexion, θ.

A collision with a nucleus of mass M and atomic number A (with m representing the mass of the electron) is described by

$$\frac{\delta V}{V} = \frac{m}{M} \tan^2 \theta = \frac{1}{1842 A} \tan^2 \theta \sim 0 \ldots \qquad \text{elastic collision.}$$

In a collision with another electron, however, we have

$$\frac{\delta V}{V} = \frac{\tan^2 \theta}{1 + \tan^2 \theta} \sim \tan^2 \theta \quad \left. \begin{array}{l} \text{inelastic collision} \\ \text{for small angles.} \end{array} \right\}$$

In single scattering, where there is only one collision during the passage through the film, the emergent electrons fall into one or other of two classes: those which have been elastically scattered by a nucleus, and those which have lost energy as a result of an inelastic collision with an electron. In the plural scattering region, on the contrary, the situation is considerably complicated by the fact that the successive collisions can belong to different categories.

19.1.4 Elastic and Inelastic Scattering at Very Small Angles

The simple models which we have described in the preceding paragraphs—electron scattering at an isolated nucleus or a free electron—only represent the real situation well for very small angles; the formulae of §19.1.3, for example, are valid in the diffraction region for angles of the order of a degree or a few degrees, but they become unreliable if θ falls below half a degree as it does in microscopy ($\theta = 5 \times 10^{-4}$ to 10^{-3} radians). In the domain of small angles, we must therefore take into account the binding between the planetary electrons and the nuclei—the theory becomes complicated, and only provides an incomplete description. It does however provide a satisfactory reply to the first of the two important questions which we are prompted to ask.

In the elementary solid angle $d\omega$ in the vicinity of $\theta = 0$, into what proportions are the retarded electrons (inelastic scattering) and the electrons which retain their initial velocity divided? The theoretical results of von Borries (1948) who, like Boersch (1947a), has a made critical study of these delicate points, are set out in Fig. 231; they are quite well confirmed experimentally. Considerably more electrons are retarded than continue without loss of velocity, and in each case, the figure gives the ratio dN/N_0 of the number dN of electrons which emerge, in the given category, into unit solid angle, to the number N_0 which are incident per unit solid angle.

Further, what is the mean loss of velocity in traversing the material, and what is the spectrum of the velocity loss at the exit? The theory has so far proved to be unable to explain the experimental results; this explains why measuring techniques in this domain are so important. That discrete losses of the order of ten volts exist has been shown experimentally, and has provided an incentive for the construction of accurate and stable velocity filters, with which this category of electrons can be eliminated; as we shall see later, the final result is an increase in the clarity of the images and diffraction patterns, by an improvement of their contrast.

19.1.5 Collision Cross-section

In the preceding paragraphs we have described the scattering and retardation phenomena in a fashion which is adapted to electron microscopy; we have, that is to say, selected certain properties and a particular vocabulary in which to describe them. However, the theory has been considerably developed, and applied to several different situations, and another way of expressing the facts is usually found to be more convenient. It seems useful to introduce the reader to this general form of expression, by defining succinctly the quantities which are found to be useful, and which differ from those which have been used earlier in this chapter.

We are led to define a new quantity, the "effective cross-section". We consider a parallel incident beam, carrying N_0 electrons *per unit area of cross-section*. We consider the number of electrons, deflected laterally through a mean angle θ, which lie within an elementary solid angle $d\omega$ which is defined by the cones with summit angles θ and $\theta + d\theta$. The quantity

$$\sigma(\theta) = \frac{1}{N_0} \frac{dN}{d\omega},$$

homogeneous with respect to area, is known as the "effective differential cross-section" relative to the angle θ. To define the total number of particles scattered into a cone of angle θ, we introduce the integrated form of this expression,

$$\Sigma_\theta = \int_0^\theta \sigma(\theta) \cdot 2\pi \sin\theta \cdot d\theta.$$

Since $d\omega = 2\pi \sin\theta \cdot d\theta$, this quantity is a measure of the total number of electrons deflected through an angle less than or equal to θ. The results of scattering experiments, therefore, will be set out by showing how σ varies with θ, and by stating whether or not it is retarded electrons which are involved.

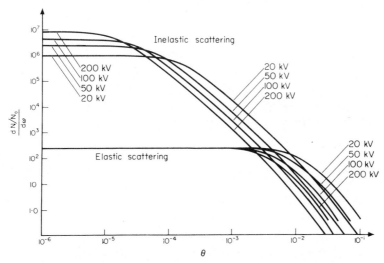

FIG. 231. The values of the effective differential cross-section in the passage through a flake of carbon (von Borries, 1949).

In Fig. 231, the distribution of the electrons which emerge from a thin film of carbon is shown in terms of the "effective cross-section"; the mass thickness μ_s is 10^{-6} g cm^{-2}, and the initial energies lie between 20 and 200 kV. The conclusion of the preceding paragraph is found expressed in a new form—for small angles, the collision cross-section is far larger for the curves which describe inelastic collisions than for those which correspond to elastic collisions. Inelastic collisions are predominant.

More recent information on this topic is to be found in an article by Marton, Simpson, Fowler and Swanson (1962).

19.1.6 Discrete Energy Losses, and the "Collective" Theory

Single scattering, which plays so important a practical role in image-formation, is still imperfectly explained from the quantum point of view; the existence of discrete energy losses, however, clearly suggests that it

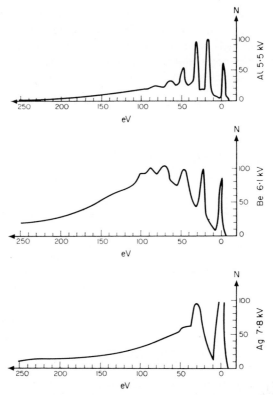

FIG. 232. Some examples of the spectra obtained by Voge and Rutheman, showing characteristic losses of the order of 10 eV.

is from this standpoint that the explanation is to be found. First, we shall summarize the clearest characteristics as they appear from several recent critical studies, and the numerous, though not wholly concordant, experimental results (see Fig. 232).

(i) The losses appear in the form of narrow or wide bands.

(ii) The energy loss associated with each band lies between a few volts and 100 V; when the bands are narrow, they can be quite numerous, and the characteristic energies are then usually integral multiples of the energy loss of the first band (the minimum loss).

(iii) It is not possible to relate the appearance of the energy-loss spectrum to the nature of the specimen unequivocally. Fine bands appear for certain metallic films (Al and Mg for example), but also for certain compounds and insulators (SiO_2, PbS, Sb_2S_3 and quartz); the spectra of some compounds are close to those of their base metals (Mg and its salts); some metals, however, have wide bands (Cu, Ag) as have certain non-metals (C).

The state of the surfaces of the films which are used is of fundamental importance, which explains the wide discrepancies between the experimental results. By evaporating an Al film within the body of the apparatus, Powell and Swann (1959) have shown that the spectrum rapidly deteriorates in a few minutes as a result of oxidization.

It is not possible to explain these energy jumps simply as the number of energy quanta which are necessary to extract a bound electron from an inner shell, as the energy in question would be too large, nor as the energy which would be required to transfer a peripheral electron into a valency or conduction band (in the sense in which this expression is used in the theory of metals) as the band would be too wide.

The bands remained without a plausible explanation until 1953, when Bohm and Pines developed a theory of the collective behaviour of electrons in plasmas. This latter is the name which is given to a medium which is electrically neutral but fluid, a completely ionized gas for example or the electrons in which the ions in a metal are immersed; the electrons are free to wander about within the metal and are in fact highly mobile, but it is difficult to understand their motion if only the mutual forces between immediately neighbouring particles are considered. The Coulomb law, which contains the factor e/r^2, shows that the more distant neighbours (which lie at a distance R) of any given electron are sufficiently numerous (they occupy a volume $4\pi R^2 \cdot dR$) to produce a field as important as that of the nearest neighbours, if they act in unison. It is certain that in fact they do not all act concordantly, but this hypothetical case gives an insight into the difficulty which has for so long delayed the calculations. A collective theory must be constructed, which deals with the behaviour of the totality of the electrons. The great achievement of Bohm and Pines was to have attacked this problem directly, and to have succeeded in solving it; starting from the individual interactions, they were able to build up the collective theory. This is a difficult procedure, and the interested reader will find a lucid guide in two articles by Pines (1954, 1955).

Here, we shall simply quote the result: the plasma can be regarded as a continuum, and hence possesses characteristic frequencies of vibration. Nevertheless, it is a very special kind of medium, a mixture of ions and electrons and we must not wholly lose sight of this granular aspect of the structure. This means that the medium possesses a frequency limit below which no vibration can be sustained. This minimum frequency has been known since the basic research of Tonks and Langmuir on ionized gases;

this frequency is the plasma frequency, and is given by

$$(2\pi f)^2 = \omega^2 = \frac{n\,e^2}{4m\,\varepsilon_0}.$$

It depends only upon the number n of electrons per cubic metre; e is the charge and m the mass of an electron, and ε_0 is the permittivity of the vacuum.

Bohm and Pines have shown that such "plasma oscillations" can exist, and consist of a longitudinal oscillation of the electrical charge density (the central force is of an electrostatic nature). The discrete losses which are observed experimentally are each the result of the emission of a plasma oscillation quantum, or "plasmon".

Bohm and Pines' original calculations required a formidable arsenal of mathematical weapons. Since their initial work was performed, however, the physical basis of the phenomenon has been brought out in various other studies (Fröhlich and Pelzer, 1955).

We consider a longitudinal exterior electric field E, frequency ω, wave number k. Through the Coulomb interaction, this excitation sets up forced oscillations among the electrons; the amplitude of the response is proportional to $\frac{1}{\varepsilon(\omega)}$ in which $\varepsilon(\omega)$ is the dielectric constant. If the frequency ω is such that $\varepsilon = 0$, the system can oscillate by itself without outside stimulus. This self-oscillation is precisely what we mean by plasma oscillation. To calculate the plasma frequency ω, we are led to study the polarization of the electron gas, or what is equivalent, to evaluate its dielectric constant $\varepsilon(\omega)$. We consider first of all a gas of free electrons, a case analogous to the situation in the upper atmosphere, which was studied by Sir Edward Appleton in the twenties. There will be an opposition between the Coulomb forces due to the applied field, and the Brownian motion of the individual electrons. But at very long wavelengths, the random motion of the electrons can be neglected, so that the screening effect is determined, or more accurately, limited by the *inertia* of the particles; the dielectric constant is thus given by:

$$\varepsilon(\omega) = 1 - \frac{n\,e^2}{m\,\omega^2\,\varepsilon_0},$$

in which n is the number of electrons per unit volume, e their charge, m their mass and ε_0 the dielectric constant of free space. The plasmons are of frequency $\omega_p = \sqrt{[n\,e^2/m\,\varepsilon_0]}$, the value of ω for which $\varepsilon(\omega_p) = 0$. When the wavelength decreases and hence the wave number increases, the calculated frequency differs more and more from this limiting value, ω_p. This aspect of the effect has been observed experimentally in the Ruthemann absorption, by a study of the angular distribution of *scattered* electrons; the energy of the discrete groups depends upon the scattering angle and hence upon k (Watanabe, 1955, 1956; Marton, Simpson and McCraw (1955, 1955a).

If the wave number k increases to such an extent that it becomes comparable with the Debye wave number k_D, the plasma oscillation disappears. k_D is the reciprocal of the screening radius† which is equal to $\omega_p/\langle V \rangle$, in which $\langle V \rangle$ is the root mean square velocity of the random motion of the electrons. This effect has been clearly observed experimentally, and displays how overwhelming is the influence of the individual motions at short distances.

We pass now to the real case of a solid, metal or insulator. A new factor appears, namely the opposition between the Coulomb forces and the binding forces associated with the ions located at the sites of the crystal lattice. We denote the mean random energy of an electron in the lattice field by $\langle \omega_{exc} \rangle$. If $\langle \omega_{exc} \rangle \ll \omega_p$, the electrostatic forces are predominant, and the plasma frequency is close to ω_p. This is true of Be, Al, Si and Ge, and for these elements we observe fine lines at the predicted positions. If, on the contrary, $\langle \omega_{exc} \rangle \simeq \omega_p$, the discrete losses are displaced, and above all, considerably widened. This is the case for many of the transition metals and compounds, among other substances. The agreement between theory and experiment is still only qualitative, but it is difficult to improve it at the present stage of development of the subject. The energy losses (in electron volts) are tabulated below for the most intense band.

I. *Good agreement*

	Be	C	Al	Si	Ti	Ge	Sn	Pb	Bi
$\hbar\omega$	19	25	16	17	18	16	14	13	14
ΔE	19	22	15	17	22	17	12	13	13

II. *More mediocre agreement*

	Fe	Ni	Zn	Ag	Cd	Au	Pt
$\hbar\omega$	31	35	13	9	11	9	30
ΔE	21	23	23	23	20	24	23

Plasmons are present in insulators, just as in metals (Mott, 1955). To take part in a plasma oscillation, an electron has no need to be free; at the frequency in question, it has no time during a single period $2\pi/\omega$ to feel

† The screening radius can be directly defined in the following way: suppose we were to alter one of the equilibrium characteristics of the plasma, in a small region around some point M; the plasma would "cure" this defect itself in that a self-modification would occur within a sphere of radius $1/k_D$ and outside this sphere, the "disease" would disappear completely. Debye was the first to consider this self-screening mechanism in the case of a liquid electrolyte.

the crystal lattice binding potential (Nozières and Pines, 1959), which on its own produces a much slower motion.

In conclusion, we should mention the existence of more complex collective modes, and in particular, the "surface" plasmons which characterize vibrations localized in the neighbourhood of the surface of the film (Stern and Ferrell, 1960, and Ritchie, 1957). The frequency predicted by the theory for these plasmons is $\omega_p/\sqrt{2}$ and this seems to be confirmed by experiment (Powell and Swann, 1959).

A very complete study of the characteristic energy losses of electrons in solids has been published by Marton, Leder and Mendlowitz (1955); this article collects together the numerous experimental results and summarizes the various theories which have been proposed to explain them. Articles on characteristic energy losses have also been published by Mendlowitz (1960a, 1960b) and Hartl and Raether (1961). In this connexion, an article by Ferrell (1956) is also relevant; see also Green (1962).

19.2 CONTRAST FORMATION IN AN AMORPHOUS
OBJECT

19.2.1 The Roles Played by the Sensitivities of the Eye and of the Photographic Plate

We consider two contiguous parts of the image: from the photometric point of view, each will be characterized by the value of the surface density of the electron current which bombards the plate or screen. It is the difference between the values of the current density which gives rise to a difference between the values of the brilliance (of a fluorescent screen) or of the blackening (of a photographic plate).

When we examine the fluorescent screen by eye, the difference between the two regions can be discriminated only if the brilliances of the two are sufficiently different. Under normal conditions, the relative difference must be of the order of 5 per cent, as the mean brilliance is not usually intense enough for the eye to be used in its optimum conditions (where the limit would be lowered to 1 per cent). The electron intensities would have to differ by at least 5 per cent for the observer to be able to focus the instrument correctly by looking at the boundary between the two zones in question.

A photographic plate records differences of contrast of this magnitude faithfully, provided the exposure is chosen correctly, that is, in such a way that the conditions which correspond to the straight section of the photographic density curve obtain for every region of the image. The emulsion could record even less marked differences in contrast, and we might hope to be able to develop the film in such a way as to make them reasonably visible to the eye by copying the primary impression onto a "hard" emulsion. The marginal gain to be had in this way is too small, however, and the cycle of operations too complicated and of too doubtful a fidelity to have seriously caught the attention of experimental workers. It must not be forgotten that it is only in the small details of the object that the microscopist is interested, which correspond to extremely tiny zones of the primary plate; statistical fluctuations produced by the grain are then appreciable, and severely limit the possibility of discriminating between the different amounts of contrast (von Borries, 1949; and Leisegang, 1954). The photographic plate is scarcely more efficient than the combination of eye and fluorescent screen.

19.2.2 Transparent Objects in Electron Optics

Normally, the objects which are used in the electron microscope are thin enough to absorb only a negligible fraction of the incident electrons. Their sole action is to scatter—elastically or otherwise—the incident particles. As there can certainly be no chromatic differentiation with a monochromatic electron beam, it is clear that the contrast which is to be seen (in Fig. 5 for example) is different in origin from the contrast which is formed in an optical microscope image.

We pointed out at the very beginning (§ 1.3) the origin of this contrast: the diaphragm which contains the aperture halts the most acutely deflected electrons, which are the more numerous as the substance has a more pronounced scattering action. A substance which contains heavy atoms (a high value of Z) appears blacker than a substance made of light atoms.

We have also to admit, however, that the contrast is not wholly lost if the stop is removed (Hillier and Ramberg, 1947). To understand this complicated combination of effects, therefore, the exact mode of image formation must be studied in each case.

19.2.3 Image Formation with a Small Aperture

We suppose that a "contrast stop" of small diameter has been placed beneath the object. The useful aperture, α, of the instrument is, therefore, well-defined, and it is always desirable that this should be the optimum aperture already defined (§ 16.2). Equally, we shall suppose that the illumination is parallel and we consider a point on the object the effective differential cross-section of which is $\sigma(\theta)$, which corresponds to an element of area dS. The number of electrons impinging upon this element of the surface is $N_0\, dS$, but only

$$N = N_0 \left(2\pi \int_0^\alpha \sigma(\theta) \sin\theta \cdot d\theta \right) dS$$

pass through the diaphragm subsequently to be distributed over the image of area $G^2 \, dS$ (where G is the magnification). The current density at the image will then be

$$i = N_0 \frac{2\pi}{G^2} \int_0^\alpha \sigma(\theta) \sin\theta \cdot d\theta.$$

As soon as the effective cross-section is known, therefore, we can calculate the contrast, but the cross-section is only known for certain in the two extreme cases of a very thin film (the usual case) and a very thick film (an exceptional case, of theoretical interest only. See Marton and Schiff, 1941; and Hall, 1949 and 1951). In both cases, only the mass thickness is relevant, and we can only determine the thickness of the object if its nature is known; in biology, we have to take into account the possible modifications which may be produced in the object by the bombardment. Two basic characteristics of the microscopic contrast become apparent: the resolving power and the contrast are linked, since they are both dependent upon the angular aperture, and two widely different details will have the same contrast with respect to the background if they have the same effective cross-section.

It is clear that the method of calculation outlined above can only give an order of magnitude. The illumination is not parallel, as often the image can only be made tolerably bright by opening the condenser beyond the optimum value (the illuminating angle is of the order of 5×10^{-4}). The contrast is then lessened, as an electron which is incident obliquely may be scattered by a detail in such a way that it passes through the diaphragm whereas with parallel illumination it would have fallen outside the opening. We have also to take into account the retarded electrons which "lap" over the image as a result of chromatic aberration. Finally, the collision cross-sections are imperfectly known.

19.2.4 Image Formation with a Wide Aperture

A contrast diaphragm with a small opening rapidly gets dirty. Charges collect on the impurities as the beam passes, and the ellipticity aberration worsens. Attempts have often been made to remedy these deficiencies by suppressing the diaphragm, and if this is done, we find that the contrast between two large adjacent regions is weakened but that the contrast between a small detail and the background is hardly affected.

It is to the aberrations that we must ascribe this feature. We consider two objects, one of which, A, scatters markedly, while the other, B, scatters hardly at all; the beam which emerges from A is wide in angular aperture and will produce a large aberration patch at the image plane as a result of the wide apertures—the brightness will not be very high, therefore. The beam which emerges from B, on the other hand, will only produce a small aberration patch which in consequence will be comparatively intense in brilliance. This qualitative explanation is illustrated in Fig. 233.

This contrast is dependent to a considerable extent upon the focusing, as it is linked to the effect of the aberrations. Although we have represented the images as being uniformly illuminated in Fig. 233, this is only for the

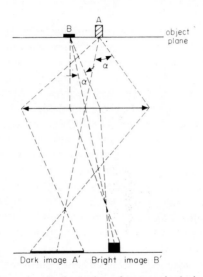

FIG. 233. Diagram illustrating the formation of contrast in the image of an object; the electrons are spread out over the aberration spot, the size of which depends upon the degree of scattering produced by the object.

sake of simplicity. In reality, the distribution is complicated, and depends upon the focal plane. The contrast in the image is accentuated by a slight defocusing, if only by producing a bright fringe (a negative fringe) which adheres to its border; the image seems clearer, and often, a slightly defocused image is preferable to an image rigorously in focus despite the slightly higher resolution which should in theory be possible with the latter. The effect of overfocusing is identical, and in fact the exact focus corresponds to minimum contrast, which is experimentally the best criterion for identifying it. A slight overfocusing blurs the image, however, as a dark (positive) fringe surrounds the dark details; the increase in contrast is accompanied, in this case, by an appreciable deterioration of the resolution. In conclusion, we should point out that the choice between a precisely focused image, poor in contrast but sure in interpretation, and a slightly defocused image in which new details often appear which must, however, be accepted somewhat circumspectly, must be made according to the type of object which is being studied.

19.2.5 Improving the Contrast

(i) *Shadowing*

It is most desirable in many cases, therefore, to find a way of increasing the contrast between the background and the details in which we are interested.

We have already mentioned the technique with which this can be done. The object is covered with a thin layer of heavy metal (10 Å thick for gold, 4 Å for uranium) evaporated on at grazing incidence. Details with a small mass thickness which alone could not produce sufficient contrast become visible by virtue of their "overlying shadow" which is not covered with metal and therefore appears bright against the dark background which is the image of the part of the object which is covered with metal.

This way of improving the contrast was introduced by Müller (1942). Further details of this shadow technique are to be found in the articles by Williams and Wyckoff (1944), Kahler and Lloyd (1950) and Koenig and Helwig (1950). The main problem is to find a metal which is deposited as a continuous layer and remains thus during observation. The majority of metals recrystallize, and under the electron bombardment these microcrystals collect into grains and hence create an artificial structure. Uranium and certain germanium and platinum alloys are free of this drawback. Examples of the image of an object shadowed in this way are to be found in Figs. 252 and 256.

(ii) *Dark field operation (strioscopy) and Foucault contrast*

This represents the translation into electron optics of the celebrated experiment performed by Cotton and Mouton with the ultra-violet microscope. Figure 234a–(i) shows schematically an optical arrangement with which the image of the object can be formed by scattered and diffracted electrons alone. The condenser contains an annular aperture with the aid of which we can illuminate the object with a hollow conical beam; a second diaphragm which contains an axial opening of small diameter is placed in the image focal plane of the objective, and removes the direct beam while allowing certain of the electrons which have been deviated from their initial trajectories to continue uninterrupted. The image of the object is then seen as a white region against a black background, and the brightness of the image increases with the scattering power of the object (Fig. 234b).

In another arrangement, which is simpler to construct (Fert and Faget, 1948) a condenser diaphragm in the form of a narrow slit is used (three to five microns wide) while in the plane of the crossover of the objective, a screen is placed parallel to the slit; this screen consists of a very fine metallized thread (0·1 μ) which interrupts the direct beam completely [Fig. 234a–(j)].

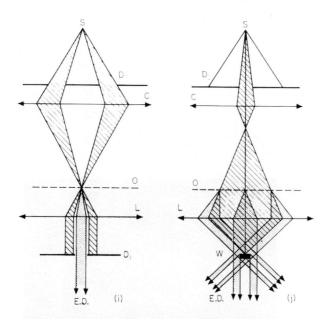

FIG. 234a. Two ways of observing against a dark background: (i) with an annular condenser diaphragm D_0; the diaphragm with a centrally placed hole, D_1, halts the direct beam and allows the scattered electrons ($E.D.$) to pass through; (j) with a condenser diaphragm D_0 in the form of a thin slit on the axis; the direct beam is halted by a metal thread parallel to D_0—a proportion of the scattered electrons $E.D.$ is transmitted.

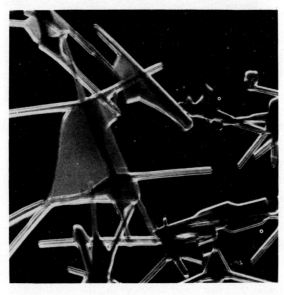

FIG. 234b. Dark field image of zinc oxide crystals (Photo: Fert).

Other types of system may be used, although they give considerably dimmer images: for example, the contrast diaphragm beneath the objective may be made eccentric, or if the object is monocrystalline, one or more of the spots of the diffraction diagram may be isolated with this same diaphragm.

This technique which was introduced by Boersch (1936–53) has been developed by Hall (1948) to obtain a resolution of 50 Å with a convenient arrangement in which the obscuring disc is placed between the object and

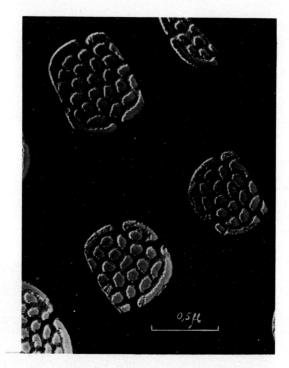

FIG. 234c. Photograph of a diatom, with Foucault contrast (Photo: Fert).

the objective. Ramberg and Hillier (1948) have calculated the theoretical resolution, and even when chromatic aberration provides a limit, it ought to be as good as in the bright field situation. We should mention too the beautiful experiments performed by Möllenstedt (1949, 1950, 1952a, 1952b), who obtained dark field images in which the images formed by the electrons which had been scattered and slowed up were displaced with respect to the image formed by those which had been scattered elastically.

In the course of a thorough examination of electron scattering, Boersch (1947) showed that just as in light optics, the resolution in dark field conditions should be better than the resolution which would be obtained if the same instrument were used with a bright field. Against a supporting layer of collodion, 100 Å thick, we ought to be able to distinguish atoms heavier than aluminium. At present, practical problems are the reason for the limit to the resolution, and in particular, the need to use stops of very small diameter at the objective in order to diminish the harmful effect of chromatic aberration.

The contrast of certain details which scatter particularly strongly can be considerably increased by removing almost all the electrons in the direct beam and half of those which have been scattered or diffracted by means of a stop in the form of a thin half-plane (a thin sheet of graphite, for example). This is "Foucault contrast" (the Schlieren method), which is characterized by an asymmetrical light distribution; the image displays marked shadowing effects, as if the object had been illuminated obliquely (Fig. 234c).

(iii) Phase contrast

In 1934, Zernike conceived a different way of observing the image from the one used in strioscopy. A thin sheet, which is known as the "phase plate", covers the whole diffraction pattern of the aperture; here, however, the plate is not opaque but instead, introduces an optical path difference between the principal wave and the diffracted waves which is chosen to be equal to $\pm \lambda/4$ (Zernike, 1942). This enables us to convert the variations of phase in the beam which are created by a transparent object into variations of the contrast at the image. For a long time, a means of transposing this remarkable technique into electron optics has been sought, so successfully has it been applied in light optics. To see how this may be done, we consider the fact that certain studies of electron diffraction (the famous Davisson and Germer experiment, for example, and the calculations made by Hillier and Ramberg, 1947, of the Fresnel diffraction effects) have shown that the potential within a substance is higher than the (relativistically corrected) potential V^* of the incident beam by an amount ΔV of the order of a few volts; the refractive index of the substance with respect to that of the vacuum is thus:

$$n = \sqrt{1 + \frac{\Delta V}{V^*}} \simeq 1 + \frac{1}{2} \frac{\Delta V}{V^*}.$$

As the refractive index is greater than unity, the wavelength in the medium is shorter than the wavelength *in vacuo*. A wave which passes through a thickness d of a material with refractive index n is retarded by a distance equal to

$$\triangle = (n - 1)\,d = \frac{\Delta V}{2V} \frac{1 + \dfrac{e\,V}{m_0\,c^2}}{1 + \dfrac{e\,V}{2m_0\,c^2}}\,d$$

(in which V is the potential applied to the gun) with respect to a wave travelling in the vacuum.

If the path difference $\triangle = n\dfrac{\lambda}{4}$, the phase difference is $\triangle q = n\dfrac{\pi}{2}$; with collodion, for example, d would be 100 Å for $\triangle q = \dfrac{\pi}{2}$ and $V = 60\ kV$.

The first attempts at using a phase plate were made by Agar, Rewell and Scott (1949); a hole about a micron in diameter was cut in the centre of a thin formvar plate, which was placed at the crossover of the objective. The phase of the electrons which have been diffracted by the object is altered with respect to the phase of the direct beam, as the former pass through the formvar whereas the latter pass freely through the central opening. These attempts were disappointing; it was thought for a time that no accentuation of the contrast could be hoped for from such a system, the blame being cast upon the system itself which was assumed to destroy the phase coherence of the beam, and hence to render any kind of phase contrast impossible. This seemed to be confirmed by the difficulties which were encountered in attempts at electron interferometry with separate beams in which one of the beams was retarded in phase with respect to the other by a material film (Marton, 1953).

Faced with these difficulties, therefore, a method of altering the phase without introducing a real plate has been sought. Locquin (1955) has been able to demonstrate that the contrast of very thin biological specimens without supporting films can be sharply increased (see Fig. 234d) by placing an insulating point in the plane of the aperture diaphragm of the objective; this point becomes charged under the influence of the beam. The mechanism by which this contrast is formed remains without a conclusive explanation, and is still the object of controversy, as no way of explaining it in terms of phase contrast in Zernike's sense of the phrase seems possible. It seems to be due rather to some kind of interference effect, similar to those which can nowadays be straightforwardly produced with the aid of an electrostatic biprism, the principle of which we shall now explain.

A divergent beam of electrons is emitted by a line source S, which is made as narrow as possible to improve the coherence (Fig. 234e). The beam passes through an electrostatic system which comprises a diaphragm containing a slit parallel to the source and an extremely fine metallized thread (less than a micron in diameter) which is stretched along the axis of the slit. This thread may be held at an electrostatic potential, Φ, which is positive with respect to the edges of the slit. If $\Phi = 0$, the two beams which are separated by the thread do not subsequently intersect, and only the shadow of the thread is seen framed within the ordinary diffraction fringes which resemble the fringes produced by the edge of a screen (the lenses which follow the biprism provide the necessary magnification). If Φ is increased, however, the two beams are deflected in opposite directions, approach one another, and begin partially to overlap; they appear to have originated

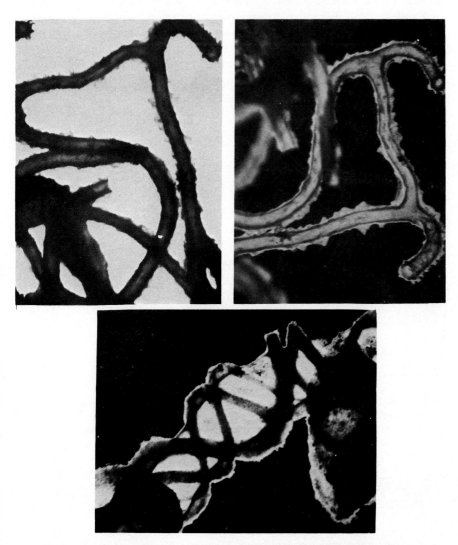

Fig. 234d. Attempts at phase contrast (Locquin, 1955) photographs of myxo-mycetis fibres. *Top left:* ordinary photograph with a light background. *Top right:* photograph at the same magnification, after introducing an insulating point into the aperture diaphragm—the contrast is improved and fresh detail appears in the internal region of the fibres. *Below:* photograph at high magnification, showing spiral structures which are invisible with normal methods of observation.

in two virtual sources, S_1 and S_2, separated by a distance $2l$. In the zone common to both diffraction fringe systems, therefore, interference fringes appear, just as in the Fresnel biprism experiment (already familiar in light optics); the fringes are progressively more closely spaced as $2l$ is made

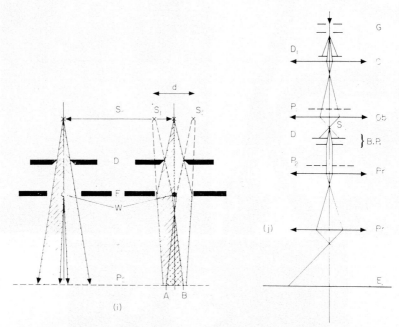

FIG. 234e. Diagram of an interference microscope. (i) The biprism with the wire not excited (left) and excited (right). S_0: linear source; D: diaphragm; F: slit and W: wire, parallel to S_0; P_2: plane of observation of the fringes which are formed in the region AB, common to the two beams which seem to originate in the virtual sources S_1 and S_2, the separation d of which varies with the excitation of the wire. (j) General diagram of the optical system. G: gun; D_1: slit-shaped diaphragm; C: condenser; Ob: objective; S_0: focus of Ob (cross-over); $B.P.$: biprism; Pr_1 and Pr_2: projective lenses; E: screen; P_2: focal plane of Pr_1; P_1: object position; either a wire for dark field observation or a phase plate can be introduced into S_0.

larger. The total number of fringes visible in extremely carefully conducted experiments may reach several thousands (Möllenstedt and Düker, 1956, Faget and Fert, 1956, 1957, Faget, Ferré and Fert, 1958); we can thus describe the coherence of the beam by attributing to it a wave-train several thousand wavelengths long, or about a hundred Ångströms. The evolution of a fringe system obtained in this way is shown in Fig. 234f.

If a thin carbon film 100 Å thick is introduced into the path of the beam just beyond the biprism, however, the fringes remain visible; *the coherence of the beam, therefore, is not destroyed* (Faget and Fert, 1956). Genuine

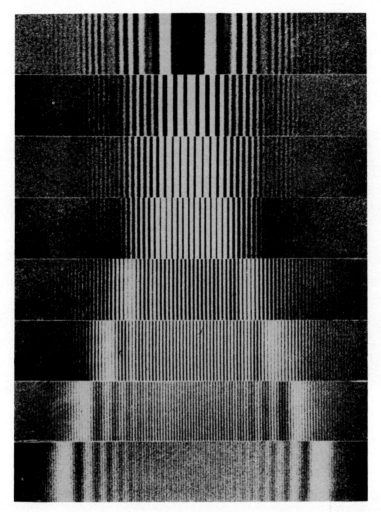

Fig. 234f. Interference fringes obtained by Möllenstedt with the aid of a device similar to a Fresnel biprism. The phase difference is produced by the electric field of a fine wire, the potential of which increases as we move from top to bottom from zero to seven volts; more and more fringes become visible (19,400 eV).

interference microscopy has thus been achieved, and we can consider afresh the possibility of producing real phase contrast.

In this connexion, we shall now give a brief description of interference microscopy techniques (Möllenstedt and Buhl, 1957; Buhl 1960; Faget and Fert, 1957). The arrangement is as follows: the condenser contains a slit diaphragm and a biprism is placed beneath the objective. The object which is to be examined is placed in an ordinary specimen-holder, so that on the screen, both the image of the object and the shadow of the thread are to be seen. When the latter is given a positive potential, fringes appear in the zone common to both parts of the beam. At each point, the order of the interference is defined by the geometrical path difference produced by the combined effects of the biprism and the difference in optical thickness between the two corresponding points of the object. If the latter is homogeneous

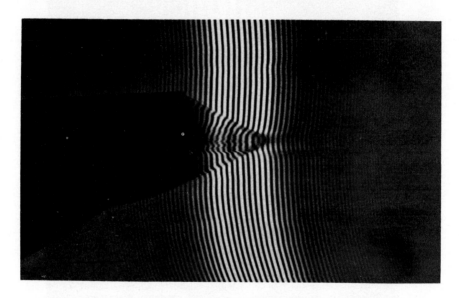

FIG. 234g. Interference microscopy: phase difference produced by a graphite flake which has been inserted into the path of one of the electron beams. The displacement of the fringes depends on both the thickness and the internal potential of the flake (Photo: Fert).

but not uniformly thick, the fringes are sinuous as the difference between any one fringe and the "average" fringe represents the corresponding difference in optical thickness (Fig. 234g). If, on the other hand, the object is of constant thickness, but consists of several constituents, the fringes produced by two neighbouring regions, different in nature, will again be shifted with respect to one another, and will thus indicate the variations of internal potential $\triangle V$. By measuring the irregularities of the fringe system, we obtain a measure of $\triangle V$. To make this measurement in practice, the object is left unsupported, so that $\triangle V$ is obtained directly with respect to the vacuum (Fert and Faget, 1958); the main difficulty lies in the precise measurement of the thickness of the object.

Let us now reconsider phase contrast, with a phase plate. A slight improvement in the contrast has been obtained by Kanaya and Kawakatsu

(1958) with the aid of a collodion or carbon plate, in which numerous holes have been pierced. The direct beam passes through the substance, while certain spots of the diffraction diagram pass unimpeded through the holes. The analogy with the optical arrangement is not always very rigorous, however, and it is only very recently that Faget, Fagot and Fert (1960) have managed to produce the exactly analogous situation. A straight-edged phase plate, accurately parallel to the slit source of the condenser, is placed in the image focal plane of the intermediate lens; to avoid conta-mination the phase plate is held at a temperature of 200°C. To produce the phase plate, a very fine thread from a spider's web (between 0·1 and 0·4 μ in diameter) is stuck on a thin collodion film, and then covered by evaporation with a thin layer of carbon. The thread is then torn away. In the corresponding zone of the plate is a region thinner than the remainder, and we check that the path difference produced by this difference in the thickness is in fact $\lambda/4$ by interference microscopy. The thin zone of the plate must cover the image of the slit source very exactly, so that the dif-fracted electrons all pass through the thicker region. An example of the im-provement in contrast which can be obtained in this way is shown in Fig. 234h.

These experiments confirm that the coherence of the beam is not destroyed by a passage through a thin object. The influence of this passage on the resolution of the instrument is as yet only imperfectly understood, however.

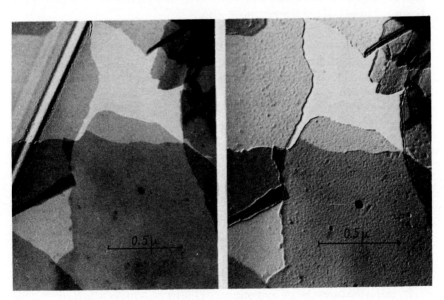

FIG. 234h. The improvement of contrast obtainable with a phase plate. *Left:* the image before the insertion of the phase plate. *Right:* the same image, after the phase plate has been introduced.

19.2.6 Contrast in Crystals

When the object is crystalline, we can observe the phenomenon of diffraction at the various lattice planes of the structure; just as in ordinary optics, this phenomenon is characterized by the presence of privileged directions along which the waves are in phase and give a diffracted wave. In every other direction, the resultant intensity is zero, as the diffracted wavelets mutually interfere destructively. We shall thus observe intensity maxima for certain relative orientations of the lattice, the incident beam and the diffracted beam. Let us suppose that a parallel beam of electrons falls on a thin crystal which has lattice-spacing d at one of the Bragg angles of incidence which are defined by

$$2d \sin\theta = n\,\lambda.$$

With $\lambda = 0{\cdot}05$ Å, $d = 2{\cdot}5$ Å, the first maximum will be observed in a direction which is inclined with respect to the incident direction at an angle $2\,\theta = 2 \times 10^{-2}$ radians. This angle is in general adequate to prevent the diffracted beam from passing through the objective diaphragm when the illumination is parallel to the axis. Generally speaking, every part of the crystal which is placed in such a way that a Bragg reflexion falls outside the objective diaphragm will appear dark; this explains the presence of dark fringes, which are more often than not sinuous, in the images of thin crystals such as molybdenum oxide. When the orientation of the illuminating beam is altered, these fringes move also.

If, on the other hand, a lattice plane is so oriented that the diffracted beam passes through the objective diaphragm (Fig. 235) and is focused, a bright image of the dark part of the object is seen on the screen (Boersch, 1943); this image is displaced with respect to the ordinary image, and this duplication can easily be seen with needle-shaped crystals (such as those of zinc oxide), and an intense illumination. These "Bragg images" can be distinguished from the paraxial image by the attendant spherical aberration, and by measuring the separation d between the two "images" we can calculate the spherical aberration constant C_s from the relation

$$d = C_s(2\theta)^3.$$

During focusing, the light and the dark images move with respect to each other—coincidence between the two corresponds to under-focusing.

For a long time, these "images" were regarded simply as curiosities. A few years ago, however, it was realized that they could provide very interesting information about the structure of the specimen, and an original technique for examining periodic structures has been developed.

By using the direct beam and one of the diffraction spots which correspond to some particular direction of the lattice planes, we obtain an image which can be considered as the projection of the structure and which contains only the first term of the illumination (represented mathematically in the form of an infinite Fourier series), apart from the fundamental. We thus obtain lines separated by the same distance as those lattice planes at which the diffracted waves which have been accepted were diffracted.

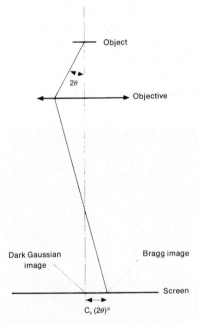

FIG. 235. The formation of a "Bragg image", displaced with respect to the main image by spherical aberration.

The lines to be seen in Fig. 227c are such a case, for example—this figure may be said to be an "image" of the $(20\bar{1})$ planes of a copper phthalocyanine crystal with a mean separation of 9·9 Å.

Images of this type were obtained for the first time by Menter (1956) who was studying phthalocyanine crystals.

If, therefore, we eliminate all the diffraction spots save one by means of a contrast diaphragm 30 to 50 μ across, we can study the lattice structure of a crystal and in particular, the faults in the lattice; this procedure is limited, however, by the resolution of the instrument, and this *direct* method of observation is only applicable when the meshes of the lattice are comparatively great (longer than 6 or 7 Å). The shortest interval which has been observed is that of the (020) planes of molybdenum trioxide (6·9 Å: Bassett and Menter, 1957). This implies that with the instruments available at present, metal lattices are the finest which can be made visible in this way.

The lattices of the large organic molecules can, on the other hand, be observed by diffraction contrast.

An indirect observation technique is now available, however, in which we use the moiré fringes which are obtained when two lattices are superimposed. For some time, fringes had been observed in the images of certain crystals, which were often periodic formed figures of varying degrees of complexity, and could not be explained in terms of simple Bragg diffraction. Several studies have been devoted to them (Boersch, 1942, 1943; Kinder, 1943; Heidenreich and Sturkey, 1945; Pernoux, 1952, Möllenstedt, 1953; and Rang, 1953, 1953a). Overall surveys are to be **found in Pernoux's thesis (1953) and in an article by Menter (1958).**

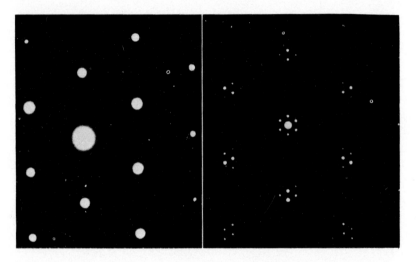

Fig. 236. Diffraction diagram of a single monocrystal (left) and of two identical superimposed monocrystals (right) (Photo: Bassett, Menter and Pashley, 1958a).

The fringes which are observed are the interference figures produced by the interaction of the direct wave and the waves which have undergone **successive Bragg reflexions at two systems of planes (see below). This can** easily be proved by examining the diffraction diagram which corresponds to the zone of the object in which the moiré fringes originate; we obtain a principal diagram of points, which correspond to the upper layer which is the first upon which the beam falls, but each spot is surrounded by a subsidiary diagram produced by the second layer (each beam which has been diffracted at the first layer behaves as a principal beam for the second **layer, and is thus partially diffracted a second time (Fig. 236; Bassett,** Menter and Pashley, 1958a).

The simplest case to which we have been referring in the previous paragraph corresponds to the superposition of two plane lattices with the same orientation but slightly different mesh-lengths (Fig. 237a). At the image, a new periodicity

$$D = \frac{d_1 \, d_2}{d_1 - d_2}$$

is obtained. Alternatively, the periodicity d of the two lattices may be the same, but they may be inclined to one another at an angle ε (Fig. 237a);

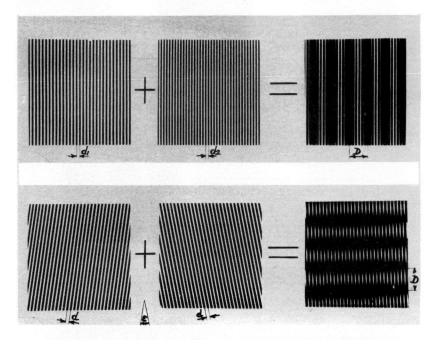

FIG. 237a. Optical analogue, showing the formation mechanism of moiré fringes by superposition of *above*—two different parallel networks, with different periodicities; *below*—two identical networks, slightly inclined with respect to one another (Menter, 1958).

we now obtain a more complicated figure, in which only the perpendicular fringes are visible when d is very small, and we now find:

$$D = \frac{d}{\varepsilon} \, .$$

In these two simple cases, therefore, we have a "moiré magnification" of $\dfrac{d_2}{d_1 + d_2}$ and $\dfrac{1}{\varepsilon}$ respectively.

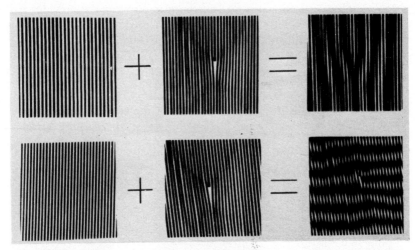

FIG. 237b. Optical analogue, showing two different ways a single dislocation may appear in a moiré fringe system.

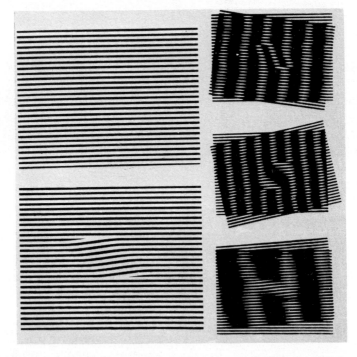

FIG. 237c. The differing appearance of a pair of neighbouring dislocations in a moiré fringe system when the relative orientation of the two real networks (left) varies (Bollmann, 1960).

These two fringe systems can clearly yield combinations of varying degrees of complexity. With the aid of the moiré pattern, a quantitative study of the lattice defects can be undertaken; if for example a dislocation is present in one of the lattices (see Fig. 237b), it appears considerably

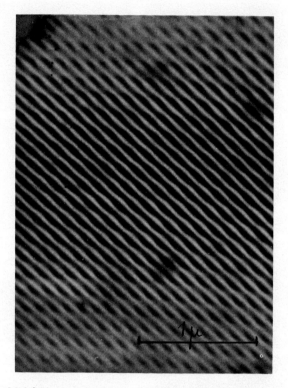

FIG. 237d. Moiré fringes in the image of a thin graphite film (Photo: Fert). Notice the triple fringe system.

enlarged in the image. Simple optical models can be used to help interpret the irregularities, often of widely different shapes, which are observed in the fringe system. Figure 237c shows the figures which are obtained when the two fringe systems (i) and (j) are superimposed, and the angle ε varied (Bollmann, 1960). In Fig. 237d, an example of moiré fringes obtained with flakes of graphite is to be found.

Again, suppose that two identical systems remain together, but are slightly curved; we may obtain dark serpentine fringes, often closed, with widely different appearances according to the inclination of the whole combination with respect to the illuminating beam (Rang, 1958). The oddest patterns of all are obtained when both systems are curved, with different curvatures, so that an empty space is left between them, shaped like a meniscus or a bubble. The figures then have a centre of symmetry and are illustrated in Fig. 237e (see Pernoux, 1953).

FIG. 237e. The figures at the centres of symmetry, created by bumps or hollows in a lead iodide crystal (Pernoux).

It is in terms of this type of diffraction phenomenon that the fringe systems which we observed in crystals can be explained, and they may have various symmetries, depending upon the relative orientations of crystal and beam (see, for example, Dupouy and Perrier, 1960); here also is to be found the explanation of the structures which appear at very high magnifications in grains of colloidal gold or microcrystals of silver deposited on mica (Matthews, 1960) or on molybdenite (Bassett, 1960), and we can thus study in particular the epitactic growth of the crystals of an evaporated substance on a suitable support. Figure 237f shows such fringes, obtained with magnesium and molybdenum oxide crystals.

Over the last four years, numerous articles have been devoted to the theory of moiré patterns, and to their applications in crystallography. We mention in particular those of Menter (1958), Rang (1958), Dowell, Farrant and Rees (1956, 1958), Stabenow, Suito and Uyeda (1958), Bassett, Menter and Pashley (1958a,b) and Hashimoto, Naiki and Mannami (1958), all of which were given at the International Conference on Electron Microscopy which was held in Berlin in 1958.

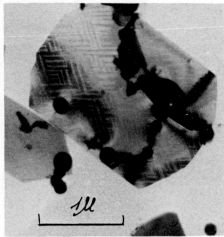

Fig. 237f. Fringes obtained in molybdenum oxide crystals (left) and molybdenum (right) (Photo: Dupouy and Perrier).

A preparation technique has been developed with which moiré patterns can be produced at will, starting with superimposed thin crystalline layers, which may be of either similar or different natures (Pashley, Menter and Bassett, 1957) and have known orientations. A thorough study of the resolution which can ideally be attained with this method shows that we could hope to observe systems with periodicities of the order of 1 or 2 Å.

19.2.7 Contrast Produced by Defocusing

In § 16.5.3, we described the way in which the Fresnel diffraction fringes at the edge of a screen (either around an opaque particle, or, on the contrary, in the peripheral zone of a hole—in a collodion membrane, for example) can be used to help focus a system and correct the ellipticity astigmatism.

If the image of the edge of the screen is over-focused, the diffraction diagram is *real* and lies *behind the object*; if it is under-focused, the diagram is *virtual* and is situated *in front of the object*. Whereas in ordinary optics the contrast of the fringes is to all intents and purposes identical in the

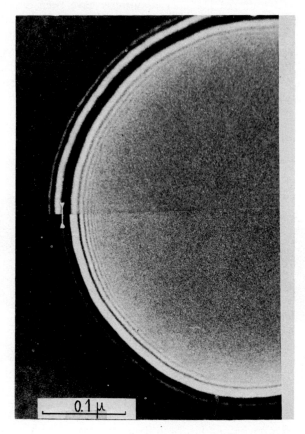

FIG. 238a. Images of the edge of a hole in a carbon film. In the lower regions, the image is *under-focused*; in the upper part, it is *over-focused*. Notice the improved contrast at the edge of the black zone in the case of under-focusing. Between the two images, the defocusing is 50 μ.

two cases, in electron optics the fringes of the virtual diaphragm are far more sharply contrasted. The first fringe of the under-focused image is a very bright line, which lies outside the dark image; the contrast of the edge of the image is accentuated with respect to the general background. The first fringe of the over-focused image is a dark fringe, which is slightly separated from the image itself (Fig. 238a). This lack of symmetry between the contrast in the two cases—which often leads the experimental worker to settle for a slightly under-focused image—can be understood if we remember that on the electron scale, a screen never has a clearly-defined edge ($\lambda = 0.04$ Å for 86 kV electrons). If we take the transition zone, across which the thickness is decreasing, into account, the results of a calculation of the luminous intensity in the fringe system agree well with experiment (Fert and Laffitte, 1961).

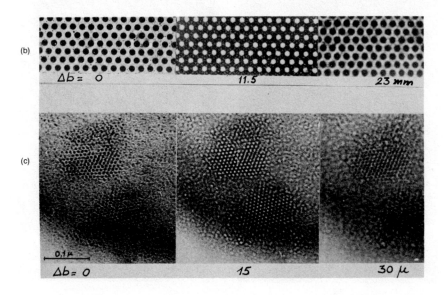

FIG. 238 b and c. Variation of the contrast with defocusing, for a periodic object.

(b) Optical model $\triangle b_1 = 23$ mm. (c) Electron image $\triangle b_1 = 30\,\mu$. For $\triangle b = \dfrac{\triangle b_1}{2}$,

the pattern has the same appearance save that the contrast is reversed.

If the image of a periodic system is formed with a sufficiently coherent illumination, it is again apparent that the contrast varies very rapidly with the degree of focusing. If, in the simple case of a periodic object consisting of lines or square meshes of periodicity a, the image is longitudinally defocused by an amount $\triangle b$, the figure which is obtained on the screen is identical to the original figure whenever $\triangle b = \triangle b_1 = \pm p \dfrac{2a^2}{\lambda}$ (in which p is an integer). If the lattice is close-packed in two dimensions (hexagonal) the period of the defocusing $\triangle b_1 = \pm 1.5\, p \dfrac{a^2}{\lambda}$ (Fagot and Fert, 1960). This phenomenon is illustrated in Fig. 238b in optical microscopy. The object is a small-scale photograph of a periodic system of opaque discs on a light background; the periodicity is $a = 91\,\mu$. With

$\lambda = 5461$ Å, the original form reappears after a defocusing of $1 \cdot 5\, a^2/\lambda = 23$ mm, while for values of $\triangle b$ between zero and 23 mm, numerous very different figures appear, which give no hint as to the structure of the object,

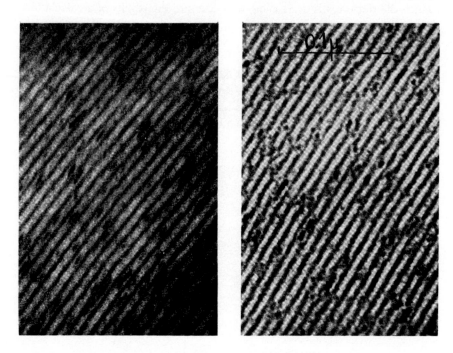

FIG. 238 d. The effect of coherence on the contrast, when the object is periodic (myeline fibres, with a periodicity of about 85 Å).

Left: incoherent illumination: $2\alpha_c = 6 \cdot 10^{-3}, 2\alpha_0 = 10^{-2}$.

Right: coherent illumination: $2\alpha_c = 0 \cdot 7 \times 10^{-3}, 2\alpha_0 = 10^{-2}$.

$2\alpha_c$ represents the angular aperture of the beam illuminating a point on the object, and $2\alpha_0$ the numerical aperture of the objective.

only its periodicity. For $\triangle b = 11 \cdot 5$ mm, the appearance of the figure has that of the original object, save that the contrast has been reversed. This experiment has been performed afresh, in electron microscopy, with coherent illumination, by Fagot and Fert (1960) and Fagot, Ferré and Fert (1961). Different test objects have been tried. A biological section of regularly distributed grains forms a pattern which resembles the optical model mentioned above; for $a = 55$ Å, and $\lambda = 0 \cdot 037$ Å we find

$$\triangle b_1 = 1 \cdot 5 \frac{a^2}{\lambda} = 30 \,\mu.$$ A mean defocusing of $15\,\mu$ gives rise to contrast reversal (Fig. 238 c).

If the object has a periodic structure which consists of parallel lines (myeline fibres for example, see Fig. 238 d), we re-encounter the pattern of the original structure when the defocusing is equal to $\triangle b_1$; for $\triangle b = \frac{1}{2} \triangle b_1$, however, the periodicity is doubled (the parallel lines are twice as numerous). This phenomenon has also been mentioned by Kamiya et al. (1958), in the images which they obtained with crystalline specimens of the mineral serpentine.

In conclusion, we would mention that it is in coherent illumination that all these phenomena appear, that is, when the angular aperture, α_c, of the beam with which the object is illuminated is small with respect to the numerical aparture, α_0, of the objective ($\alpha_c < 10^{-3}$, for example, and $\alpha_0 = 10^{-2}$). The angular aperture is reduced with the aid of a condenser diaphragm, which in interference microscopy may be a very narrow slit. The resolution of the microscope is of the same order of magnitude whether the illumination be coherent or incoherent, but the contrast is better in coherent illumination for the reasons already discussed. Nevertheless, the interpretation of the image is liable to become a most delicate affair.

19.2.8 Improvement of the Contrast, the Very Low Voltage Microscope

In contrast to the strainings towards very high tensions, which are necessary above all for the examination of thick biological sections, several attempts have been made during the last few years to use very low energy electrons in microscopy, with a view to reinforcing the contrast of thin unshadowed preparations. We now know how to produce ultra-thin sections, and this is an essential condition if electrons of a few kilovolts which have such a low penetrating power are to be used. The resolution of the instrument in theory decreases when the potential V falls, but is not nevertheless catastrophically reduced by a large reduction of the accelerating potential. From § 16.2.4, we know that

$$d_{th} \simeq C_s^{1/4} V^{-3/8},$$

so that if we work with a fixed focal length, $d_{th} \simeq V^{-3/8}$ whatever the potential V; at low values of the potential, however, the focal length of the lenses, and hence C_s, may be reduced, according to the rule:

$$f \simeq V^{1/2} \quad \text{or} \quad C_s \simeq V^{1/2}.$$

The resolution, therefore, behaves thus:

$$d_{th} \simeq V^{-1/4}.$$

If V alters from 60 kV to 6 kV, the limit of resolution will only be multiplied by about two. In these conditions, Nixon (1958) hopes to be able to obtain a resolving power of 15 Å.

From the point of view of the contrast, the work of von Borries and Ruska (1940b) shows that for substances with neighbouring atomic masses, it is advantageous to use low tensions to increase the contrast. A detailed study of the scattering of electrons by matter shows that if we are to examine a preparation which rests upon a support of given thickness e, there is an optimum potential for which the contrast is greatest; if for example $e = 100$ Å for a carbon support,

$$V_{opt} = 3 \cdot 8 \text{ kV} \qquad \text{(Nixon, 1958)}.$$

The work of Nixon (1958), Wilska (1960) and van Dorsten and Premsela (1960) has demonstrated that all the accompanying problems can be satisfactorily solved. It is such difficulties as the reinforcement of the brightness of the gun and the brilliance of the screen at low tensions, the use of sufficiently sensitive emulsions such as the Gevaert Scientia GE 56 (van Dorsten and Premsela, 1960) and the obtaining of an effective accentuation of the contrast which require solution.

If we compare the results of this procedure with the reinforcement obtained if an objective diaphragm is used, we find that a reduction of the accelerating potential from 40 kV to 20 kV improves the contrast to the same extent as a thirty-three fold reduction in the diameter of the opening in the diaphragm; the latter operation is impractical both because of the ensuing contamination of the hole and of the increase in the size of the diffraction disc.

19.3 VELOCITY ANALYSERS

The discrete energy losses which we mentioned in § 19.1.6 will provide information about the structure of solids when their origin has been more clearly elucidated, and this will be a new and valuable spectroscopic technique. For this reason, many devices known as "velocity analysers" have been developed over the last few years, with which these velocity losses—of the order of 10 eV, we must remember—can be measured.

19.3.1 Magnetic Prism Analysers (cf. Chapter 22)

The apparatus which Voges and Ruthemann (1939) originally used was later considerably improved by Ruthemann (1941, 1942 and 1948). It consists of an evacuated chamber, in the form of a half-ring, which is placed between two coaxial solenoids which produce a magnetic field inside the chamber perpendicular to the plane of the ring. At one extremity is the electron source, a straight tungsten filament, which is situated in front of a narrow slit (0·1 mm), parallel to the filament. The filament is almost completely screened, except for a wide slit which faces the filament—the screen is held at the same potential (Fig. 239). The accelerating potential can be varied between 2 and 8 kV, and is applied between the filament and the slit. The flat electron beam, the aperture of which is a few degrees, falls upon one of the objects placed on the specimen-holder two or three

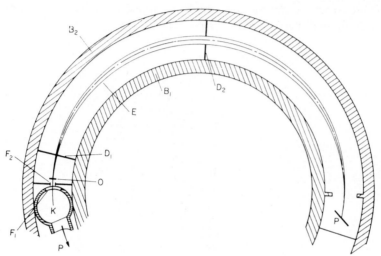

Fig. 239. The Ruthemann (1941) and Lang (1948) analyser. K: tungsten filament; F: accelerating slit; O: specimen holder; D_1, D_2: diaphragms; P: photographic plate; E: vacuum chamber; B_1, B_2: magnetic coils.

millimetres from the slit; a remote control is provided, so that one or other of the specimens can be brought into the path of the electron beam from outside.

The velocity analysis is performed after the beam has been deflected through a semi-circle by a magnetic prism. In Chapter 22, the principle upon which such a prism acts is to be found, together with a calculation of the resolving power. To improve this latter, Voges and Ruthemann use a magnetic field which decreases very slowly from the centre towards the periphery. In this way, the spherical aberration is corrected. A field of this kind is produced by two long air-filled coaxial coils with radii R_1 and R_2 carrying currents which flow in opposite senses, and by two thick turns of radii R_1 and R_2 which lie one on each side of the chamber, and which carry currents which flow in the opposite sense to that of the nearest coil. The total field is constant to about one part in 3×10^3 along the mean trajectory and falls off slightly towards the exterior by about 5 per cent. After an extremely careful stabilization of the supply currents (a few parts in 10^4) and of the accelerating potential, Ruthemann attained a resolution of 1 V in 1200; it would doubtless be possible to improve the focusing by taking advantage of recent work on β-ray spectrographs (cf. Chapter 25). The difficulties which have to be overcome are less serious here than for β-rays, as the dispersion spectrum is very narrow and the mean energy small.

Apertures are placed along the path of the beam, and at the far extremity of the vacuum chamber the beam impinges on a mobile photographic plate on which several spectra can be recorded, and which correspond to a maximum dispersion of 6 or 7 per cent.

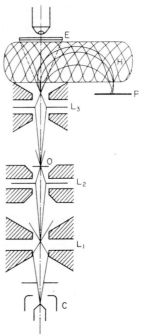

FIG. 240. The Hillier and Baker (1944) analyser. C: gun; O: object; L_1, L_2, L_3: magnetic lenses; E: fluorescent screen; H: uniform magnetic field; P: photographic plate.

Hillier and Baker (1944) have taken great pains to find out the position and the extent of the region of the preparation through which the electrons pass. Two magnetic reducing lenses (Fig. 240) form an electron probe in the vicinity of the specimen; its diameter can be as small as 20 mμ. This point source is used first of all to throw a highly magnified image of the object on the screen by a shadow effect.

The observer selects the region of the image in which he is interested, and brings it to the optic axis. The probe is then shifted into the object plane (which means that the magnification is infinite and the image blurred); the region of the object which is to be analysed is reduced to the diameter of

the probe. The third lens focuses the divergent beam into a still further reduced image of the probe at the entry to the region in which a homogeneous
magnetic field has been established; the latter deflects the beam through
180° and converges it onto a photographic plate. The beam is no longer
laminar, as it was in the preceding paragraph, but is axially symmetric.
Stops are placed within the last two lenses to reduce the aperture with the
object of diminishing the aberrations and increasing the resolving power
of the analyser. In principle, the resolution should reach 1 volt in 10,000
but it proves to be useful to increase the probe diameter to 200 mμ to avoid
contaminating the object. The resolving power is in practice about 1
in 1000.

A far more highly perfected device is at present being used by Marton,
Simpson and McCraw (1955a) to study the distribution of the retarded
electrons as a function of the scattering angle. A very fine electron beam passes
through the specimen—the scattered electrons are then caught by a magnetic
field which deflects them through 180° and concentrates them on a fluorescent
screen where the brightness is recorded by a photomultiplier. By placing
a small diaphragm in front of the analysing field, and making the whole
combination of electro-magnet and receiving system rotate about an axis
passing through the centre of the object and the screen (Fig. 241), we
can collect the electrons which have been scattered through any angle
whatsoever. For each position of the system, the magnetic field can be
varied, and hence electrons which have lost different amounts of energy
can be brought onto the screen. The actual system is remarkable in that
it is highly automatic, and hence the angle, the loss and the intensity can
all be recorded within the space of a few minutes. The resolution is about
one part in 1000 for an accelerating voltage of 20 kV.

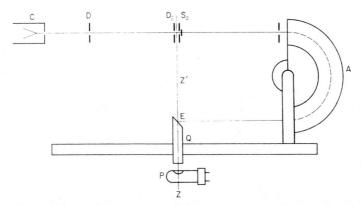

FIG. 241. Analyser used by Marton, Simpson and McCraw (1955a). C: gun; D_1,
D_2: diaphragms; Sp: specimen; E: fluorescent screen; A: magnetic field analyser;
Q: quartz rod; P: photomultiplier; ZZ': axis of rotation.

19.3.2 An Analyser with a Decelerating Electrostatic Field

Boersch (1954) has recently perfected the method of analysis which is due to Lenard. No optical system is required, and the best resolving powers are obtained; Boersch obtained the velocity spectrum of the thermal emission from various different cathodes, and his measurements were repeatable to within 4×10^{-3} V for a high tension of some tens of kilovolts—a resolution close to one part in 10^7.

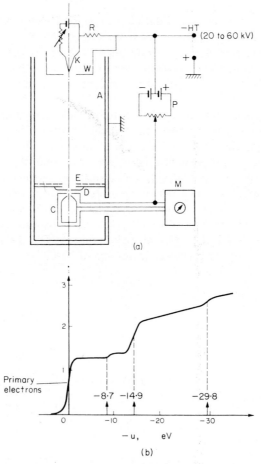

FIG. 242. Boersch's analyser (1954).

(a) The principle of the instrument, K, W, A: electron gun; E: screen; D: diaphragm; C: Faraday cage; M: electrometer valve; P: polarization for the Faraday cage.

(b) An example of the spectrum obtained after passage through an aluminium foil.

The apparatus is illustrated in Fig. 242a. The electron beam which is emitted by a tungsten filament is focused by a triode gun and the image of the source is formed on a fluorescent screen in which a small hole has been cut. The electrons which pass through this diaphragm encounter an intense retarding field which fills the space between the screen and a Faraday cage which is held at a potential close to that of the cathode. Only those

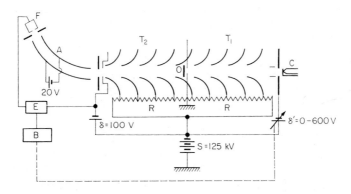

FIG. 243. The Blackstock, Birkhoff and Slater (1955) high energy analyser. C: gun; T_1: accelerating tube; T_2: decelerating tube; O: object; A: electrostatic analyser; F: collector; E: electrometer; B: automatic spectrum recorder.

electrons with sufficient kinetic energy can reach the receiver. The potential of the cage, $\Phi \pm \triangle \Phi$ is slowly and continuously varied about Φ, the value of the negative potential applied to the filament. The current i at the receiver changes from its full value to zero. A galvanometer measures i directly, and the spot is directed onto a photographic plate which is displaced by an amount which is proportional to the variation of the potential $\triangle \Phi$; in this way, the velocity spectrum is recorded directly, whether it be the spectrum of the emitting source, or that of an absorbent film placed between the gun and the screen.

The diagram which is obtained in this way represents the *area* under the ordinary spectrum, the *integrated spectrum*; intensity jumps are observed rather than bands. Figure 242b represents the spectrum which is recorded when electrons pass through a layer of aluminium—the positions of the points of inflexion of the diagram correspond to the values of the characteristic losses.

The immense improvement in resolving power and the precision of the measurements owe a great deal to the meticulous adjustment of the optical system of the receiver. When the cage potential is slightly more negative than that of the cathode, this system is effectively a convergent retarding lens; only a fine beam of electrons close to the axis can pass through this system as over a certain region, radius r_0 about the axis, the potential is approximately equal to that of the cathode, while for $r > r_0$, it is more negative and the electrons are reflected. If all the incident electrons which are to be analysed lie within a cylinder of radius $r \leqq r_0$, however, they will all meet the same potential barrier. The secondary electrons emitted in the cage, on the other hand, must remain there. These two conditions can be satisfied by making a judicious choice of the dimensions of the retarding electrode and the stops; a detailed discussion of this interesting device is to be found in the article by Boersch.

A variant of this type of analyser has recently been built with a view to studying the losses at very high energies (Blackstock, Birkhoff and Slater, 1955). It consists (Fig. 243) of the electron gun of an oscillograph which gives a parallel beam, a long accelerating tube constructed in sections in the same way as an electrostatic high voltage generator tube within which the field is uniform, a target through which the electrons pass with high energies, a retarding tube similar to the other tube, and finally the actual analyser itself. The analysis can thus be made on electrons with energies of a few hundreds of volts either with the Boersch device, or with an electrostatic analyser analogous to Ruthemann's magnetic analyser.

In conclusion, we should mention another type of filter-analyser, described by Simpson and Marton (1961); the electrons are slowed down in an ordinary three-electrode immersion objective after passing through the anode which is provided with a stop D_1. The potential of the grid is adjusted to give an extremely reduced image of D_1 in the plane of the plane cathode K, in the centre of which a small hole has been cut; the potential of K is close to that of the filament. The electrons are then again accelerated in an objective lens symmetrical to its predecessor, and with the same cathode K. The anode diaphragm D_2 is the image of D_1 produced by the complete system, and the emergent beam is then identical to the incident beam. At D_2, a fluorescent screen is placed, followed by a photomultiplier. The "sharpness of the cut-off" can be defined as the voltage $\triangle \Phi$ necessary to reduce the intensity of the beam from 90 to 10 per cent of the current transmitted with the lens positive. This quantity depends upon the diameter d of D_2—if $d = 1$ mm, $\dfrac{\triangle \Phi}{\Phi} = 5 \times 10^{-4}$ and if $d = 0.32$ mm, $\dfrac{\triangle \Phi}{\Phi} = 5 \times 10^{-5}$. If we use the lens as a velocity analyser, the order of magnitude of the energy losses which we can measure is appreciable. This quantity depends upon the maximum energy E_m of the electrons in the retarding plane, for which the optical properties are conserved. For $d = 1$ mm, E_m may reach 1 per cent of Φ; for $d = 0.32$ mm E_m would be about 0.5Φ.

19.3.3 Analysers with an Electrostatic Lens

For some years now, we have been aware of the abnormal way in which an electrostatic einzel lens behaves when the potential of the central electrode is more negative than that of the cathode. This is the behaviour which we designated transgaussian in § 8.2. Möllenstedt (1949) has found a means of taking advantage of the high chromatic sensitivity of this type of lens in the transgaussian region.

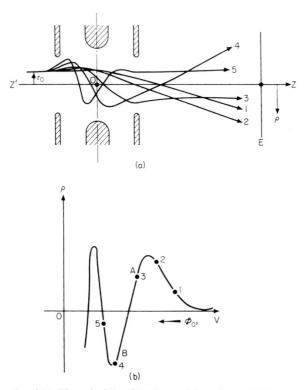

Fig. 244. The principle of Möllenstedt's analyser (1949).
(a) The appearance of the trajectories when the central electrode is made steadily more negative.
(b) The ordinate of the trajectory at the fluorescent screen when the potential at the centre of the central electrode tends towards the cathode potential which is chosen as origin.

We consider a very fine monochromatic electron beam, rigorously parallel to the axis, which approaches the lens at a distance r_0 from the axis. When the bias of the central electrode is slowly varied, becoming more and more negative, the convergence of the lens first of all increases so that the trace of the beam at the screen E (Fig. 244) moves further away from the axis. The convergence then passes through a maximum and the ray returns towards the axis, intersects it, and moves away from it again, then returns and subsequently oscillates more and more rapidly; finally, the lens is transformed into a mirror. The displacement of the spot from *A* to *B* during the first oscillation corresponds to only a very slight change in the potential of the electrode, and in the vicinity of the axis the displacement is to all intents and purposes proportional to the change in potential. Conversely, if we set the potential of the central electrode at the value which corresponds to telescopic behaviour of the lens (when the emergent ray is parallel to the incident ray), and slowly vary the accelerating potential of the incident electrons, this effect is most sensitive and the trace of the emergent beam is displaced across the screen. After calibration, even the velocity dispersion in an electron beam can be discerned. In reality, it is a laminar electron beam and a cylindrically symmetrical analysing lens which we use, so as to obtain a band spectrum on the screen. It might seem tempting to use a higher order transgaussian trajectory, as the chromatic sensitivity of the lens increases very rapidly with the order of the emergent ray; the difference in velocity between two successive oscillations is too narrow for the needs of the analysis, however—what we should like in fact is a virtually linear dispersion over a zone 100 or 200 V wide in the vicinity of the accelerating potential. Further, the resolving power of the instrument would not be increased as the size of the spot increases as well as the dispersion.

The essential features of an electrostatic analyser are, therefore (see Fig. 245):

(i) An electron source as monochromatic as possible. The standard tungsten filament can be replaced by an oxide coated cathode, as the velocity dispersions are respectively 0·3 to 0·5 V and 0·1 V.

(ii) A circular or slit objective which produces a beam with a given angular aperture in front of the analysing lens.

(iii) A specimen-holder which is placed directly behind the objective; the specimen-holder can be retracted, and the analyser can then be calibrated simply by varying the initial velocity of the beam in ten-volt steps with an auxiliary group of batteries.

(iv) A slit with sides as nearly parallel as possible, between two and five microns wide. An adjustment with which the slit can be moved laterally *in vacuo* to adjust its distance from the axis is provided.

(v) The analysing lens, which is placed immediately after the slit. This is a slit lens, the edges of which must be strictly parallel to the edges of the analysing slit if the trace of the beam on the screen is to be undistorted. The outer electrodes of the lens are in reality circular holes, to give a slight longitudinal magnification to the trace.

(vi) The receiver system, which consists of a retractable fluorescent screen which is used for making adjustments, and a photographic plate which is equipped with a special type of shutter with the aid of which the plate can be exposed twice—one half records a calibration spectrum while the other half records the unknown spectrum.

(vii) A very carefully smoothed H.T. supply; to first order, the behaviour of the instrument is independent of fluctuations in the supply, in the same

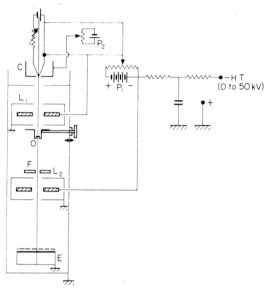

FIG. 245. Basic features of an electrostatic analyser. C: gun; E: screen and photographic plate; O: objective; F: slit; L_1: objective; L_2: analysing slit lens; P_1: supplementary gun polarization, for fine adjustment; P_2: Wehnelt polarization.

way as an electrostatic microscope, since the polarization potentials change in the same ratio. Here, however, the energy loss which is to be measured remains constant and does not vary with the fluctuations of the potential. The result is a broadening of the lines due to the action of the residual alternating potential which must, therefore, be smoothed out with extreme care.

The calibration spectrum is obtained by taking successive photographs on the same plate of the bands which correspond first to the accelerating potential Φ and then to potentials $\Phi - k\Phi$ (which correspond to slower electrons); the latter are obtained by applying a series of carefully measured potentials to the cathode, in steps of 10 V for example.

Figure 246 represents a spectrum obtained in this way (Gauthé, 1955) and in the upper part, the spectrum produced by a beam which has passed through a thin film of silver. The value of the characteristic loss can be read off immediately. In a slightly improved form (Klein, 1954),

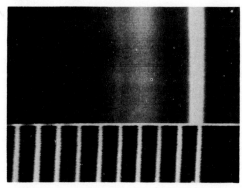

FIG. 246. An example of the type of spectrum obtained with the electrostatic analyser (Gauthé, 1955). The electrons have passed through a thin silver layer. The calibration bands correspond to 8 V intervals.

the resolving power of this instrument can be 1 V in 35,000 with a tungsten cathode and a one or two micron slit, and 1 V in 70,000 with an oxide-coated cathode (Möllenstedt, 1952a, 1952b).

The most marked defect of this analyser lies in the widening of the image of the slit which is observed when we move away from the axis, while the dispersion only increases slowly; the consequence of this effect is a pronounced reduction in resolving power for bands far from the origin. Möllenstedt and Dietrich (1955) were able to remedy this drawback, bringing all the bands successively to the centre, by compensating for the velocity loss produced by the specimen. The fluorescent screen is provided with a narrow slit which is placed in the path of the rays which emerge parallel to the axis (Fig. 247) and which subsequently produce the band which corresponds to electrons which are not slowed up. On the normal fixed potential Φ_a ($-U$ in the figure) of the analysing lens, they superimpose a correction potential which varies linearly with time $\triangle\Phi(t)$ (or $\triangle U$). The emergent rays which correspond to the various energies which are produced by the retarding process sweep across the screen and successively cross the slit F_2. If the photographic plate which is placed beneath F_2 is moved with the same velocity as the trace of the beam, every point of the spectrum will be recorded with the same degree of definition. The slit F_2 can then be widened to increase the brightness; a still better arrangement is obtained if the photographic plate is replaced by a transparent fluorescent screen followed by the photoelectric cell of an electron multiplier.

Lippert (1955) has made a theoretical study of the optical properties of einzel lenses working in transgaussian conditions, and of the final resolving power of this type of analyser as a function of the thickness of the central electrode and of the inter-electrode distances. If the size and angular aperture of the incident beam are sufficiently reduced (although at the expense of the luminosity), it is possible to attain a resolution of the order of one part in 10^6 with the lens which has the optimum dimensions. An analyser of this type, working at a maximum potential of 70 kV, has been described by Kushnir *et al.* (1960). The incident particles may be either electrons or lithium ions. In the spectra which are obtained, the losses which are

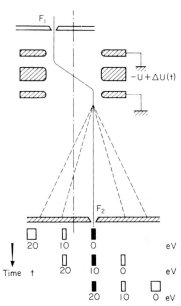

FIG. 247. An improved arrangement, with which a resolving power which is constant over the whole extent of the spectrum can be obtained (Möllenstedt and Dietrich, 1955).

found with Li^+ ions are of the same order of magnitude as those which electrons display on passing through such gases as helium, argon, nitrogen and oxygen.

Another very high performance version has been described by Dietrich (1958 a, b, c); a great deal of practical advice is to be found in his articles. (See also Metherell and Whelan, 1965, 1966.)

19.4 VELOCITY FILTERS

The slowing up of the electrons by the sample results in a reduction of the contrast at the image. With a velocity analyser, however, this can be overcome, as we can separate the retarded electrons from the remainder and collect them before they reach the image. This is the task of the "velocity filters" which are used in microscopy and diffraction.

19.4.1 Grid Lenses

The principle of the filter is simple: all that need be done is to insert a potential barrier into the path of the beam which is sufficiently high for only the unretarded electrons to be able to get past, and to accelerate the beam again before forming the image. Putting this principle into practice is, however, a matter of great difficulty. The barrier must be of the same height over the whole extent of the beam; the electrons, which have very low velocities in this region, are highly sensitive to space charge effects and to parasitic electric or magnetic fields. Finally, the barrier behaves as a lens, and we must be sure that this additional lens neither upsets the ordinary focusing nor introduces new aberrations.

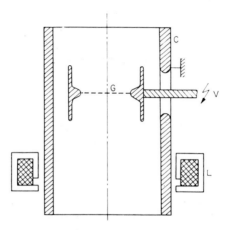

FIG. 248 a. Filter with a grid, with which the contrast in an electron diffraction photograph can be heightened (Boersch, 1949).

The simplest solution is to place a sheet of mesh which is held at the same potential as the cathode behind the first lens of the microscope or diffraction camera (Boersch, 1949, 1953, see Fig. 248a). The shadow of this grid need not be visible at the image, but the secondary electrons which are produced by the impacts of the ions which are created by the reflected electrons will perturb the final image. The best results are found in electron diffraction. When the filter halts any electron which has been retarded by more than 4V, the continuous background of the diagrams vanishes almost completely. An interesting fact is observed—the Kikuchi lines entirely disappear under these conditions, a clear demonstration of the fact that they are a consequence of retarded electrons only.

A form of mesh filter has recently been described by Beaufils (1959); it consists of two closely inter-related parts (Fig. 248b). The upper part is a retarding immersion objective, the cathode of which is held at the same potential as the microscope filament, and consists of a plane grid (mesh-length: 25μ). The focus of the objective coincides with the crossover

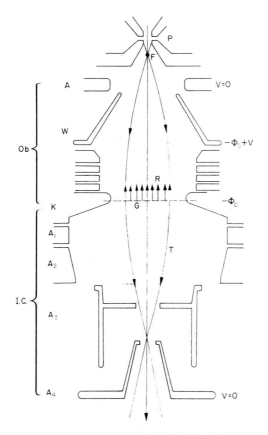

FIG. 248 b. The grid filter (Beaufils). P: intermediate projective; F: crossover, common to P and to the decelerating immersion objective Ob; A: anode; W: Wehnelt; K: cathode with several electrodes. to ensure that the equipotentials be plane and parallel; G: grid (at the source potential); $I.C.$: electrostatic image-converter (K: cathode; A_1, A_2, A_3, A_4: accelerating electrodes).

of the projection lens which is placed immediately above. Under these conditions, the electrons arrive normally at the grid. They then pass through the second part of the system which consists of a particular type of immersion objective which resembles those employed in image converters (§ 14.2.3, Fig. 174) and which produces a twice magnified image on the final fluorescent screen. The useful surface of the grid corresponds to a circle 3 cm in diameter. When the grid is connected to the cathode, the potential at the centre of the holes is equal to $(\triangle V + 4)$ volts where $\triangle V$ is the contact potential between the metal of the grid and that of the cathode.

19.4.2 The Filter Lens

In an ordinary unipotential lens, the potential barrier on the axis is as high as 90 per cent of the cathode potential, and it is very difficult to raise it even a few volts further as the aberrations increase rapidly. We can,

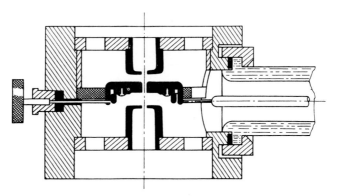

FIG. 249 a. Filter with a lens, with which the contrast in an electron image can be improved (Boersch, 1953).

however, lower the potential of the central electrode or reduce the diameter of the central hole. Boersch (1953) chose the latter solution, and placed a thin film of platinum in which a small hole a few tenths of a millimetre in diameter had been pierced, in the mid-plane of the lens. This film was gripped between two 3 mm diaphragms (Fig. 249 a) which act on the potential distribution like immersion lenses. The quantities must be determined extremely precisely, as it is necessary that the lens shall work in the transgaussian region at the third maximum of the convergence—at this point, the spherical aberration and the distortion are minimized. The working point corresponds to an axial potential of 3 V for an H.T. of 27 kV. As the aberration constants are higher than those of the other lenses, the filter lens can only be used as an intermediate lens or as a projective lens, after the objective has reduced the aperture of the beam considerably. As all the useful electrons are to surmount the same potential barrier, a field diaphragm of a few hundredths of a millimetre must be introduced to avoid edge effects.

The filter increases the contrast in the image considerably, especially for thick objects (such as biological sections, for example). This technique has been gradually improved over several years, in particular by Forst (1958) who employed a very thick central electrode instead of a thin diaphragm.

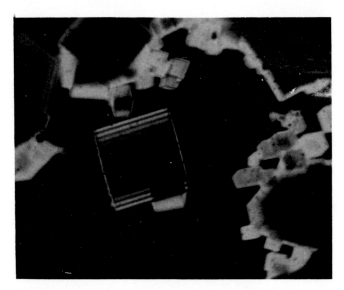

FIG. 249b. Filtered images of magnesium oxide crystals. *Above*: Elastically scattered electrons ($\triangle E = 0$). *Below*: Electrons which have suffered the first characteristic loss ($\triangle E = 12\cdot5$ eV).

19.4.3 The Magnetic Filter

A velocity analyser allows us to separate in space beams of electrons which have lost different amounts of energy on passing through the object; at the exit of the analysing unit (a magnetic lens or prism), a beam with known characteristics can be selected. Using this technique, and a stigmatic magnetic prism, Castaing and Henry (1962) have been able to obtain images of the specimen, produced first by the unperturbed but filtered beam (all the electrons which have made inelastic collisions are eliminated) and then by the electrons which have suffered a characteristic energy loss (Fig. 249 b).

CHAPTER 20

SPECIMEN PREPARATION†

20.1 THE PROBLEM OF PREPARATION

Electron images are formed in conditions very different from those which are encountered in optical image formation; so marked is the difference that when, towards 1939, the electron microscope had reached a sufficiently high degree of perfection to be appreciated and employed by microscopists, its usefulness was still in doubt. Opinions were no less pessimistic as regards examining a surface (it is difficult to thin down a metal film sufficiently for it to be observable by transmission and the reflexion image is very distorted) or a biological section (the vacuum and the electron bombardment destroy the sections; it is difficult to obtain a thin enough section; before shadowing was discovered, it was impossible to see a particle less than two or three hundred Å in diameter). It was not enough to build a good instrument—methods of specimen preparation specially adapted for the electron microscope had also to be discovered. Although many problems are still without a totally satisfactory solution and the methods employed are in constant evolution, anyone who nowadays wishes to use an electron microscope has a vast battery of preparation techniques at his disposal, which are capable of providing a basis for the examination of the majority of specimens. What seemed to be virtually an absolute limit in 1940 was in reality only a complex mass of problems. Ingenious methods of overcoming these have been developed, which both contribute to the conditions in which good electrons microscope images are formed and leave the structure of the specimen untouched.

Each new application of the electron microscope requires, generally speaking, a new technique to be developed or one or more existing techniques to be adapted or combined. For this reason, we shall describe no more than the basic ideas, and refer the reader to the bibliography for further details. The recent advances in electron microscopy are due above all to the development of preparation techniques and to the use of reliable and straight-

† Very detailed discussion of many aspects of this subject is to be found in Magnan's *Traité de Microscopie Électronique*, Hermann. Paris, 1961.

forward microscopes, rather than to the introduction of ultra-high quality instruments. To obtain useful results, it is in no sense indispensable to be able to work continually close to the theoretical resolution limit, any more than it is inevitably necessary to use an immersion objective in light optics.

20.2 SUPPORTING THE OBJECT

20.2.1 Objects which Can Be Used Without a Support

A certain number of objects, mainly metal oxides (such as those of Mo, Zn, Cd and Mg) are capable of maintaining themselves on the rim of the specimen-holder with no other support than their cohesive forces. They hang from the rim of the opening in a diaphragm. This is the case with zinc oxide, mentioned in Chapter 1 (Fig. 2). Such specimens (in particular, the globules of carbon which are obtained by the combustion of a drop of benzine) are excellent as test objects with which the quality of an instrument can be judged; since the different particles lie in different planes, every photograph contains a region in focus.

20.2.2 Supporting Grids

In an optical microscope, the object is placed on a glass plate. In an electron microscope, the problem with which we are faced is to find a support which is strong enough to carry the object during the examination and which is transparent to the particle being used in the instrument without possessing a structure which could produce contrast. If the object is large, a grid can be used; this method is used for thin sections, wires, some diatoms, and generally speaking for any object larger than the mesh-length of the grid. These grids are either constructed by electrolytic deposition or by weaving the wires, but the latter have the disadvantage of an undulating weft. A method of obtaining grids with a high transparency coefficient and a very short mesh-length photographically has recently been described (Challice and Sutton, 1952). With such grids as these, it ought to be possible to examine with ease a large number of kinds of very small object without any other support, even down to the larger bacteria. An organic supporting film liberally pierced with holes can also be used (Jaffe, 1948). Grids are commercially available with which the coordinates of the various squares of the network can be determined; this means that it is possible to mark some particular part of the object, and make a comparative study in the optical microscope and the electron microscope for example (Bishop, 1949a; Hyam and Nutting, 1952; and Nankivell, 1953).

20.2.3 Organic Supporting Films

The most interesting objects are too small to be examined by the above technique, and must be supported on a transparent film between 150 and 200 Å thick which can be obtained by evaporating a few drops of a very dilute solution of varnish in a suitable solvent. This film is then placed on

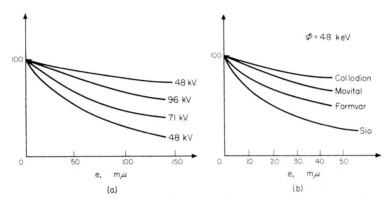

FIG. 250. Curves showing the electron transparencies of the various substances used to support the object. (a) The influence of thickness and potential. (b) The influence of thickness and the nature of the substance.

a grid which is in turn supported by the specimen-holder. For further details of this preparation technique, the reader is referred to a general examination by Drummond (1950), and to the articles by Schuster (1946) and Leroux (1953). The thickness of the film can be obtained by measuring the area of the membrane into which a known volume of the solution is transformed.

The table on p. 695 summarizes some recent attempts. In Fig. 250 the transparency of the supporting film to electrons is shown as a function of energy and thickness (Weber and Fragstein, 1954).

20.2.4 Mineral Supporting Films

Organic supporting films are not very strong and tear under the action of electrostatic forces; they are easily melted under the effect of bombardment. For this reason, mineral films which are obtained by evaporation in a vacuum are found to be preferable. The installation required for vacuum evaporation is almost identical to the apparatus required for shadowing.

(i) *The apparatus*

This is shown in Fig. 251. The vaporized atoms move in a straight line
and settle in a thin film on a surface which lies in the path of the flow.
A vacuum of about 10^{-4} mm Hg is necessary. A wise precaution is to hold
the oven at a temperature slightly below the temperature at which the

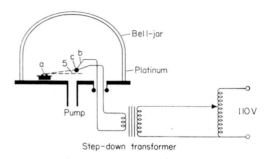

FIG. 251. Diagram of an installation for vacuum evaporation with which mineral
supporting films can be produced and specimens "shadowed". As shown in the
drawing above, the device is ready for shadowing; the atoms from *c* which is
heated by a filament *b* impinge almost tangentially upon the preparation *a*.

metal evaporates for about a minute, in order to outgas completely the
various parts which are to be held at a high temperature; the preparation
is screened off during this period.

The evaporation oven can be constructed from a tantalum gutter (5 mm
wide and a few tenths thick) or a heated filament which is bent into a helical
or conical form, and which may be smeared with aluminium oxide, or
beryllium oxide for substances which evaporate at a high temperature
(Olsen, Smith and Crittenden, 1945; and Dunoyer, 1951). In the case of
uranium or platinum, Williams and Backus (1949) recommend that the
filament should be doubly folded into the shape of a hairpin. The nature
of the heating filament in any given case (W, Ta, Mo, Cr, Chromel) depends
upon the material which is to be evaporated (Caldwell, 1944). It is sometimes
possible to avoid corroding the filament too rapidly by plating it with
platinum; this involves immersing it in platinum chloride, and subsequently
heating it gently in air. As only small amounts of metal are to be evaporated
it is pointless to use filaments larger than 0·5 mm in diameter (heated with
12 to 20 A) and we thus reduce the thermal damage which may be in-
flicted on the object. If the metal to be evaporated forms an alloy with
the filament, or is not particularly volatile, a diameter of 0·75 mm and
a heating current of 50 A may be necessary.

Organic varnish	Concentration %	Solvent	Evaporation over	Properties	Reference
Parlodion Collodion Nitrocellulose	1 to 3	Amyl acetate Butyl acetate	Water	Not very resistant	
Formvar 15/95 (polyvinyl formol)	0·2 to 0·4	Dioxane Ethylene dichloride	Glass (spreads poorly over water)	Resistant but not very flat	Schaeffer and Harker (1942)
Alkathene		Warm xylene	Water or glass	Not very flat; to be used when we need a film which is insoluble in an organic solvent	Ellis (1947)
Chlorinated rubber	2 to 3	Toluene Amyl acetate	Water or glass	Insoluble in alcohol; resistant to bases and acids	Davoine and Pernoux (1951)
Methyl-cellulose		Water			

The best ovens operate with either induction heating (Otis, 1933; and Picard and Joy, 1951) or heating by electron bombardment (Brochard, Giacomo and Jacquinot, 1951). These devices, which are noticeably more complicated, allow us to lay down a finer deposit.

(ii) *The materials evaporated*

The principal materials used are:

(a) Silicon monoxide, SiO, which can easily be evaporated on a very thin collodion film which is then burnt or dissolved. The film, which has no structure, can only be used at thicknesses less than 20 Å (Hall, 1948). SiO is preferable to silica which is less easy to evaporate.

(b) Beryllium, evaporated onto a glycerine surface (Hast, 1948). A 20 Å layer is sufficiently resistant, but tends to crystallize; Kaye (1949) remedied this by using an Al–Be alloy (40–60).

(c) Silicon, evaporated in a crucible of beryllium oxide (this is very solid, and chemically inert).

(d) Carbon films are excellent supports. They can be obtained by vacuum discharge (König, 1948, 1951) or with an auxiliary ion gun (Möllenstedt and Hubig, 1954) as well as by evaporation in a vacuum (Bradley, 1954). The films which are used are about 40 Å thick.

(iii) *The thickness which is evaporated*

It is difficult to estimate the thickness of the evaporated layer, as it is not easy to hold the conditions constant and to obtain perfectly reproducible results. Some workers use a known quantity of the substance, which is, for example, measured by using a known length of wire (Howard and Berntson, 1952) which is wholly evaporated; others cover the base with a large volume of the substance, and determine the mass M which is evaporated as a function of time and heating current (Robillard, 1949; and Bishop, 1949a, 1949b).

The thickness of the film can be calculated if we know the evaporation law of the oven (see the various ways of performing the calculation in Holland and Steckelmacher, 1952). With a point filament emitting isotropically in all directions, and a small plane surface inclined at an angle α to the direction of evaporation, the thickness is

$$\varepsilon = \frac{M \sin \alpha}{4\pi \, d^2 \, a} \, ,$$

in which M is the mass evaporated, d the distance, and a the density.

This formula does not give very precise values, as the density of the layer may be very different from the density of a macroscopic quantity of the same substance; this, however, is not very important as it is only the mass thickness $\mu_c = a\,\varepsilon$ which appears. Williams and Backus (1949) suggest that the calculated thickness should be multiplied by $^3/_4$ to take into account the efficiency of evaporation which does not usually exceed 75 per cent. Further details are to be found in the articles by Rang (1952), Schoen and Davis (1951) and Nandy (1954).

20.3 DIRECT OBSERVATION OF THE OBJECT

20.3.1 Thin Biological Sections

If the object is small in size (powdered oxide, for example) all that need be done is to distribute a few grains over a support. If the object that we wish to examine is thick, however, we must cut a thin section which will be transparent to electrons. Further, the large depth of field of the instrument will produce a superposition of the images of various planes of the object at the screen, so that interpretation of the details in the image of a relatively thick object will be a very delicate operation.

Sections which are to be used in an electron microscope must be about 100 Å thick; ordinary microtomes cannot be used therefore (thin sections for optical examination are a few microns thick) and we shall now say a few words about the refinements which have proved to be indispensable.

(i) *The double-sectioning method (Sjöstrand, 1951)*

This method gives very fine sections (a few hundred Ångströms thick) with which a very high resolution (20 to 50 Å) can be attained. A thick section of the object is first of all cut (0·5 to 1 μ); this section is then enveloped in collodion, and the majority is removed with a special microtome (see Fig. 253). The part which remains is then examined—it can be extremely thin, and need not be at all deformed by the knife-blade. Unfortunately, this is a most delicate technique.

(ii) *Modified standard microtomes*

The present tendency is to use ordinary microtomes which have been modified in such a way as to reduce their travel; in particular, this can be done by thermal expansion (Newman *et al.*, 1949, 1949a; Latta and Hartmann, 1950; and Eden *et al.*, 1950). Sjöstrand (1953, 1953a, 1954 and 1955) has built an extremely high precision microtome with which sections between 100 and 200 Å thick are normal, and the finest sections can be as thin as 70 Å. The object is fixed eccentrically at the end of a stout rotatable rod, and only passes the knife-blade once in each cutting cycle. The same result (with a single cut) is obtained in Haanstra's microtome, where

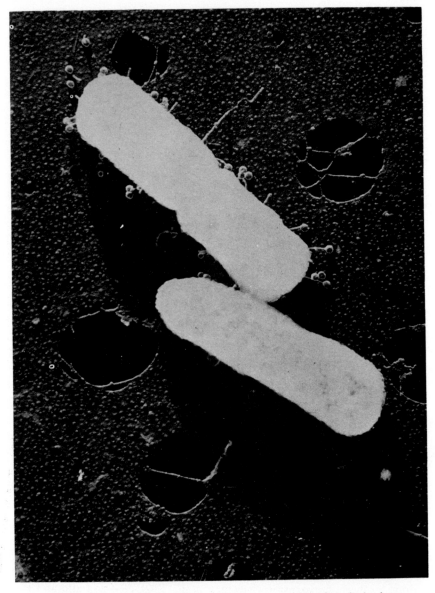

Fig. 252. An example of a biological object examined after shadowing.

Biological Interpretation: The particles of bacteriophage attach themselves to the cell at the top of the figure by their tails; this cell is "receptive" (an *E. coli* F cell); they do not, on the other hand, attach themselves to the cell at the bottom which is "not receptive" (an H Fr cell). Further, a connecting bridge between the two cells can be seen, which allows genetic material to be transferred from one type of cell to the other.

the object oscillates to and fro; as the object returns under the knife, the support of the latter contracts slightly by magnetostriction, so that the blade is slightly separated from the object (Sjöstrand, 1954).

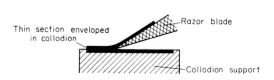

Fig. 253. Method of obtaining the double section.

The cutting-edge of the knife can be an ordinary razor-blade (von Ardenne, 1948; and Bernhard and Gautier, 1951); alternatively, a splinter of diamond, synthetic sapphire, or tungsten carbide can be used, or more simply, a glass splinter prepared in the way described by Latta and Hartmann, (1950), or Lépine and Croissant (1953). Although many workers prefer to use steel knives sharpened in the laboratory (Richards, 1950; Hillier, 1951; and Sjöstrand, 1953) the present tendency would seem to be a return to glass knives, which consist of the edge of a freshly broken wedge. A trough full of liquid is attached to the knife support, in such a way that the cutting-edge comes into contact with the meniscus. The sections are laid out on the liquid, and are thus gathered automatically.

Finally, we should mention that it seems that the methods of cutting biological sections have become refined enough for materials to be examined on the molecular scale; the examination can now be made with a resolution of 20 Å, but an improvement of this down to 10 Å would advance our understanding of the fundamental cellular processes considerably.

20.3.2 Thin Metal Sections

One method of examining metals consists of thinning down a metal disc until the thickness is small enough for the transparency to be adequate. The first efforts of this kind were made by Heidenreich (1949) who set out with a metal disc the same size as the grid of the specimen-holder, a few tenths of a millimetre thick; the two faces of this disc were then successively thinned down at the centre by electrolytic polishing until a few small holes started to appear. In favourable cases, certain zones close to the holes are thin enough to be examined by transmitted electrons. The method has been simplified by Saulnier and Mirand (1958), who employ an electrolytic polishing device which is available commercially; the specimen is first of all thinned down to a few hundredths of a millimetre, and then polished electrolytically on the two faces successively, in order to get rid of beaten-in layers. Finally, it is further thinned down, using the same bath at high intensity for a few tens of seconds. Fragments of the specimen are torn away and remain in the bath where they are collected; some of these are between 0·3 and 0·5 mm in diameter and a few hundred Å thick, and can be used directly.

In another method, the electrolytic polishing is followed by a final thinning down by cathode sputtering, produced by bombarding the two faces of the specimen with ions (Castaing and Laborie, 1953, 1954; Castaing and Lenoir, 1954), or alternatively, the whole thinning-down process is

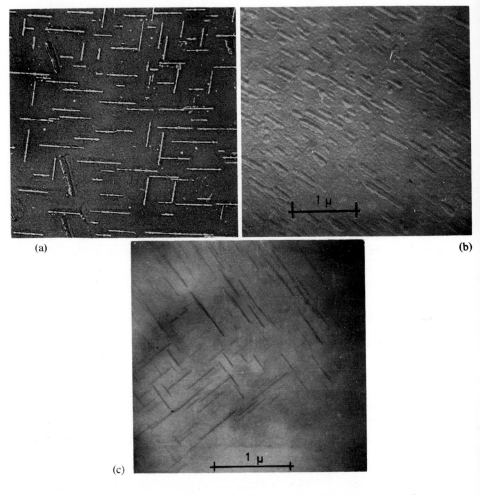

(a)   (b)

(c)

FIG. 254. A sample of the alloy Al–Cu (4%), aged at 200°. (a) Oxide replica (Castaing). (b) Evaporated aluminium replica (Baillie). (c) Direct transmission after thinning down (Castaing).

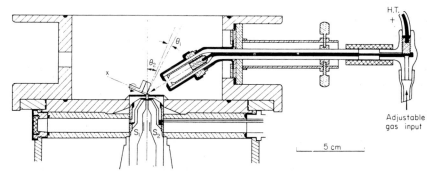

FIG. 255a. Diagram of the reflexion microscope (Fert, Marty and Saporte, 1955). The electrons fall on the object at an angle θ; the ion gun to be seen on the right is used for cleaning the surface; $S_1 S_2$ is a stigmator.

FIG. 255b. Microphotograph of an electrolytic copper surface: the two magnifications differ by a factor 10, the illumination angle is 2° and the observation angle 6° (Fert, 1954).

conducted with ions (Castaing, 1955a, b). The surface of the specimen is perfectly clean, and any source of artificially added substance is eliminated. The full resolution of the electron microscope can therefore be fruitfully employed in examining the metal; finally, the various phases which are apparent in the micrograph can be identified by localized electron diffraction (Castaing, 1955a, b, 1956). The thinning down can be performed within the microscope itself, and it is possible to watch the process continuously while it is actually in progress (Castaing, 1954a, b). Specimens which are nonmetallic and poor conductors, and which cannot be polished electrolytically, can be thinned down and examined at a very high resolution by the same process. It seems likely that the standard methods of observing replicas will have to be replaced by ionic thinning down in cases where a very high resolution is required.

20.3.3 Observation of Surfaces by Reflexion

It is possible to build the electron optical counterpart of the metallographic microscope in which the surface of the specimen is examined by reflexion. The first attempts to do this were made by Ruska (1933, 1940) and later by von Borries (1940). The illumination fell upon the specimen at grazing incidence, and good images were obtained. The method was then abandoned

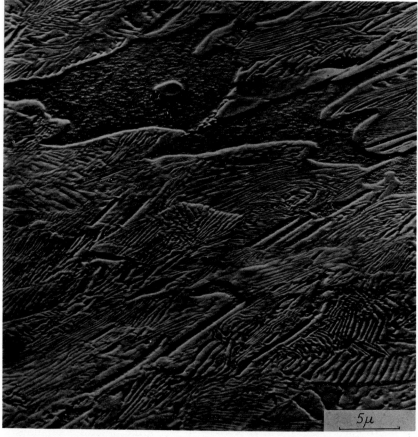

FIG. 255 c. Perlitic casting. The magnifications differ by a factor 2·5, the illumination angle is 2° and the observation angle 23° (Fert, 1955).

for a considerable time, but has now returned to favour, and a number of instruments have been specially constructed for examination by reflexion (Menter, 1952; Cosslett, 1952; and Fert, 1952, 1952a and 1954). The resolving power is limited by the chromatic aberration of the objective and by the velocity dispersion of the scattered electrons; Haine (1953) estimated his limit to be about 70 Å, but various other workers have reached only 300 Å. The angle at which the illuminating beam is incident must be very acute (a few degrees); the instrument possesses the special property of being highly sensitive to variations of relief in the specimen—differences between the various levels which are smaller than 20 Å are perceptible, provided their lateral extent is greater than the limit of resolution.

The majority of reflexion microscopes work with a scattered beam which emerges at a very acute angle as well. The scattered intensity is then appreciable, and the image is bright. The image does possess, however, the serious drawback of being considerably contracted in one direction, and since it is not easy to recognize the surface of the object at so acute an angle, the image is difficult to interpret. Fert and his collaborators (1955) avoided this problem by using a much larger angle of emergence (about 24°) but at the expense of a considerably longer exposure which is then about half a minute; in this way, he obtained excellent images with a resolving power of 300 to 400 Å, easy to interpret in that the object is comparatively normal (the ratio between the magnifications in the two directions is about 2·5; see Fig. 255c).

20.3.4 Observation of Living Biological Sections

When biological sections are introduced into the vacuum which is maintained within a microscope, the cells dry out—only dead tissue can be examined. Further the penetrating power of electrons accelerated through a few tens of kilovolts is so small that in practice one is obliged to use extremely thin films, at most a few hundred Ångströms thick. If living matter is to be studied, therefore, we must employ a special specimen-holder within which the preparation experiences normal conditions of pressure and humidity, and work with very high energy electrons, which are capable of passing through living cells of the order of a micron in thickness, the substance on which the preparation is supported, and the layer of humid air which surrounds it, without absorption and also without excessive scattering. The experimental work of Stojanowa (1958) has shown that the intensity of a beam of 80 kV electrons, incident on two collodion membranes 100 Å thick, strengthened by the evaporation of a layer of carbon, together with 0·01 mm of air at a pressure of 600 mm Hg, is reduced by three-quarters.

Recently, Dupouy and Perrier (1960, 1961) have succeeded in transforming into reality one of the oldest of the biologists' dreams as regards the electron microscope, namely, the formation of an electron image of living cells without killing them either before or during observation. They use a special type of magnetic microscope (which has already been described, in § 18.4.3) which is operated at potentials between 600 kV and 1·5 MV. The specimen-holder which is used is illustrated in Fig. 255 S1; it has the form of an extremely short cylindrical chamber (the height of the cylinder can be varied between one and ten microns), the plane end-faces (F_1 and

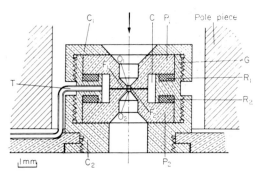

FIG. 255 S 1. Special specimen-holder for observing living cells at atmospheric pressure. F_1, F_2 are membranes of parlodion–carbon, stuck onto the plane faces of the parts P_1 and P_2 and closing off the holes O_1 and O_2; C: air reservoir; T: tube leading to the atmosphere; C_1 and C_2 are pieces which on tightening, close up the joints R_1 and R_2 so that the thickness of the central cell $F_1 F_2$ can be altered.

F_2) of which contain holes O_1 and O_2 each 0·1 mm in diameter which are in turn closed with parlodion membranes strengthened with carbon. These membranes constitute the windows which are transparent to the electrons; an extremely fine tube T provides communication with the outside atmosphere (If the volume of the reservoir of air C which surrounds the axial zone is adequate, the tube can be dispensed with when the observations last only for a short time.) When the chamber is placed in a vacuum, and the pressure is atmospheric, the windows have only to withstand a force of 8/100 g wt. The cells to be examined (bacteria) are placed on one or other of the two windows, and the chamber is then assembled under an optical microscope with the aid of which the thickness of the "strip" of air can be measured; the chamber is then placed in the gap between the pole-pieces of the objective.

The bacteria which are being studied are subjected to the action of the electrons and also to the fairly intense X-radiation which is produced by the electrons in the part of the beam which falls on the entry diaphragm of the chamber. The first experiments seemed to show that when the instrument is well adjusted, the bacteria remain alive despite the radiation. Figures 255 S2 and 255 S3 are photographs of bacteria taken at relatively

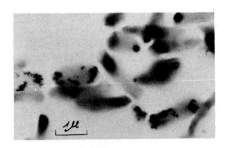

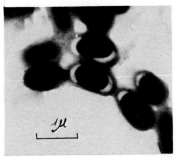

FIG. 255 S3. Photograph of anthracomorphous bacillus, showing the spores (Photo: Dupouy and Perrier).

low magnification and high tension (650 and 750 kV respectively), during the first attempts to adjust the instrument. This method ought to be capable of providing valuable information on the evolution with time of the constituents of living cells, with a resolution which is of the same order of magnitude as that of ordinary microscopes.

20.4 INDIRECT OBSERVATION OF THE OBJECT

20.4.1 Replicas

It was Mahl (1940, 1941) who thought of an artifice with the aid of which the surface of a massive object could be examined without thinning and without the oblique examination which reflexion entails; his idea therefore marks one of the major advances in the history of electron microscopy. The process consists in examining by transmission a moulding of the surface, a sufficiently thin moulding for the electrons to pass through easily, which is known as a replica or imprint of the surface. This moulding can be examined with all the resolution of which the microscope is capable—the only limit to the fineness of the details of the surface which we can see is constituted by the fidelity of the replica. We must obtain as fine and as faithful a replica of the surface being examined as possible, by covering this surface with a thin film which settles into all the irregularities, and then withdrawing this film without deforming it. After this, we can extract all the information which this replica is capable of giving with the aid of a good microscope.

We shall classify the various replica techniques under two main headings, according as the replica is obtained directly or in two stages. We shall only describe the most typical methods, without entering into the host of variations which various authors have suggested.

20.4.2 Direct Replicas

(i) *Oxide replicas*

This was the first method of replication which was used (Mahl, 1940, 1941 and 1942), and so far as fidelity is concerned, it is probably the best which is known at present. Unfortunately, it has the serious drawback that it can only conveniently be applied to aluminium (Heidenreich and Shockley, 1947; Brown and Honeycombe, 1951; Chaudron and Bussy, 1953; and Bussy, 1954) and aluminium alloys (Geissler and Keller, 1947; Castaing, 1949; and Castaing and Guinier, 1949a, 1949b). After electrolytic polishing, and chemical attack, the surface of the specimen is oxidized anodically, in such a way that a film of aluminium oxide between 200 and 400 Å thick is formed on its surface. This film has then to be detached; various ways of performing this operation without tearing the film have been proposed, and the reader will find details of these in the articles already mentioned, and in the articles by Walkenhorst (1947), Hass and Scott (1950), Gulbransen (1948), Mahl and Nielsen (1948), Tsou and Nutting (1950) and Yamaguchi (1951 and 1952).

(ii) *Plastic replicas*

After the surface of the specimen has been subjected to an attack the purpose of which is to display the variations of its internal structure as variations in the relief, it is wetted with a dilute solution of some plastic substance (a solution of 0·5 per cent formvar in dioxane, for example). When the solvent has evaporated, the specimen is covered with a thin plastic skin (less than 1000 Å thick) in intimate contact with its surface. To obtain the desired replica, we have simply to detach this plastic film. This delicate operation is the only difficulty encountered in this method, which is otherwise very rapid and well adapted for a systematic long-drawn-out study. We can withdraw the film either wet, easing it away with tweezers after immersing the specimen in a liquid (warm water, for example), or dry; in the latter case, a specimen-holder grid is placed against the surface of the specimen which is covered with the plastic film, and an adhesive ribbon is then pressed against the whole combination. The ribbon is gently drawn away, bringing with it the grid and the replica (Schaeffer, 1942; Nutting and Cosslett, 1950). This latter method is, however, more rough, and there is a risk of introducing foreign materials if the film is not very thick (Habraken, 1954).

20.4.3 Indirect Replicas

In this case, the replica is made in two stages: first, a thick mould of the specimen is taken, which is easy to withdraw; a thin replica of the face of this first imprint which had been in contact with the specimen is then taken. This second replica can then be isolated with ease, simply by dissolving the intermediate mould. We shall mention two methods which are very much in favour at present.

(i) *The evaporated aluminium replica (Hass and McFarland, 1950; and Fert and Dargent, 1952)*

The surface to be reproduced is covered with a relatively thick layer of aluminium by evaporation *in vacuo*, and this layer is removed with a band of adhesive ribbon. The whole combination is then subjected to anode oxidation. The aluminium layer is oxidized over the face which had been in contact with the specimen (we have found that the aluminium layer has to be covered with a collodion film before the adhesive ribbon is struck on if we are to avoid oxidation of the back face, and hence the formation of a double layer of oxide). The oxide coated aluminium is detached from the adhesive ribbon simply by immersing it in water, slightly acidified with nitric acid if necessary. The layer is then placed in a solution of mercuric chloride which dissolves away the aluminium but leaves the aluminium oxide replica. The replica is robust, and gives a very fine image with good contrast; this is a valuable method for many types of object.

(ii) *The carbon replica (Bradley, 1954)*

A thick film of formvar is deposited onto the surface to be studied, and then pulled away dry. A thin film of carbon (about 50 Å thick) is deposited onto the face which contains a mould of the specimen, by evaporation *in vacuo*. This is done by placing two bars of hard graphitic carbon, each sharpened to a point, in the vacuum chamber with their points just in contact. If a current (of 30 or 40 A) is passed between the points, an intense local heating is produced and hence a slight evaporation; the current is cut when the deposit has reached a suitable thickness. The formvar is then dissolved in chloroform, and a fine carbon replica remains. Despite the intermediate formvar replica, the resolution is excellent; according to Bradley (1955), this indicates that the poor resolution which is obtained

FIG. 256a. Carbon replica of a broken surface of mild steel (the magnification is 20,000; Plateau).

FIG. 256b. Evaporated aluminium replica of the surface of an Fe–Cu alloy which has undergone thermal attack (magnification: on the left, 90,000, and on the right, 20,000; Plateau).

with a normal formvar replica is not, as has been generally believed, due to the infidelity of the mould, but to the perturbations created by the electrostatic field produced by charges which attach themselves to an insulator.

The carbon replica can also be used as a direct imprint (when the carbon film is evaporated directly onto the specimen itself); in this case, however, the specimen must be dissolved to detach the replica. Bradley ob-

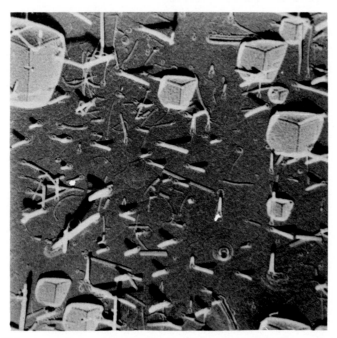

FIG. 256c. A specimen of an Al–Mg–Si alloy aged at 250°. The replica is an oxide replica shadowed with chromium (Castaing).

tained very handsome replicas of mineral crystals in this way. If the carbon film is thick enough, stereoscopic views of large crystals can be obtained. Some photographs of replicas are reproduced in Figs. 256.

20.4.4 Contrast in the Replica

The image which is obtained depends essentially on the mode of replication which has been used and if errors of interpretation are not to be made, the mechanism of contrast formation must be well understood in each individual situation. We have seen (§ 19.1.1) that the shadowing technique confers upon the image the appearance which the surface of the object would have if it were strongly illuminated at a small angle of incidence; the shadowing layer therefore creates the main contrast, and the replica method which has been used is not very important. The image is very "plastic" and can be interpreted immediately. The shadowing can, however, reduce the resolution slightly, and this is not always desirable; we shall study the contrast in a replica which is not shadowed, therefore, and we shall examine three typical cases—a plastic replica, a replica of the "evaporated" type, and a chemical replica.

(i) *The plastic replica*

This is the case of the direct unshadowed formvar replica. The lower part of the plastic layer settles more or less perfectly against the surface to be reproduced. If the relief of the object is not too pronounced, the upper

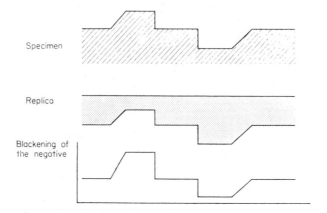

Specimen

Replica

Blackening of the negative

FIG. 257. Contrast formation in a plastic replica.

surface of the layer is virtually plane (Fig. 257). The contrast results from variations in the thickness of the plastic film, which is thin where the original surface was ridged, and thick where the surface had deep valleys— the positive, therefore, will be dark in the hollows and bright at the projections. The appearance is analogous to that of the bottom of a lake, when seen from a certain height.

(ii) *The "evaporated" type of replica*

In this category, we place all the replicas with a constant thickness, such as the silica replica, the carbon replica, and the evaporated aluminium replica, although in the last case, the mould proper is obtained chemically. The film of uniform thickness settles against the surface to be reproduced.

The contrast results from the variations in the apparent thickness which are due to changes of slope of the surface (Fig. 258). The thickness of the material through which the electron beam passes is $e \sec \theta$. The transmission varies inversely as the local slope; on the positive, a dark region corresponds to a steep incline, without any indication as to its direction. The appearance of the positive can be related to the appearance which the surface would

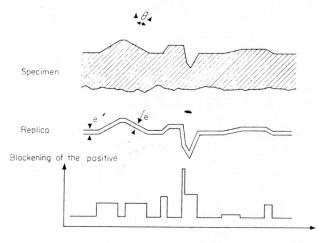

FIG. 258. Contrast formation in an evaporated replica.

have if it were observed with a metallographic microscope at small aperture; *a priori*, nothing distinguishes hollows from projections, so that a group of hemispherical projections would appear as a group of dark rings with clear centres, and would be indistinguishable from a group of hollows of the same shape.

(iii) *The chemical replica*

The metallographic oxide replica is representative of this case. If the specimen is homogeneous (the pure metal), the replica falls into the preceding category (replicas of the "evaporated" type), and the contrast is produced according to the same laws; we have a simple replica of the "shape", to deal with. If, on the other hand, the specimen is heterogeneous (for example, a precipitate), the precipitates may not be attacked in the oxidization process; they remain in their place on the film of oxide, and show up black on the positive. Often, however, the precipitates are dissolved during the oxidization (for example, anodic oxidization). The replica then possesses the property of itself displaying the structure of the specimen, without a preliminary attack being necessary. The various cases which may appear for precipitates of different sizes are illustrated in Fig. 259 (Castaing, 1949), when the anodic oxidization potential is not too high. Any very small precipitates remain included within the film of oxide, and show up black on the positive if their scattering power is higher than that of the oxide. The presence of larger precipitates is betrayed by holes or (clear) slits in the film

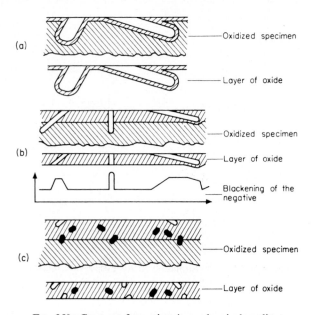

FIG. 259. Contrast formation in a chemical replica.

of oxide. Coarse precipitates are dissolved by the oxidization, and their edges are covered by the film. If the anodic oxidization potential is high even the small precipitates produce holes (punctures) in the oxide film; the effect on the positive is a number of white points which are not, unfortunately, always easily distinguishable with certainty from the accidental punctures which are statistically distributed over the surface of the replica.

20.5 THE INCREASE IN THE TEMPERATURE OF THE OBJECT, AND THE DISTURBANCE OF THE LATTER BY THE IMPACT OF THE ELECTRON BEAM

The intense electron bombardment to which an object being examined in a microscope is subjected may raise the temperature of the object considerably. Since the thermal capacity of the region being bombarded is very small, thermal equilibrium is reached very rapidly. The equilibrium temperature of any given point on the object depends upon:

(a) the energy absorbed locally; this in turn depends upon the intensity and the accelerating potential of the beam, and upon the thickness, the density and the mean atomic number of the region in question;

(b) the rate at which the calories absorbed are dispersed, by radiation and by conduction.

To calculate the equilibrium temperature rigorously is difficult; we mention the approximate calculations of von Borries (1948) who takes only the radiation into account and of von Borries and Ruska (1939) who make allowance for conduction but only in the case of a supporting film. An exact calculation would require the percentage of electrons which lose a given energy, the percentage of electrons absorbed, and the importance of secondary emission all to be known. According to von Borries, the minimum electron intensity with which the image can be focused with a resolution of 50 Å raises the temperatures to 50°C when the accelerating potential is about 100 kV and the object has a mass thickness of 10^{-6} g cm^{-2}; for a resolution of 10 Å and an object of mass thickness 10^{-5} g cm^{-2}, the temperature rises above 500°C. If the accelerating potential is increased, the light yield at the screen falls, while at the same time the energy losses in the object diminish; for a given brightness, therefore, there is an optimum potential which is close to 100 kV. Stefan's law shows that the radia-

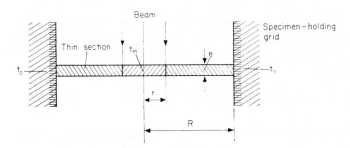

FIG. 260. A simple model for studying the heating of the object in an electron microscope.

tion losses are scarcely affected by cooling the chamber. On the other hand, should the conduction losses become predominant, the temperature of the object settles down to a value which is a given amount above the temperature of the vacuum chamber. It is advantageous, therefore, to diminish the latter temperature.

A special case in which the conduction losses predominate is provided by the examination of thin metallurgical specimens by transmission (Heidenreich, 1949; Castaing, 1954a,b). For the study of light alloys in particular, it is important that the temperature of the specimen should not exceed about fifty degrees, as otherwise the structure may be modified. It is a straightforward matter to calculate the temperature increase of a metallic object which is uniform in thickness, if we neglect radiation losses. If the object is of thickness e, and is supported along a circle of radius R by the specimen-holder which constitutes a thermal sink at the temperature t_0 and

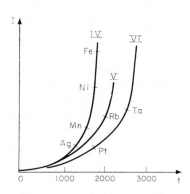

FIG. 261. Curves showing the melting point as a function of the electron intensity which is necessary to melt the layer observed in the microscope.

if the bombardment is uniform over the centre of the object across a circle of radius r (Fig. 260), while the power absorbed is W, then the maximum temperature attained at the centre is (if we assume e to be very small) given by:

$$t_m = t_0 + \frac{W}{4\pi\,JC}\left(1 + 2\log_e\frac{R}{r}\right), \qquad \text{(Castaing, 1951)}.$$

(J is the mechanical equivalent of heat, and C the thermal conductivity). If we suppose that W_0 is the electron energy which is incident upon the specimen, we can assume for the purposes of an approximate calculation that this energy is absorbed, roughly speaking, exponentially according to a law of the form $W = W_0\{1 - \exp(-\sigma e)\}$; for small values of e, this law reduces to $W = W_0\,\sigma e$. For the heating, therefore, we find

$$t_m - t_0 = \frac{W_0\,\sigma e}{4\pi\,JC}\left(1 + 2\log_e\frac{R}{r}\right).$$

As an example, we shall calculate the heating at the centre of a layer of aluminium ($C \simeq 0.5$), cooled along a circle of radius $R = 1$ mm and subjected to an electron bombardment of intensity $1\,\mu A$ and accelerating potential $50\,\text{kV}$, which is uniformly distributed over a circle of radius

$r = 0.05$ mm in the centre of the disc (these are approximately the working conditions in a R.C.A. microscope at a magnification of 20,000). The coefficient σ, which is defined by the electron absorption according to Lenard and by the retardation of the electrons according to Thompson and Whiddington, is about 1000 in this situation (Zworykin et al., 1949). The heating which is found at the centre, therefore, is about 13°. This slight heating corresponds to an electron intensity of 12 mA cm^{-2} at the object, or 3×10^{-11} A cm^{-2} at the image when the magnification is 20,000 (straightforward focusing). If the brightness of the image is increased by a factor 100 (to 3×10^{-9} A cm^{-2} which is a very brilliant image), which can be done by concentrating the electron beam into a circle with radius $r = 5\,\mu$ (condenser "at the focus"), the heating at the centre is still only about 22° (Castaing, 1956).

We see that metallic objects which are good thermal conductors remain relatively cool under the impact of the beam in ordinary illuminating conditions; this advantage has another aspect, however—the contamination is very rapid so that a given region can only be usefully examined for less than a minute. In these conditions, it is clear that the shadowing technique will be particularly valuable to reduce the thermal load of a specimen of insulator. These considerations are, however, only applicable to specimens which are sensibly uniform in thickness. The observation of thick and absorbent particles which are spread over a thin supporting film and which are in practice cooled by radiation alone can heat them to such an extent that they melt. Forestier, Haasser and Uhl (1950) have systematically studied the melting of the various elements as a function of electron intensity (Fig. 261), and the results agree with Mendeléeff's columnar classification.

Apart from simple thermal effects such as recrystallization, melting or partial sublimation, the effect of electron bombardment upon the specimen can equally well be a chemical transformation; we mention the transformation of $Ca(OH)_2$ into CaO and of MoO_3 (transparent flakes) into MoO_2 (jagged crystals; Pernoux, 1953). Organic substances too undergo fundamental transformations—according to König (1948) collodion supporting films are transformed into graphite during bombardment. It is not easy to be confident, therefore, that the object which is being observed is the same as the one which was inserted into the microscope. The simplest way of reducing these transformations to a minimum is to conduct the observation with the minimum electron bombardment, and hence at as low a magnification as is tolerable. Alternatively, we should mention the method suggested by von Ardenne (1940, 1948) in which a jet of hydrogen at low pressure is directed onto the specimen: (this should also reduce the contamination—see Castaing and Descamps, 1954); a diaphragm is placed above the region of the specimen in which we are interested, and is retracted only for taking photographs.

CHAPTER 21

THE DIFFRACTION CAMERA

21.1 ELECTRON DIFFRACTION

21.1.1 Description of the Phenomenon

The first verification of de Broglie's theory of the wave-nature of the electron is to be found in the interpretation of the precise though complicated experiments of Davisson and Germer (1927), who studied the reflection of slow electrons at a nickel crystal. Shortly afterwards, a striking confirmation appeared when first Thomson (1928–9) and later Ponte (1929) showed that it was possible to obtain diffraction diagrams with electrons which are similar to the diagrams obtained with X-rays. To do this, a very fine

FIG. 262. The diffraction diagram of zinc oxide (Lafourcade).

beam of fast monoenergetic electrons is fired through a very thin poly-crystalline layer. Figure 262 is a reproduction of the diagram obtained by transmission through polycrystalline zinc oxide; it is to be noticed that analogous diagrams can be obtained by "reflecting" the electron beam from a plane surface at grazing incidence.

21.1.2 Translation of the Bragg Theory

We shall now forget the corpuscular aspect, and henceforward, the elec-tron beam—which is supposed parallel and monoenergetic—will be charac-terized by the wavelength λ of the associated wave which we have already defined in § 15.2. It will be the relativistic version of the formula which

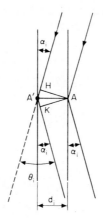

FIG. 263a. The scattering of a wave by the atoms of a crystal.

relates λ to the energy, V, of the electrons, which will be useful, in view of the extreme precision of the experiments and the high potentials which are used in certain diffraction cameras, namely

$$\lambda = \frac{h}{\sqrt{2em V}}\left(1 + \frac{e V}{2m c^2}\right)^{-1/2} = \frac{12 \cdot 25}{\sqrt{V}}(1 + 0 \cdot 9786 \times 10^{-6} V)^{-1/2},$$

where λ is in Å and V in volts. At 100 kV, the relativistic correction has reached 5 per cent; an accelerating potential of 60 kV corresponds to a wavelength of 0·05 Å.

We consider a crystal within which is a family of lattice sites two of which are drawn in Fig. 263a. The electron wave is scattered by one of the atoms A' of one of these planes, and we examine the resulting phenomena in a direction which is inclined to the incident beam at an angle θ_i. The intensity in this direction will be appreciable only if the various waves diffracted by the atoms at the receiver are in phase; otherwise, the waves will be destroyed by interference, so numerous are they.

This phase coherence must first of all be assured for all the waves diffracted in the selected direction by the atoms in any one plane. This condition leads to the standard reflexion law, which we have already taken into account in Fig. 263a. Further, the coherence must still be maintained for a pair of atoms which lie in different lattice planes distance d_i apart. This requires that $H A' + A' K$ should be a whole number of wavelengths, which implies

$$2 d_i \sin\alpha_i = 2 d_i \sin\frac{\theta_i}{2} = n \lambda.$$

Diffraction will only be found in a direction making an angle θ_i with the incident direction if electrons are incident at an angle α_i on a crystal which has lattice planes a distance d_i apart.

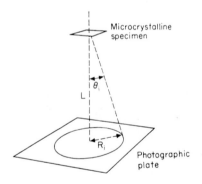

FIG. 263b. Diffraction by a polycrystalline specimen.

If the specimen is a single crystal, the diagram which is obtained consists of a regular pattern of spots which form a geometrical design the appearance of which depends both upon the nature of the crystal and the direction of the incident electrons. If the specimen is composed of a large number of small randomly orientated crystals, however, the selective reflexions give rise to diffracted beams in the form of hollow cones, of semi-angle θ_i, which produce rings of radius R_i on a screen at a distance L (Fig. 263b); this is the analogue of a Debye–Scherrer diagram for X-rays. By measuring R_i, we can calculate the corresponding value of d_i. Since the angles θ_i are very small (at most a few hundredths of a radian), we have

$$d_i = \lambda \frac{L}{R_i}.$$

If the rings which correspond to large values of d_i are to be observed, L must be large; for $d_i = 10$ Å and $\lambda = 0.05$ Å, we should have $R_i \geqq 1$ cm if $L \geqq 2$ m which would give a prohibitively large instrument. We shall see how this difficulty can be surmounted electron optically.

Another diffraction pattern, which only appears when the object is thick, is known as the "Kikuchi line". Inelastic scattering in the thick object produces a homogeneous continuous background, which appears to be traversed by bright or dark lines and bands which cross one another and form regular geometrical patterns. These patterns which are in general obtained with a highly convergent illumination, are also valuable in determining crystal structures (Wilman, 1948a, 1948b).

21.1.3 Comparison of X-rays and Electron Rays

It is of some interest to make a careful comparison between X-rays and electron rays; the essential characteristics are as follows:

(i) Wavelength

For accelerating potentials between 50 and 100 kV, the wavelengths which are associated with electrons lie between 0.054 and 0.036 Å. The X-ray wavelengths which are normally used in crystallography, and which correspond to the K_α lines of the various elements, are far longer, as the following table shows.

Element	Pt	W	Mo	Cu	Fe	Cr
λ(Å)	0.19	0.21	0.71	1.54	1.93	2.29

The line which corresponds to the shortest wavelength will be the K_γ line of uranium, $\lambda = 0.1$ Å.

(ii) Penetrating power

Electrons have a vastly smaller penetrating power than X-rays. For example, an aluminium sheet 3 cm thick only absorbs about 9/10 of the intensity of a beam of hard X-rays ($\lambda = 0.2$ Å), while $5\,\mu$ of the same material are sufficient to halt 100 kV electrons completely. In both cases, the absorption increases with the wavelength and with the atomic number of the element through which the beams pass. Transmission electron diffraction can only be observed, therefore, with specimens of a mean thickness less than $0.1\,\mu$. A smaller quantity of the substance is required; in reflexion, only surface layers can be studied effectively.

(iii) *Monochromatism*

The X-radiation which is emitted by a tube consists of a number of lines superimposed on an intense continuous background. To obtain monochromatic radiation, we must either use filters, or selective reflexion at a crystal, which produces a marked reduction in the intensity of the beam. It is very easy to obtain a monochromatic electron beam, on the other hand, in which all the power dissipated by the source corresponds to precisely the chosen wavelength.

(iv) *Power*

The efficiency of the energy yield of an X-ray tube scarcely exceeds one part in a thousand, and the radiation is very diffuse. To obtain a beam which is both very fine and very intense, therefore, is an extremely difficult task. Further, the selective reflexion coefficients are of the order of 10^{-4}. Exposures of several tens of minutes are often necessary to obtain X-ray diffraction diagrams. In electron diffraction, conversely, the intensity is much higher, and the diagram can often be recorded in a few seconds. This fact is due to two fundamental causes: firstly, the possibility of concentrating electrons into a fine and very intense pencil, with the aid of a suitable electron optical arrangement, and secondly, the fact that the interaction between matter and electrons is considerably more powerful than the interaction between matter and X-rays.

(v) *Refractive index*

In both cases, the refraction of the radiation as it passes through a crystalline medium can be neglected in a first order approximation, as the refractive index is very close to unity. For X-rays,

$$\mu - 1 = - \frac{N e^2}{2 \pi m (v - v_i)^2} \, ,$$

where N represents the number of electrons per unit volume of the substance, v the frequency of the radiation and v_i the frequency of the absorption line of the material nearest to v; $\mu - 1$ varies between 10^{-6} and 2×10^{-4} for various substances. For electrons, we must take the mean electrostatic potential Φ of the crystal lattice into account; Φ is of the order of 10 V and the refractive index is effectively given by

$$\mu = 1 + \varepsilon, \quad \text{where} \quad \varepsilon \simeq \frac{\Phi}{V},$$

and V is the accelerating potential. For the usual accelerating potentials, therefore, $\varepsilon \simeq 10^{-4}$. We have already encountered this refractive index in the section devoted to phase contrast (§ 19.4.3).

If we take this refractive index into account, the angle which is defined by the uncorrected Bragg relation

$$\sin \frac{\theta_i}{2} = \frac{n\,\lambda}{2\,d_i},$$

must be replaced by an angle φ_i, such that

$$\sin^2 \frac{\varphi_i}{2} = \sin^2 \frac{\theta_i}{2} - (\mu^2 - 1).$$

We shall see later on how this refraction can be made apparent and Φ measured.

21.1.4 The Different Diffraction Techniques

Two arrangements are in use, in one of which we work with the transmitted beam, in the other, the reflected beam.

In the first case, the beam passes through a specimen which is in the form of a very thin film (a few hundred millimicrons or a few thousand Ångströms at most). The specimen is usually polycrystalline, with the crystals randomly orientated, and we need not worry greatly whether or not the beam is perfectly normally incident upon the preparation; if we are dealing with a single crystal, however, or orientated organic molecules, the angle of incidence must be defined exactly.

In the second case, the beam falls at grazing incidence upon the specimen, which can be either massive, or a thin film resting on a massive support. The mechanism of this phenomenon is the same as that of the preceding case, as in spite of appearances, this is once again transmission diffraction. So small is the penetrating power of the electrons that they are diffracted while passing through the microscopic surface irregularities. We can only collect half-rings, as the other halves are masked by the object.

Later on, we shall examine in detail the problems raised, in particular by transmission diffraction; the techniques are less refined and the instruments simpler in the reflexion case.

21.2 THE PROPERTIES OF A DIFFRACTION CAMERA

21.2.1 Resolving Power

(i) *Definition*

The object of every piece of research in this field is to measure the radii of the rings which are obtained on a photographic plate (or the distance between the spots in the case of monocrystals) and to separate closely neighbouring rings when the specimen is homogeneous.

It is natural to adopt the conventional definition of the resolving power of a recording instrument, which is given by the expression

$$\mathscr{R} = \frac{R}{\triangle R},$$

in which $\triangle R$ is the smallest distance between two neighbouring rings which can just be discriminated on the photograph. The ratio $S = \dfrac{\triangle d_i}{d_i}$ is often used as a definition of resolving power in terms of the lattice constants. $\mathscr{R}$ can be deduced from S by taking into account the precision with which L and λ have been measured, as we shall show later. It should be noticed that $\mathscr{R}$ defined in this way is proportional to R and λ and inversely proportional to the lattice spacing d_i and to V; we cannot give an estimate of $\mathscr{R}$ without quoting the values of λ and d_i to which it corresponds.

(ii) *The general expression for the resolution*

We consider the sketch of the most simple type of diffraction camera with a lens, shown in Fig. 264. The electron beam emerges from a source S of radius ϱ_c, passes through a weak lens L_1 which forms the image S' of S on a fluorescent screen (or a photographic plate). The specimen is placed

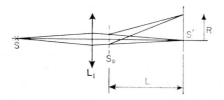

FIG. 264. The ray-paths in a simple diffraction camera.

immediately after the lens, a distance L from the screen. The diffracted beams which correspond to lattice-planes of the same family also converge onto the screen, giving very fine rings. The thickness of these rings, $\triangle R$, depends upon two factors, very different in nature:

(a) the radius of the spot on the screen ϱ_s when there is no diffraction; this depends upon the quality of the optical system, and can be reduced at will by improving this system.

(b) a source of natural broadening due to the passage through the specimen, about which the experimenter can do practically nothing.

In modern instruments, it is finally this second phenomenon which limits the resolution $\mathscr{R}$ to values below 10^3 whereas the optical system alone is capable of giving $\mathscr{R} = 10^4$. If ϱ_e is the contribution of the specimen to the broadening of the spot, and r the overall radius of the central spot

$$r = \varrho_s + \varrho_e.$$

The limiting distance between two rings which can just be separated is given by

$$\triangle R \simeq 1 \cdot 7\, r,$$

if we assume that the intensity is distributed uniformly over the spot. If we round off the coefficient, and take

$$\triangle R = 2r,$$

we find

$$\mathscr{R} = \frac{R}{2r} = \frac{\lambda L}{2r\, d_i}.$$

The distance between the specimen and the screen, L, can, as we shall see, be artificially increased by introducing electron lenses. L will then represent the "equivalent length" of the instrument, which is defined by the ratio (see Fig. 265)

$$L = \frac{R_i}{\theta_i},$$

in which R_i and θ_i are the values of R and θ which correspond to planes separated by d_i; L can be ten times the actual distance between the object and the plate.

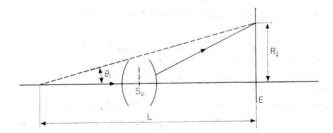

FIG. 265. Definition of the equivalent length.

The quotient $Q = \dfrac{L}{2\varrho_s}$ is often known as the "quality factor" of the instrument—it depends only upon the optical part of the system, and gives a method of comparing different models. In the ideal case of a specimen which produces a negligible broadening of the spot ($\varrho_e \ll \varrho_s$) we shall have

$$\mathscr{R} \simeq Q\frac{\lambda}{d_i}.$$

(iii) *The broadening of the spot which the specimen produces*

This broadening is due to a host of phenomena, some of which are inevitable.

The first of these is provided by the perturbation to which the diffracted beam is subjected in passing through the specimen; the electrons may be deviated from their trajectories by elastic collisions, and in instruments where the specimen is followed by a lens or several lenses, electrons which have undergone inelastic collisions will be affected by chromatic aberration. These phenomena resemble the ones which were analysed in connexion with the formation of image contrast in microscopy (cf. Chapter 19).

The width of the ring also depends upon the size of the microcrystals involved, just as the width of the band which is produced when monochromatic light is diffracted at a grating varies inversely as the number of lines of the grating; in the electron case, the width is proportional to the number of diffracting planes, and hence to the width of the lattice, or finally, to the thickness of the crystal. As the Bragg angles concerned are very small, the effect of the transverse dimension t of the crystals is by far the most important. This enlargement term is effectively equal to

$$\delta \simeq 0.9 \, \frac{L\lambda}{t},$$

and this formula is identical to the corresponding X-ray expression. Although this relation has been obtained for the case of a single crystal or for microcrystals all of the same size, it is equally valid as a statistical mean when the crystals are unequal in size, provided their dimensions are spread in a Gaussian distribution about t.

Finally, the refraction of the diffracted beams which is due to the fact that the refractive index is slightly different from unity may produce duplication of the rings, which has the appearance of broadening if the resolution is inadequate.

There is virtually no way of reducing the effects of these phenomena save by an increase in the H.T. There are other defects, however, which can be considerably lessened by good experimental technique: for example, dispersal of the electric charges which have accumulated on the specimen, and a satisfactory distribution of microcrystals in the preparation. During the bombardment, the object becomes charged due to the inadequacy of the secondary emission which is produced by fast electrons; if the specimen is a good insulator, the effect is even more pronounced. These charges can often be eliminated simply by mounting the specimen on a conducting grid in the transmission arrangement, but with reflexion the specimen is often massive, and a special discharging device has to be incorporated into the instrument; an auxiliary gun sprays slow electrons (200 to 1000 V) over the whole surface of the object which is discharged by the abundant secondary emission which results, as the multiplication coefficient is much greater than unity, at low potentials. The ionization of the residual gas also has a favourable effect. These charges, which vary both in number and in distribution with time, deviate the diffracted rays by a mechanism exactly the same as the electron scattering at the lattice. We must also prevent the photographic film from becoming charged by placing it on a metal support, although these charges are less harmful since their effect occurs at the end of the trajectory.

Further, the width of the rings depends a great deal upon the preparative technique. In reflexion, the surface must be perfectly polished so that the dimensions of the residual irregularities shall be of the same order as the depth of penetration of the electrons. In transmission, the thickness must be uniform; if the specimen is in powder form, the distribution must be as perfect as possible so that there are no opaque aggregates, and so that every orientation is present, and hence complete diagrams will be obtained. Examining the specimen in the microscope is a valuable help in obtaining suitable specimens.

In conclusion, we should point out that for different substances, the actual broadening varies widely with the degree of perfection of the crystals. For example, magnesium oxide rings are about three times finer than the rings produced by a layer of aluminium evaporated onto the same support with crystals of the same size (Hillier and Baker, 1946).

(iv) *The radius of the spot produced by the lens when the beam is perfectly monoenergetic*

We call ϱ_c the radius of the source S used, that is the radius of the cross-over of the gun, and consider the most usual arrangement (Fig. 266); here the specimen is placed after the lens. The spherical aberration of this

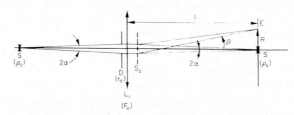

FIG. 266. Aberrations: the object follows the lens.

lens is the only aberration which affects the radius of the final spot. If C_s is the constant of spherical aberration, the radius of the spherical aberration spot is

$$\triangle = C_s \alpha_i^3 = C_s \left(\frac{r_a}{l}\right)^3,$$

in which α_i is the semi-aperture of the beam which produces the rings, r_a is the radius of the opening in the diaphragm, and l is the distance from the diaphragm to the screen, which is the same as the distance from the lens to the screen. l is usually in the vicinity of $2 f_0$, so that

$$C_s \alpha_i^3 = \frac{C_s}{2 f_0} \frac{r_a^3}{(2 f_0)^2} \cdot$$

$C_s/2 f_0$ is of the order of 25 for a magnetic lens when $2 f_0 = 20$ cm, and about 10 cm in diameter when we use an unshielded lens. The linear magnification G of L_1 is always close to unity, so that

$$\varrho_s = G \varrho_c + C_s \alpha_i^3 \cong \varrho_c + C_s \alpha_i^3.$$

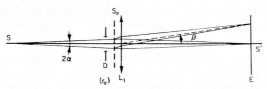

FIG. 267. Aberrations: the object precedes the lens.

In some instruments, the specimen is placed before the lens L_1 (see Fig. 267). This is not a very good design, because extra aberrations appear, namely coma and astigmatism. The coma term, which is the more important, is given by

$$K \beta r_a^2,$$

in which β is the angle at which the ring in question would be seen from the image nodal point of L_1, and K is the coma constant. This term varies proportionally with the radius R, therefore, and hence this arrangement should be rejected.

We can always reduce the angle α_i to reduce the spherical aberration to a negligible value. When f_0 is of the order of 10 or 20 cm, $G = 1$ and $r_a = 0 \cdot 1$ mm, the spherical aberration disc is of the order of a fraction of a micron in diameter; if the distance between the lens and the screen is 20 to 40 cm, the effect is negligible, and

$$\triangle R \simeq 2\rho \simeq 2\rho_c.$$

ϱ_c must be reduced to the uttermost; to do this, the source S is obtained either by powerfully reducing the image of the crossover obtained with a simple triode gun, or more simply by placing a stop of very small diameter immediately after the anode of the triode gun. With the former, ϱ_c can easily be reduced to below 10μ. For $L = 50$ cm, $\lambda = 0.05$ Å ($V = 60$ kV), and $d_i = 1$ Å, we should have

$$R_i = \frac{\lambda L}{d_i} = 2.5 \text{ cm}$$

when $2 \varrho_c = 5 \mu$, so that if we take the complete optical arrangement into consideration,

$$\mathscr{R} = 5000 \quad \text{and} \quad Q = \frac{L}{2\varrho} = 10^5.$$

These conditions do not lead to a prohibitively large loss of brightness, and the exposure time remains of the order of a minute.

We can, therefore, find a good enough optical system for the resolution to be limited by the structure of the object itself. In certain problems, we wish to measure precisely the contribution of the object to the enlargement of the spot; in such a situation, we prolong the exposure and work in conditions such that the intrinsic diameter of the spot is negligible, which can give $\mathscr{R} \simeq 10,000$.

(v) *The radius of the spot provided by the lens when chromatic aberration is present*

The accelerating potential may contain slight ripple; or it may display small, fast fluctuations $\triangle V$ due, for example, to microdischarges. These defects result in two new enlargement terms.

The wavelength of the associated wave varies slightly, which produces an increase $\triangle R_1$ in R:

$$\frac{\triangle R_1}{R} = \frac{\triangle \lambda}{\lambda} = \frac{1}{2} \frac{\triangle V}{V}.$$

To obtain $\dfrac{\triangle R_1}{R} \simeq 10^{-4}$, the H.T. must be stabilized and the residual alternating component suppressed in such a way that $\dfrac{\triangle V}{V} \leq 5 \times 10^{-5}$, which is at the limit of present-day possibilities. In the more usual situation, where a resolution of 100 to 200 suffices, it is easy to construct an H.T. supply such that $\dfrac{\triangle V}{V}$ is of the order of 10^{-3}, and hence, such that this first source of enlargement can be neglected. (We refer the reader to § 18.1 where information about H.T. sources is to be found.)

The magnification G of the lens varies with the electron velocity; for a magnetic lens, f is proportional to V. A variation in f results in a broadening $\triangle R_2$. It can easily be verified that this term is far smaller than its predecessor.

In the case where the object lies before the lens, the two terms are of the same order of magnitude but of opposite signs. We shall later examine the attempts which have been made to obtain an arrangement which will be achromatic even for large variations of V.

21.2.2 Fidelity

As any ring is defined by the corresponding diffraction angle $\theta_i = \dfrac{R_i}{L}$, measurement of the radius R_i of the ring must give the value of θ_i unambiguously, and hence the lattice-spacing d_i also:

$$d_i = \frac{\lambda}{\theta_i} = \frac{\lambda L}{R_i}.$$

We must ensure that the quantities λ and L are stable over a period of time, and that the given experimental conditions can be attained exactly.

This fidelity is affected by a number of factors:

(a) variations in the potential, V, with time, and the immensely difficult problem of obtaining the same value of the potential V, in two different experiments separated by a fairly long interval; the latter, which is the more serious, may be the result of inadequately accurate measurement of the high tension, or of uncertainty about the value of λ.

(b) variations in the equivalent length L which, as we shall see, depends upon the convergence of the lenses and of the optical system which follows the specimen in many instruments. Here again, the lens convergence must remain constant with the passage of time, and take the same value in two successive experiments if we apply the same magnetizing current. Nevertheless, hysteresis of the magnetic framework can produce slight variations. Similarly, an unexcited lens has a slight convergence as a result of the remanent field, the value of which may fluctuate with time.

In simple diffraction cameras, this objection does not occur in transmission, since L and the distance from the object to the screen are equal (the object is placed immediately behind the lens). In reflexion, however, there is doubt about the position of the point of impact of the electron beam which is nearly tangential to the surface being examined. Considerable effort has gone into attaining this long-term fidelity, but although the excitation currents in magnetic lenses can be measured to within about 1 per cent it is very difficult to obtain measurements which are reproducible to better than 1 per cent. The fidelity is considerably better during a series of consecutive experiments throughout which the various parameters can be treated as constants. We can get round the problem by verifying the "constancy" of C, defined by $d_i = C/R_i$, before each important piece of research is undertaken, by looking at a known material which is perfectly stable in time.

The layers of evaporated gold and aluminium which have often served as test specimens in the past are in fact a very poor choice, as the lattice-spacing varies with the conditions under which the layer is prepared. Thin films of sodium chloride or lithium fluoride evaporated on collodion are to be preferred. We can also use an object-support designed to hold two specimens, and superimpose a part of each of the two diagrams which correspond to a known specimen and to the specimen being studied. An efficient though laborious solution is to supply two identical diffraction cameras working in parallel with the same potential, and to take the photo-graphs of the known and the unknown specimens simultaneously, one in each of the two instruments. At each moment, therefore, we can deter-mine the "constancy" of the instrument in question (Yamagushi, 1953).

In certain situations, where we are studying the structural modifications which a mechanical or thermal treatment produces on the specimen by examining the variations which result in the radii of the rings, a precise calibration of the apparatus is unnecessary as the form of the original diagram is known at the beginning.

21.2.3 The Accuracy of the Measurements

(i) *The accuracy of the measurements made on the recorded diagram*

This depends upon three quantities.

(a) The error in measuring R on the plate; this can be reduced to a negli-gible amount after a suitable enlargement of the photograph. On some diagrams with poor contrast, in particular for the rings close to the over-exposed central zone, a difficulty in determining R arises. The distribution of brightness in the photograph is of the form shown in Fig. 268, in con-

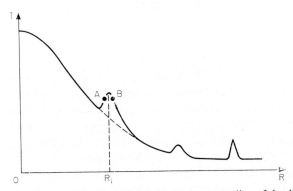

FIG. 268. The distribution of luminous intensity across a recording of the diagram, with an intense continuous background.

sequence of the very intense continuous background; the radius of the ring no longer corresponds to the distance between the highest point of the peak and the axis, and the error increases as the radius of the ring is reduced. In this situation, we take the average of the distances from the points of inflexion A and B to the centre as the value of the radius. There is an advantage in moving the first rings away from the central zone, by using a projective lens which magnifies the diagram sufficiently. Experiment shows that the over-exposure of this central zone is due in fact to parasitic rays, and is hardly magnified at all by the lens.

(b) The error in L, which is appreciable in reflexion; with a good instrument, this error can be reduced to a few tenths of a millimetre, say to $\dfrac{\Delta L}{L} \leqq 10^{-3}$.

(c) The error in V which can be eliminated by bracketing the measurement on the unknown sample between two auxiliary measurements on a test-object. If this is done carefully, the length of the lattice-spacing can be measured to better than 1 per cent, with a good instrument; the ultimate precision which is obtainable in practice in favourable cases lies in the region of 0·5 per cent.

(ii) Measurement of the thickness of the rings

With this measurement, a supplementary factor intervenes—the thickness is a function of the exposure time, since the blackening of the photographic emulsion varies rapidly as a function of the incident charge Q. Figure 187 of Chapter 15 is an example of a photographic density curve. The exposure time must be measured carefully, and we must work on the linear part of the photographic density curve; the width of the ring is given directly. Here again the presence of a continuous background can introduce errors (Lafourcade, 1954) and make the measurement uncertain, and we have to employ a special procedure to reinforce the contrast of the rings (Brockway and Bartell, 1954).

(iii) Electron intensity measurements

The intensities of the various rings or spots can be measured from photographs, but the absolute accuracy is poor, especially when the peaks are superimposed on a strong continuous background (5 or 10 per cent). It is virtually impossible to compare the intensities of the diffracted beams with that of the direct beam using this method, since the central zone of the diffraction diagrams is violently over-exposed. Furthermore, the continuous background caused by electrons scattered inelastically at the object is often troublesome.

The electron current can be recorded at every point of the diagram either by moving a Faraday cage provided with a very small entry diaphragm across the region where the diagram is produced or, alternatively, by shifting the actual diagram by means of deflexion coils over a slit or hole cut in the observation screen. The current can then be measured with a Faraday cage or a crystal scintillator followed by a photomultiplier, and the current density along a diameter (or a chord) of the diagram is then recorded. Here again, the continuous background causes difficulties. It is possible, however, to reduce this parasitic current to a very large extent by placing a filter lens between the diaphragm and the measuring device, which will allow only elastically scattered electrons to pass (Denbigh and Grigson, 1965). If the blocking potential is varied, it becomes clear that the rings (or spots) consist of two classes of electrons: elastically scattered electrons and a certain fraction of electrons that have lost energy inelastically (and in particular, those that have undergone one of the characteristic losses). Despite this inelastic scattering, the latter have therefore retained some degree of coherence. Filtering thus enables us to reduce the width of the peaks and to increase the resolution and accuracy of the traces on the photographs; in particular, we reduce the error in determining the size of the micro-crystals by diminishing the uncertainty in the width of the peaks.

21.2.4 The Interest in Using Very High Tension

In conclusion, we shall summarize the reasons why very high tension is of interest in diffraction.

With a very high tension, we can, first of all, examine thicker specimens (the majority of substances in powder form which are used in industry have grains larger than a micron in size).

The intensity of the continuous background of the diagram is decreased as fewer electrons are scattered.

As a result of the high penetrating power of the electrons, the specimen can be examined at even more closely grazing incidence. We can thus examine monomolecular layers.

Finally, the enlargement term of the rings which depends upon the transverse dimension of the crystals and which is proportional to λ decreases when V is increased, as does the broadening due to refraction.

21.3 SOME PRACTICAL EXAMPLES

21.3.1 General Summary

After thirty years' evolution, specialized instruments using this technique have been developed in all the important fields of research.

(i) *The search for high resolving power*

The earliest diffraction cameras had no lenses and were faithful replicas of their X-ray counterparts. An extremely fine electron beam was obtained with the aid of an electron source (hot or cold) and a long collimator containing a series of diaphragms with very small holes (a few hundredths of a millimetre); the beam passed through the specimen, after which the diffracted beams struck the plate a distance L away.

The search for ever higher resolutions led to the use of smaller holes which produced an inconvenient diminution of the brightness, and exposures which were far too long. No other way of maintaining the degree of accuracy was available, except a considerable tightening of the tolerances on the variations of V. The development of electron optics has provided a way out of this dilemma; the designers have introduced a focusing lens with which the current density incident upon the plate can be considerably increased, and a real image of the source S can be formed upon it. The source at first consisted of a small hole in a diaphragm placed directly behind the gun, but later a powerful condenser behaving as a reducing lens was introduced, and with this a still finer source was obtained.

(ii) *Direct magnification of the rings and accentuation of the contrast*

The resolving power can also be increased by lengthening L; further, high values of L are necessary if the rings of small radius which correspond to large values of d_i are to be examined conveniently (these large values are encountered in the interesting case of organic crystals). We can, however, scarcely make the instrument longer than 60 cm without considerably slowing down the pumping time whenever the specimen is changed, and making the magnetic screening difficult. By placing a supplementary lens L_2 after the specimen, however, we can eliminate this inconvenience, and give L a value which can sometimes be longer than two metres. In Fig. 269, the lens L_1 forms an image of S and hence forms a diagram d_1 near the object focal plane of L_2; the latter forms a magnified image, d_2, of d_1 on the screen. In these conditions, the equivalent length is far greater than the actual object–screen distance. By altering the convergences of L_1 and L_2 it is clear that L can vary between wide limits.

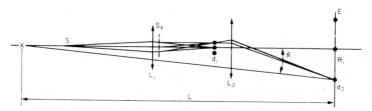

FIG. 269. Enlargment of the diagram by an electron lens.

This arrangement does not improve the resolution, however, since L_2 increases the radius of the spot as well as the radius of the rings—the resolution $\mathscr{R}$ obtained is governed by L_1. The essential advantage is that the rings close to the centre of the over-exposed central halo, which is produced by the direct beam and the electrons scattered by the object, are moved away outwards. Many of the scattered electrons have suffered considerable losses of velocity and as a result of the chromatic aberration of L_2, they are spread out, which reduces the density of the continuous background at the centre; on the other hand, the diameter of the over-exposed zone is practically independent of the size of the spot, and even if the latter is made larger the zone remains the same size—the reasons for this are poorly understood.

A large proportion of the retarded electrons can be removed with the aid of a velocity filter (cf. § 19.4) which increases the contrast on the diagram considerably. This filter lens also increases the resolution, as it reduces the enlargement term due to the object itself by removing the electrons which have been retarded after selective reflexion, and which would not be focused at exactly the same point as electrons which had not been slowed down.

(iii) *The microscopy of the specimen*

The region of the specimen which is illuminated can be made very small, and we can for example pick out a single crystallite in a heterogeneous body. We can thus make a chemical and a crystallographic analysis of the specimen. It is extremely useful to be able to chart accurately the crystallites of which the specimen is composed, or to note the appearance of the particles by which the diagram is being formed, in order better to recognize the various components of a mixture. We must find a way of allying the two techniques of microscopy and diffraction. This can be done in two ways. Either we build a high resolution diffraction camera, equipped with the necessary accessories for examining the object by shadow microscopy, or we use an electron microscope which can be transformed into a high quality diffraction camera; this type of instrument can only rival the other type described above if the traditional position of the object in the optical system is changed, and if the structure is modified until it is closely similar to that of a diffraction camera (cf. Duncumb, 1962).

This solution has the advantage of allowing us to obtain genuine micro-diffraction, that is, to obtain the diagram of a microscopic zone of the specimen with a good resolving power. This allows us to identify the different phases in the case of an object which is finely heterogeneous (a thinned down metal alloy, for example).

21.3.2 High Resolution Diffraction Cameras

Some typical models are described in the articles by Hillier and Baker (1946), Picard, Smith and Reisner (1949), Allen, Mackle and Sutton (1951), Le Poole (1947) and Lees (1954). The more specialized instruments described by Le Poole (1954) and Corbet *et al.* (1955) will be examined later. The pumping systems, the electrical supplies for the gun and the lenses, and the design of the specimen-holder and of the camera with which the image is recorded are closely similar to the corresponding parts of a microscope, which have already been described; we shall only elaborate on the general arrangement of the optical system and its possibilities, and the features peculiar to the diffraction. We should mention that the potentials employed very often exceed 50 kV, which suggests that magnetic lenses should be used. The reader is referred to Chapters 17 and 18.

(i) *The electron source*

The working potential is selected between 50 and 100 kV—only a few special instruments operate at very high voltages, 100 to 400 kV or even 1 MeV (Papoular, 1955). The source is normally a heated filament, but may also, more rarely, be a cold discharge. The gun resembles a three-electrode microscope gun, with autobias to increase the stability of the output. The "point" source can then be obtained in two ways.

In simple diffraction cameras, an anode diaphragm with a hole of 50 to 80 μ is sufficient for extremely fine rings to be obtained, provided that it is followed by a weak lens with magnification close to unity (Lafourcade, 1954).

To increase the resolution, one or two powerful reducing lenses are placed in front of the gun, and with these a genuine electron probe about a micron in diameter can be obtained; this reduction of the diameter results in a considerable increase of the angular aperture of the emergent beam; suitably placed stops eliminate any electrons which have been too widely deviated and which might well strike the walls and the pole-pieces of the lenses, and after reflexion, alter the appearance of the diagram. The diaphragms do, however, reduce the brightness considerably, and a gun capable of a larger output (100 μA) becomes a necessity.

To give some orders of magnitude, we quote the results obtained by Castaing (1951). With a suitable optical system, we can easily attain an intensity of 10^{-7} A in an electron probe one micron in diameter, when the angular aperture of the beam is of the order of 10^{-2}. If we reduce this aperture to 10^{-3} by placing a diaphragm in front of the diffraction lens, the intensity which reaches the central spot on the screen falls to about 10^{-9} A. When the object is present, the maximum diameter of the spot is 100 μ and the thickness of the rings is of the same order. The current density in the rings (or spots) again far exceeds 10^{-11} A cm^{-2}, which is the value which corresponds to a highly magnified electron microscope image, with an exposure time between a half and one minute.

(ii) *The optical system itself*

Several different lens systems have been proposed, but they can all be traced back to two basic combinations:

(a) a so-called "diffraction" lens L_1 forms the image of the source directly on the screen. This lens is very weak (focal length between 10 and 25 cm) and is often placed mid-way between the source and the screen. The object is placed after the lens, both so that the process of focusing does not affect the size of the rings, and to reduce the aberrations. The aperture is reduced by a factor of several thousand, by a diaphragm which precedes L_1, and the distance between the specimen and the screen, L, can vary between 25 and 70 cm (see Fig. 267).

(b) alternatively, the lens L_1 is followed by a projective lens L_2, which gives a magnified image d_2 of the small diffraction diagram d_1 which is produced by L_1 (Fig. 269). The size of d_1 depends upon the distance between the object and L_2; if this is too short, L_2 must be a strongly convergent lens, but the aperture of the beams must be reduced to prevent the spherical aberration from being too large. If the diagram is large, other aberrations also intervene, namely coma and astigmatism. To obtain the optimum resolution, we must find a compromise. L_2, however, must always be perfectly corrected for distortion, and it is preferable that it should consist of two lenses in tandem wound in opposite senses, to annul the rotation of the diagram.

The first system, which is the simpler, always gives the better resolution (Hillier and Baker, 1946). When the final spot is a few microns across, the resolution can in theory reach 10,000 (Tanaka and Hashimoto, 1953). A diaphragm, with an aperture of at most 50 μ, is also required. In the majority of cases, however, a diaphragm of about 0·1 mm, which gives higher brightness and leads to an exposure time of 10 or 20 sec, will be adequate. The second arrangement, as we have already mentioned, gives us a means of varying L. We have

$$L = \frac{ab - b^2 - af}{f},$$

with the notation of Fig. 270b. By altering f and readjusting L_1 for the focusing, we can make L vary between 0 and 2 m, for example, for values of a between 25 and 50 cm.

(iii) *Shadow microscopy*

On considerably increasing the convergence of L_1, a very fine electron probe is obtained in the vicinity of the specimen (Fig. 270a). When L_1 is the only lens, a highly magnified shadow of the zone which is illuminated

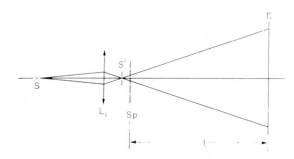

FIG. 270a. Shadow microscopy; the principle.

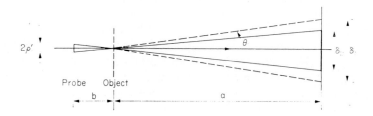

FIG. 270b. Shadow microscopy: the limit of the useful magnification.

is obtained directly, and the magnification is given by the ratio between the distances between the object and the screen and the object and the probe. G could represent an infinite magnification if the probe were a point, but the effect of the penumbra which is produced by the finite extent of the beam limits the useful magnification. If we neglect the spherical aberration, the resolution at high values of the magnification is bounded by the presence of the diffracted rays. At every point on the object, the incident ray is transformed into a conical pencil of semi-aperture θ_i (for θ_i, we can take the angle which corresponds to the last very intense ring on the diffraction diagram of the object; the lattice spacing is then d_i). The size of the diffraction spot is $\delta_i = 2a\,\theta_i = \dfrac{2a\lambda}{d_i}$, where a represents the distance between the object and the screen (Fig. 270 b). It is necessary that the size of the penumbra $\delta_2 = 2\varrho_s\dfrac{a}{b}$ should be less than δ_1 (b represents the distance between the probe and the object, and ϱ_s the radius of the probe). This means that

$$2\varrho_s\frac{a}{b} \leqq \frac{2a\lambda}{d_i},$$

or

$$G \leqq \frac{a\lambda}{\varrho_s d_i}.$$

For $a = 20$ cm, $d_i = 2$ Å, and $\lambda = 0.05$ Å, for example, we obtain $G \leqq \dfrac{5 \times 10^3}{\varrho_s}$ (ϱ_s measured in microns).

If we include the effect of spherical aberration, which distorts the image, we find that the limit is of the same order. In the best instruments, the resolving power scarcely exceeds 200 Å at 50 kV with a magnification of 10,000 and a probe about 100 Å in radius. For most cases, this is adequate, the image being simply a means of control. When we have a two-lens system available, it is better to move the probe further away from the object and obtain a high magnification by exciting L_2.

(iv) Shadow microdiffraction

We dispense with L_2 and vary the excitation of L_1 continuously between the values which correspond to Figs. 267 and 270; if a diaphragm with a very small aperture (20 to 50 μ) precedes the lens, we pass through a series of intermediate stages (Fig. 271) which are used to obtain the diagram of very small regions of the specimen, and in particular, of individual crystallites. The size of the spot at the screen increases in proportion as the illuminated area decreases, however, and the resolving power falls extremely rapidly. With the notation of Fig. 271, we have

$$d_0 = 2r_a\left(1 - \frac{l'}{v}\right) + 2\varrho_c m_1',$$

$$d_d' = 2r_a\left(\frac{l + l'}{v} - 1\right) + 2\varrho_s,$$

in which m_1' is the magnification of L_1, ϱ_c is the radius of the source S, ϱ_s is the radius of the spot focused on E, and r_a is the radius of the aperture diaphragm.

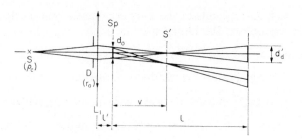

FIG. 271. Shadow microdiffraction.

When ϱ_s is very small, and the values of both ϱ_c and ϱ_s are a few tens of microns, it is easy to obtain an illuminated area of a few tenths of a micron with a resolving power of the order of 10 or 100 (Hillier and Baker, 1946); this is often sufficient to examine spot diagrams. The central spot encloses the direct shadow image of the illuminated zone, and the diffraction spots are images of the same zone against a dark background.

(v) *The specimen-holder and other accessories*

If a diffraction camera is to operate both in transmission and in reflexion, the specimen-holder must be highly refined; it must be mobile in every direction and rotatable in several ways so that we can adjust the angle of incidence of the beam easily by tilting the object. There are still more complicated specimen-holders which allow the specimen to be heated or cooled, over a temperature range of $- 180°C$ to $1000°C$. This is necessary in metallurgical studies.

The observation chamber has, therefore, to be large enough to contain the numerous devices with the aid of which the object can be subjected to various mechanical and thermal treatments:

(a) the mechanical apparatus which is required for cutting, grinding and polishing under vacuum;

(b) an ion gun with which the surface can be slowly etched and progressively cleaned; this is nowadays a standard technique (Trillat, 1954; Fert, 1954);

(c) ovens, for preparing thin films of various substances by evaporation;

(d) a gun to provide slow electrons for discharging the specimen. A gun of this kind has to provide current densities of the order of 0.4 to 5 mA cm^{-2}, without contaminating the surface of the specimen.

(vi) *Recording the diagram*

The plate cameras in everyday use resemble those which are used in microscopy, but recently, a recording technique has been developed with which the evolution of the specimen can be filmed during any kind of treatment (Boettcher and Thun, 1954; and Trillat, 1954). The Debye diagram is formed on a screen in which a very fine diametrically placed slit has been cut, beneath which a photographic film slowly passes. A ring is recorded as a pair of straight lines; as soon as the structure of the specimen alters, the distance between the lines varies and new lines appear, the intensity, width and position of which vary with time, depending on the crystallographic modifications of the specimen.

The electron density distribution in the diagram (see §21.2.3) can also be recorded by invoking all the resources of modern electronics for measuring low currents (of the order of 10^{-15} A with scintillators and low-noise particle photomultipliers with no windows, followed by counters). Either the measuring device is moved across the diagram (Lennander, 1952) or the diagram is shifted electrically across a diaphragm in front of the measuring instruments (Goodman, 1961). If this scanning is very rapid, the current density curve can be fed into an oscilloscope directly; this is the principle underlying the scanning diffraction devices (see §21.3.3), which thus allow one to observe very rapid changes in the structure of the specimen.

(vii) *Recent efforts: pulsed diffraction cameras*

Attempts have also been made to render the lenses of a diffraction camera nearly achromatic; in this way, we should hope to obtain high resolution without having to eliminate the fluctuations in the H.T. supply which could be tolerated up to a relative value of 5 or 6 per cent. Lenses which are insensitive to big variations in potential have been described in the literature (Cosslett and Jones, 1950; Le Poole and Salvat, 1950; Bernard and Salvat, 1954; and Le Poole, 1954). In the simple case of a system with a single lens, the principle is as follows—the specimen is illuminated in parallel light with the aid of a suitable system of condensers, and the lens is placed after the object (Fig. 272a). If the accelerating potential increases by $\triangle V$, the diffraction angle α_i is decreased, and the corresponding rays intercept the lens closer to the axis. The increase $\triangle V$ will also produce an increase of the focal length of L_1, so that the angle β through which the beam is deviated by L_1 will be smaller. For certain values of f, a and b, these opposed effects exactly compensate one another. We have the relations:

$$\beta = \frac{h}{f} = \frac{a\alpha}{f}, \qquad R_2 = R_1 - \beta b = \alpha(a + b) - \frac{\alpha a b}{f},$$

but

$$\alpha = \frac{\lambda}{2d} = \frac{C_1}{V^{1/2}} \quad \text{and} \quad f = \frac{C_2}{V}.$$

That the effects balance out is expressed by $\dfrac{dR_2}{dV} = 0$, so that the achromatism condition is

$$f = \frac{3ab}{a+b}.$$

But $R_2 = \dfrac{2}{3} R_1$ and the sensitivity is reduced.

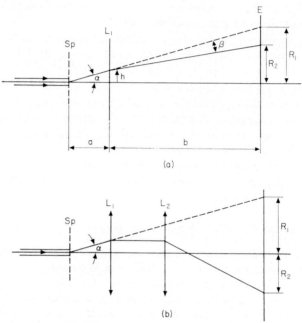

FIG. 272. An achromatic diffraction camera. The principle of (a) the single-lens situation and (b) the two-lens case.

We can establish relations which are analogous, though more complicated in form, for the two-lens case. These are given in detail in the article by Corbet *et al.* (1955). By a judicious choice of the geometrical parameters of the arrangement of these lenses, we can not only suppress the rotation of the diagram, but also make the distance L partially independent of the position of the object and cancel the distortion for the mean operating potential V. Since the distortion reappears for potentials on either side of V, the lenses must be chosen in such a way that this aberration is very small. The distortion, which has the effect of making the outer rings too large (cushion distortion), also broadens these rings when we are working with a variable potential. A magnetic system which fulfils these specifications and, in addition, makes the radii of the rings to some extent independent of the variations in the excitation of the lenses has been studied by Le Poole (1954).

Another achromatic two-lens arrangement (Fig. 272 b) is equipped with an instrument (Corbet *et al.*, 1955) with which the specimen can be studied simultaneously by microscopy. A diffraction camera of this kind is illustrated schematically in Fig. 273. There are three diffraction lenses with which either direct or magnified diagrams can be obtained. The equivalent length can vary between 65 cm and 450 cm, while the distance between the object and the screen is 60 cm. With the two condensers, the diameter and the parallelism of the incident beam can be adjusted; by altering the convergence of these condensers, and thanks to the possibilities introduced by the two lenses L_4 and L_5, shadow microscopy is possible with a magnification up to 5000. The gun is supplied with pure alternating potential, but the grid is

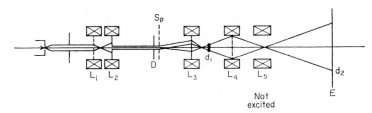

FIG. 273. An example of a diffraction camera which uses an alternating supply.

autobiased in such a way that only a small fraction of each period in the neighbourhood of the maximum (100 kV) reaches the gun. The pulses are sufficiently narrow for the potential variation during each pulse, ΔV, to be of the order of 5 or 6 per cent. The currents are measured to about one part in 10^3; the equivalent length is independent of the position of the object provided the axial displacement of the latter is not more than ± 5 mm. In these conditions, the resolution is of the order of 1000, and the accuracy with which d_i can be measured reaches 0·5 per cent, which is equal to the accuracy of the best ordinary instruments.

21.3.3 The Scanning Electron Diffraction System

This apparatus, which was suggested by Grigson (1961, 1962), contains no optical elements beyond the specimen; with a system of deflexion coils, the diagram can, however, be moved rapidly (the maximum velocity corresponds to 50×10^{-3} s) across a hole 20 or 50 μ in diameter cut in the fluorescent screen, behind which is a photomultiplier. The electron density distribution along a diameter is transmitted to an oscilloscope and can be photographed. With an auxiliary Faraday cage, the curves so obtained can be calibrated.

In a considerably improved version, Denbigh and Grigson (1965) have placed a retarding field filter behind the slit. If the filter is supplied with

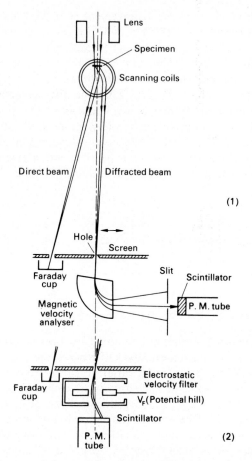

FIG. 273a. Diagram of the scanning electron diffraction unit (1) with magnetic
energy analyser; (2) with electrostatic energy filter.

a steady retarding voltage that can be adjusted in magnitude, electrons that
have lost large amounts of energy can be progressively eliminated so that
finally, only elastically scattered electrons are collected. If, furthermore,
the voltage of the high tension electrode of the filter is modulated with a
weak alternating potential, we obtain a small modulated component in the
transmitted current, corresponding to electrons lying within a band of
energies of width $\triangle V$ centred on a given energy V (which can be shifted
across the whole spectrum by altering the steady voltage, V_F, at the filter).
The diagram corresponding to a given energy will then be obtained by
using an a.c. amplifier at the receiver. This electrostatic filter, which
is not simple to use in practice, can also be replaced by a magnetic filter
(Grigson, 1965), consisting of a deflecting magnet followed by a thin slit

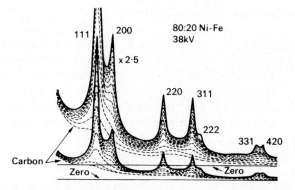

FIG. 273b. Growth of an 80–20 Ni–Fe film on carbon. Successive contours recur at thickness increments of approximately 10 Å. Sweep frequency 0·025 c/s; 38 kV electrons, two intensity scales (Grigson, 1965).

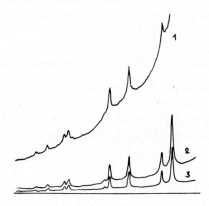

FIG. 273c. Diagram from a thin film of aluminium: improvement of contrast and resolution by velocity filtering (1) diagram without filtering; (2) all electrons that have lost more than 18 V are eliminated (so that electrons that have suffered the 15 V characteristic loss are transmitted); (3) the loss barrier is now raised to 3 V: the continuous background has virtually disappeared (Denbigh and Grigson, 1965).

and a photomultiplier. Just as in the preceding case, only those electrons corresponding to a predetermined energy V in the whole diagram will be collected at the photomultiplier.

With this scanning electron diffraction instrument, rapid alterations in structure can be followed, or the continuous growth of a thin film evaporated inside the instrument. Figure 273a illustrates the design of the device and Fig. 273b is an example of the diagram obtained during the growth of a Ni–Fe film on a carbon layer; Fig. 273c shows an example of contrast enhancement by electrostatic filtering.

21.3.4 Diffraction in an Electron Microscope

The high quality of the optical system and the electrical supply in electron microscopes might lead us to hope for a good resolving power when the instrument is used as a diffraction camera. The demands which are made upon the optical system in microscopy and in diffraction are rather different, however, and it often awkward to transform a very good microscope into a good diffraction camera. The introduction of supplementary lenses which are used in diffraction and not in microscopy, and *vice versa*, has proved to be indispensable in instruments which are to serve both ends (and which are known as "universal").

At first, the microscope was transformed into a diffraction camera by moving the specimen, but the present tendency is to transfer from one mode of operation to the other without touching the specimen. Nevertheless, if we require pictures with very high resolution ($\mathscr{R} \geqq 1000$ for example), the only solution is to fall back upon the arrangements peculiar to diffraction cameras, and to a specialized instrument.

(i) *Transformation of a microscope into a diffraction camera*

In a two-lens microscope (von Borries and Ruska, 1940, and Picard and Reisner, 1946), a diffraction camera can be produced by removing the pole-pieces of the lenses and introducing an extremely small diaphragm into the condenser, all without disturbing the object. The projective lens L_2, excited

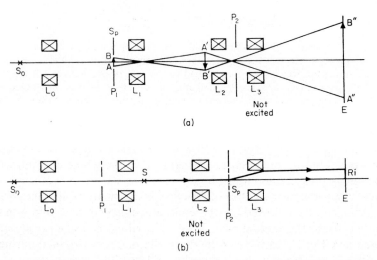

(a)

(b)

Fig. 274. Diffraction in an electron microscope with a mobile specimen-holder. (a) The instrument as a microscope. (b) The instrument as a diffraction camera.

alone, may serve as a diffraction lens, but the combination of L_1 and L_2 is equally practicable. The necessity of removing the pole-pieces arises from the fact that we must avoid intercepting the outermost diffracted beams.

In other instruments, the projective lens L_2 is followed by a unit which is designed specially for diffraction work, and contains a second specimen-holder P_2 and a weak diffraction lens L_3. When the second specimen-holder is not in place, the instrument behaves as a microscope (Fig. 274a) and the lens L_3 is switched off. To obtain a diagram, the object has to be withdrawn from the instrument and placed on P_2. The objective L_2 then produces a fine source, while L_3 gives the desired diagram on the screen (Fig. 274b). Shadow microscopy too is possible (Hillier, Baker and Zworykin, 1942; and Rühle, 1950).

(ii) *The direct passage from microscopy to diffraction; high resolution micro-diffraction*

In the majority of recent microscopes, we can pass from one mode of observation to the other with comparatively few operations, namely, variation of the excitations of the lenses and of the apertures in the stops, which can be effected from outside. Van Dorsten *et al.* (1950) describe an instrument which has a condenser L_0 and four lenses. The microscope proper works with three lenses (Fig. 275). As we have explained in § 15.5, L_1 is the objective and L_3 and L_4 are projective lenses; L_2, which is weak, is

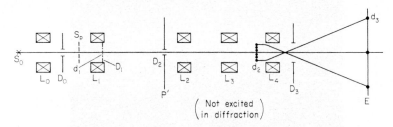

FIG. 275. Diffraction in a microscope, in which one can pass directly from one type of observation to the other.

used only in diffraction, to image the diagram produced by the objective L_1 in front of L_4. The 40 μ contrast diaphragm D_1 which is situated behind L_1 is replaced by a 1 mm diaphragm which does not intercept the diffracted rays. In the plane P', where the intermediate image of the object is formed, a zone of any desired area can be isolated with the aid of a variable stop D_2. If L_1 has a magnification of 100, we can select a 1 μ zone of the prepa-ration with a 100 μ aperture. From Fig. 277a, where the electron paths through the system are shown, we can see that the rays which pro-duce the diagram really are those which correspond to the portion of the

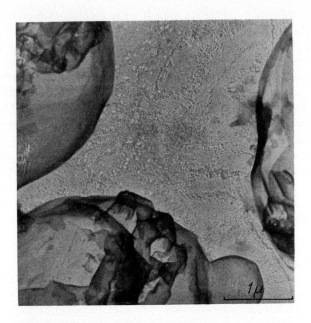

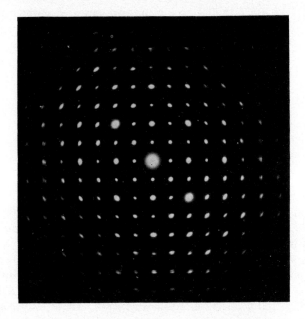

FIG. 276. Microdiffraction (Photos: Fert). *Above*: the image of the zone being studied. *Below*: the corresponding diagram.

object defined by this aperture (Fert and Ito, 1955). An example of this remarkable technique is given in Fig. 276. Unlike the shadow micro-diffraction case, there is no loss of resolution when the illuminated area is reduced. The diaphragm D_2 usually consists of two interlinked parts which describe a square hole (Fig. 277b). The area can then be varied simply by moving the two parts relative to one another. A high degree of mecha-nical precision is always necessary.

To obtain high resolution, the condenser must be readjusted. In general, the passage from microscopy to diffraction is possible in all three-lens microscopes. The third lens L_3 works at a fixed excitation in conditions of minimum distortion, and the convergence of the weak second lens L_2 can be varied between wide limits (Grivet and Regenstreif, 1950; Tanaka and Hashimoto, 1953). Instead of forming a second image of the object still further magnified with L_2, we simply reduce the convergence considerably in order that the image d_2 of the diagram d_1, which is formed by the objec-tive close to its image focal plane, shall be formed near the object focus of L_3. This mode of operation has been described in § 15.5. Under these conditions, the magnification of L_2 is always less than or equal to unity, and d_2 is small enough to be magnified entirely by L_2 working normally. The aperture of the diffracted beams is, however, larger than the aperture of the beams which originate in points on the second intermediate image of the object in microscopy, and the various aberrations (spherical aberration,

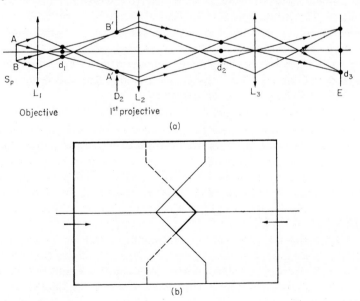

FIG. 277. Microdiffraction. (a) The ray paths. (b) An example of a field dia-phragm.

coma, ...) are accentuated. We have thus to make every effort to obtain an extremely fine source S in front of the object, either by introducing a very small diaphragm if the object is illuminated directly by the beam which is emitted by the gun, or by a very different adjustment of the condenser if such a lens already exists.

(iii) *The reflexion diffraction camera used in conjunction with an emission microscope*

This combination has the same advantage in reflexion as micro-diffraction in transmission, namely, that the spot can be very exactly localized within an extremely narrow and well-defined region of the object, together with the added advantage that both the diffraction diagram and the surface being studied can be examined *simultaneously* on two separate fluorescent screens.

The instrument is shown diagrammatically in Fig. 277 S 1. The electron beam of the diffractograph falls on the surface under examination at almost grazing incidence, and produces the picture which corresponds to the illuminated region on the lower fluorescent screen. The object forms the cathode of an electrostatic immersion objective with its axis horizontal, which produces a magnified image of the surface on a second fluorescent screen by means of the electrons which are emitted by the surface either by thermionic emission or by secondary electron emission from an auxiliary ion beam. The diffraction camera beam is stopped right down and thus illuminates only a very small area of the surface; at the same time, a few slow secondary electrons are liberated, which are collected by the immersion objective and give rise to a bright spot on the image of the surface. By shifting the object, we can check the position of the spot and bring it exactly onto the detail which is to be examined.

A device of this type has been built by Arnal (1958 a, b). The accompanying thermionic emission microscope is founded on the instrument described in § 13.4; at a distance of 50 cm from the object, it gives an image magnified 70 times. The object is earthed in order to avoid disturbing too radically the diffraction camera beam which crosses the cathode–grid gap of the immersion objective. The anode of the latter, together with all the interior of the surrounding chamber and the fluorescent screen, is held at an accelerating potential of 30 kV.

The diffraction camera beam is only slightly disturbed by the relatively small accelerating field of this type of objective; to reduce both the cross-section at the level of the plane in which the object lies, and the elongation (due to the grazing incidence) of the zone which is illuminated, the beam is stopped down by means of a slit, $100 \times 2\,\mu$. The spot at the object can be reduced to $8 \times 32\,\mu$ in this way without the brightness of the diagram becoming insufficient.

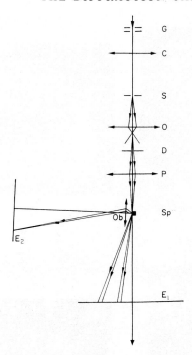

FIG. 277 S1. Diagram of the optical system of a composite instrument, diffraction camera and emission microscope (Arnal). G: gun; C: condenser; O: objective; P: projective; S: slit; D: diaphragm; Sp: specimen; E_1: fluorescent screen and camera (diffraction diagrams); Ob: immersion objective; E_2: fluorescent screen (image of the surface).

This is a most valuable instrument for the metallurgist, and it has in particular been used to study the phase transformations in titanium, near 950°C, after a lengthy heating at 900°C. The crystalline structure of a specimen is almost always perfect in these conditions, and we can examine a diagram which displays Kikuchi lines. The size of the spot is much less than that of the titanium crystals, and can easily be directed onto the zones which are in the process of transformation (Fig. 277 S2). Similarly (Sorel, 1959, 1962), we can follow the progress of surface oxidization, or analyse the epitactic coating on a massive specimen.

21.4 SOME APPLICATIONS

We shall simply review these briefly; for further information the reader is referred to the bibliography established by Rymer (1953).

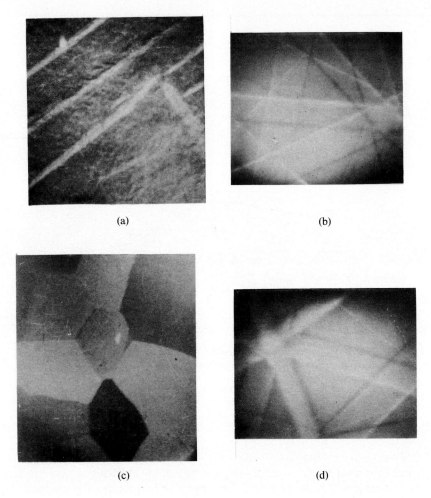

(a)   (b)

(c)   (d)

FIG. 277 S 2. Examples of the application of the above instrument. (a) The surface
of a specimen of α-titanium; the bright spot (top left) shows the position of the
spot of the diffraction camera ($G = 70$). (b) The diagram corresponding to the
crystal struck by the beam of the diffraction camera (Kikuchi lines), [021] axis.
(c) The surface of a specimen of β-titanium ($G = 70$). (d) Diagram corresponding
to the central crystal for the preceding specimen, [311] axis (Photo: Mrs. Sorel).

21.4.1 Geometrical Crystallography

(a) Determination of the lattice constants, with the possibility of identifying the sample. The latter may be a very thin layer at the surface of an object being observed by reflexion (a layer of oxide or of lubricant, or a polishing layer for example).

(b) Measurement of the changes of the lattice constants during a mechanical deformation of the specimen.

(c) Information about the type of deformation which has occurred (slip with rotation, for example) can be obtained from the diagrams of mono-crystals (Wilman, 1951).

(d) The size of the crystals can be calculated from measurements on the thickness of the rings (Lafourcade, 1954), and the form of individual crystallites can be obtained from the shape of the spots on the diagram.

(e) Kikuchi reflexion diagrams are produced only when the crystal lattices are perfect; this gives a means of estimating the "finish" of a surface, and its degree of order.

(f) Determination of the structure of organic crystals, and of the positions of the large molecules in a crystal—the latter may display preferred orientations.

21.4.2 The Study of the Internal Potentials of Crystals

The mean electrostatic potential within a crystal differs from the electrostatic potential of the surrounding vacuum by about 10 or 15 V. This potential difference produces a refraction of the electron rays, the refractive index being slightly different from unity. This refraction can easily be seen when we work with reflexion, as the diffracted beams may be displaced several millimetres away from the position which the simple theory predicts. In transmission, with an instrument capable of very high resolution, it has been possible to detect a duplication of the rings as the diffracted rays are refracted either one side or the other of the Bragg angle, according to the orientation of the faces of the microcrystal at which the rays enter and leave. The distance between the two "components" of a ring does not in general exceed $100\,\mu$ for $R \sim 1$ or 2 cm. By studying this separation, we can obtain data about the value of the internal potential.

Onto the mean potential is superimposed a spatially varying potential which is closely related to the crystal lattice. The amplitude of this variation and the thickness of the crystals can be determined by making measurements on the interference fringes produced by neighbouring diffracted rays. We can obtain either fringes of equal inclination, by illuminating the crystal with a convergent beam (shadow micro-diffraction), or fringes of equal thickness which appear when the crystal is irradiated with a parallel beam. The reasons for the existence of these fringes have been explained in Chapter 19.

CHAPTER 22

ELECTRON PROBE DEVICES

22.1 ELECTRON PROBES AND MICRO-PROBES

The name "electron probe" is given to the real image of a real electron source, and in particular, to the crossover which lies at the exit of an electron gun. The simplest example is the spot of a cathode ray oscillograph, but, except in extremely unusual circumstances, the diameter of this spot is not less than a few tenths of a millimetre.

The progress which has been made in electron optics during the last twenty years has been such that we can now obtain very small probes indeed, which may be as little as a few hundred Ångströms in diameter; these are often known as "micro-probes". By operating an electron microscope in the opposite sense, we could clearly obtain a probe with a diameter of the same order of magnitude as the resolution of the instrument. The current in such a probe would, however, be inadmissibly tiny (10^{-14} to 10^{-16} A, for example). The search for high brightness in the probe is always accompanied by efforts to compress the greatest possible intensity into the smallest possible diameter; the reasons for this will be apparent when we come to consider applications later in this chapter.

In the first section, which is devoted to a type of instrument which is particularly useful in metallurgy, the micro-probe analyser, we shall describe the general method by which a micro-probe may be obtained, the problems encountered, and the inherent limitations. Next, we shall give a brief description of two other types of micro-probe apparatus: the X-ray projection microscope and the scanning microscope. Finally, we shall say a few words about some recent techniques in which the kinetic energy carried by an intense electron beam is used to work metals; a very small intense probe may be used for drilling, and probes of several kilowatts for welding or melting the refractory metals, under vacuum.

22.2 MICRO-PROBE ANALYSERS

22.2.1 The General Principle

An electron probe device, with which an almost point-by-point chemical analysis of a massive metal specimen can be carried out without destruction of the latter, has been studied and built by Castaing (Castaing and Guinier,

1949, 1950 and 1951; Castaing, 1951, 1954 and 1955). The impact of electrons onto any kind of anticathode results in X-radiation which contains the X-rays which are characteristic of the various elements from which the anticathode is formed. A spectrographic analysis of the X-rays which are emitted therefore gives us information about the chemical composition

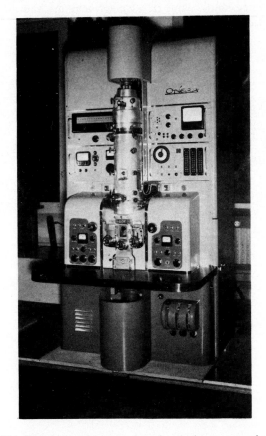

FIG. 278 S1. Castaing's electron probe microanalyser: general appearance.

of the extremely small volume which the probe has irradiated. For massive specimens, the resolution of the instrument is limited by the depth d to which the electrons penetrate, and there is no point in producing a probe smaller than d, in diameter. (When the specimen is thin, an electron microscope preparation for example, we try to obtain a probe the diameter of which is equal to the thickness of the preparation—a few hundred Å.) For electrons, accelerated through some tens of kilovolts, we shall have

$d \simeq 1\,\mu$. We shall describe in detail Castaing's electron probe, the exterior appearance of which is shown in Fig. 278 S1. The instrument contains four main parts:

(a) the electron optical system which produces the probe;

(b) a metallographic reflexion microscope, with which the specimen can be examined and which can be sighted on the point to be analysed;

(c) the specimen-holder with its displacement mechanism;

(d) X-ray spectrographs for quantitative analysis.

22.2.2 Electron Optics: Obtaining the Probe

The optical system must be capable of concentrating as high an intensity as possible into a circle less than a micron in diameter; in theory, we have simply to reduce the image of a heated cathode to the required degree with the aid of two very convergent lenses. Figure 278 S2 is a diagram of the

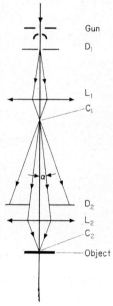

Fig. 278 S2. Formation of a micro-probe—the ray paths.

optical system which is employed. An ordinary triode gun emits electrons which are accelerated through a potential V which can be varied between 10 and 40 kV, and which are then incident on a first diaphragm, some tens of microns in diameter; this diaphragm lies in the plane in which the crossover (or "circle of greatest contraction") of the beam is formed.

D_1 acts as entry pupil, and creates a boundary to the useful surface S of the cathode, for which we can define a brightness B such that

$$B = R \frac{i_0 \, eV}{\pi \, kT};$$

i_0 is the emissive power of the cathode, eV is the acceleration energy, R is the efficiency of the gun and k is Boltzmann's constant. The diameter of D_1 is much smaller than that of the crossover, in order that we should be able to obtain as high a uniform brightness as possible. Measurements which have been made with $V = 30$ kV, $T = 2700°$K, and $i_0 = 2$ A cm^{-2} give $B = 58,000$ A cm^{-2} sterad^{-1}, or an efficiency of $R \simeq 70$ per cent. If the diaphragm D_1 is not used, the efficiency falls to about 10 per cent. The first lens L_1, a condenser, produces a very reduced image C_1 of the cathode, but the aperture of the beam is considerably increased; a second diaphragm D_2 which is placed at the entry to the second lens L_2, the objective, reduces this aperture to a value α of about 10^{-2} radians. An image C_2 of C_1 which is still further reduced is obtained in the neighbourhood of the focal plane of L_2. Were the aperture α infinitely small, the diameter of C_2 would be the Gaussian diameter d_0. To increase the intensity, however, L_2 must accept a value of α as large as possible, and the aberrations will thus increase the diameter of the probe. The dominant aberration is the spherical aberration of L_2 (L_1 is used at a very small aperture, of the order of 10^{-3}); if we denote the spherical aberration constant of L_2 by C_s, we can write

$$d_{real} \simeq d_0 + C_s \alpha^3.$$

The brightness B of the successive images of the cathode remains constant, and we can show that the total intensity I in the probe of diameter d, which varies as α^2, reaches a maximum if $\alpha = (d/4 \, C_s)^{1/3}$, that is, when the diameter of the aberration disc is equal to one-quarter of the final diameter d. For this optimum value of α, the intensity I is given by the expression:

$$I = \frac{9\pi}{64} \frac{e\,V}{k\,T} R \, i_0 \frac{d^{8/3}}{(4 \, C_s)^{2/3}}.$$

For $C_s = 3.6$ cm and $d = 1.2 \, \mu$, we ought to obtain a probe current of the order of $1.07 \, \mu$A with $V = 30$ kV.

Experiment shows, however, that another aberration which increases the diameter d must be taken into account; this is the aberration which is due to the residual ellipticity of the openings in the electrodes of L_2, which results in an astigmatism which we cannot neglect. Fortunately, this aberration can be wholly corrected by a device placed outside the lens, as Bertein foresaw as early as 1946. Using a weak multipole lens, which is placed beneath L_2, and employing a new method of eliminating astigmatism, Castaing has been able to produce a probe with a diameter which is close to the minimum value predicted by the theory, and to obtain a current I of $0.77 \, \mu$A in a $1.2 \, \mu$ probe.

To increase I, it will be advantageous to employ lenses with very little spherical aberration. This is one of the reasons why magnetic lenses are regarded as superior to electrostatic ones for this purpose. In this type of analyser, however, designed as it is for solid specimens, the lens L_2 must have its focus external, and will hence have a rather long focal length; the spherical aberration constant is therefore rather high. Very asymmetric lenses are employed, in which the exit opening is extremely small (Mulvey, 1958), as this is the arrangement which allows us to work as close as possible to the specimen (the latter, which may be ferromagnetic, must be outside the field). The contrary is true in analysers for thin specimens working in transmission conditions, as the specimen can now be immersed within the lens, just as in microscopy, and we can thus use very short focal length lenses (Duncumb and Melford, 1959). Further improvement is possible by introducing spherical aberration correctors, which resemble those which have been proposed in electron microscopy (Archard, 1959. See Chapter 7). This last solution has not as yet been incorporated into any instruments of this type on account of the difficulty of the adjustments to be made.

22.2.3 Observation of the Object and of the Point of Impact of the Probe

A genuine metallographic microscope is closely linked to the electron optical system and must satisfy several requirements: it must be placed far enough from the specimen for the scattered electrons—which would rapidly contaminate the glass surface—not to be received, the magnification and the numerical aperture must be adequate, an axial hole must be pierced to allow the incident electron beam to pass, and there must be no insulating surface in the way of the beam as electrostatic charge would perturb the probe.

All these conditions are satisfied in Castaing's apparatus by a special mirror objective, which consists of two concentric spherical mirrors. The focal length is 12 mm, the aperture 0·48, and the resolving power about $0.7\,\mu$. By using a system with two mirrors and two semi-transparent plates inclined at 45° to the optic axis we can throw the light from a lamp placed outside the vacuum enclosure onto the object, and capture the rays reflected by the surface of the object in an eyepiece (which is provided with a graticule) which is also placed outside (Fig. 278 S 3). The total magnification is 400.

With this microscope, the object can be observed while the instrument is in action, and the point of impact of the probe rigorously localized. This point should coincide with a point of intersection of the cross-hairs of the graticule. It is visible on the surface of the object as a black point (which is due to contamination by carbon molecules), which appears after a few minutes bombardment.

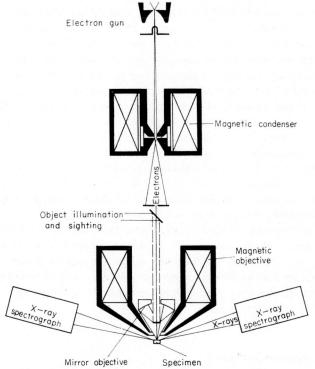

FIG. 278 S3. The principle of the electron probe microanalyser.

22.2.4 The Specimen-holder

The specimen must be cylindrical with diameter between four and eight millimetres, and about the same height. The specimen-holder is such that the perfectly straight edge of a tantalum plate can be brought onto the axis of the probe; the shadow of this edge on the fluorescent screen which lies at the bottom of the vacuum chamber is a guide to the correction of the astigmatism of L_2. Further, forty examples of pure elements, which are placed on one side or the other of the sample being analysed, can be placed successively under the probe without having to modify any of the adjustments.

The movements, parallel to and across the axis, are effected with the aid of screws of extremely fine thread, with graduated heads.

22.2.5 The Spectrographs

The X-radiation which is emitted by the object is analysed with the aid of two spectrographs, each provided with a curved crystal and a counter, which are placed within the vacuum. The crystal and the counter rotate on the same focusing circle 25 cm in radius, and the radiation reflected by the crystal is concentrated onto the centre of the window of the counter. At each instant, the angle at which the X-radiation is incident on the crystal—and hence the reflected wavelength—is indicated on a series of drums graduated in degrees and minutes, with an accuracy of better than a minute of arc. One of the spectrographs contains a quartz plate and a Geiger–Müller counter, and is used to detect radiation with a wavelength between 0·6 and 4·5 Å; with this spectrograph, all the elements between chlorine and molybdenum can be analysed using the K lines, and all the elements heavier than molybdenum with the L line. The other spectrometer detects soft X-rays (4 to 10 Å), and makes it possible to analyse light alloys (of aluminium or magnesium). It contains a curved mica reflector, 50 cm in radius, and a special counter with a Mylar entry window; a gas current flows continually (90 per cent Ar, 10 per cent CH_4).

The spectrograph chambers are separated from the probe region by exit windows which consist of beryllium plates 0·3 mm thick; these windows can, however, be retracted when the whole system has been evacuated, so that as little radiation as possible is absorbed between the object and the counter. The arrangement of the spectrographs on either side of the specimen is shown in Fig. 278 S4.

22.2.6 Chemical Analysis

It is possible, therefore, after choosing an interesting region of the object, to bring up the probe, and to measure very accurately the wavelength and the intensity of the lines emitted by a small region one or two microns in diameter around the point at which the probe is incident. We first make a qualitative analysis, scanning the spectrographs over the whole possible range of wavelengths, and after this we can set about making a quantitative analysis. We note the intensity of a line characteristic of some particular element which has been identified in the zone being analysed, and then place the representative of this element in the pure state under the probe. To a first approximation, the ratio of the two measurements gives the concentration of the element at the point in question. To make a second, more accurate, approximation, various corrections are necessary. The greater part of the X-radiation is created in the volume being analysed, but before reaching the spectrograph, it undergoes a certain amount of absorption within the specimen itself; if the latter contains highly absorbent substances, the measurement may be falsified. A correction must be applied to the measured intensities, therefore, the "absorption correction", if we are to obtain the proper values of the intensity with which to form the ratio. The necessary correction curves have been obtained empirically by Castaing and Descamps (1958), and show that for the light elements it is better to

FIG. 278 S4. The main tube of the microanalyser. *a*: adjustment of the height of the object (focusing of the eyepiece). *b*: displacement of the object. *c*: window through which the object is introduced. *d*: levers, for retracting the X-ray exit windows. *e*: means of introducing film for recording the Kossel lines. *f*, *g*: hoods for the evacuated spectrographs. *h*: opening for introducing film (diffraction). *k*: eyepiece.

use a small accelerating potential (10 to 15 kV). It is for these elements that the analysis is least accurate.

These corrections assume that the emergent angle of the useful rays is well defined and equal to 18°; this requires the specimen to be perfectly plane, and hence forbids the use of electrolytic polishing; a very careful mechanical polishing is to be recommended. A further correction, for the fluorescent radiation emitted by the specimen, can also be made, as certain atoms of the specimen may be excited by the primary X-radiation. Correction formulae are available (Castaing, 1951; Castaing and Descamps, 1955) with which this source of error can be eliminated.

22.2.7 Some Applications of the Micro-probe

The most usual field of application of the probe is metallurgy, but we might equally well analyse any type of specimen, conductor or insulator (provided the latter is first coated with a thin conducting layer so that the charge in the incident beam can flow away). Examples are to be found in the numerous articles by Castaing and his associates, and in an article by Mulvey (1958); particular examples are the chemical analysis of steels, the deposits produced by rubbing one metal against another, and inclusions in alloys.

The problems of intermetallic diffusion can be studied most elegantly with a micro-analyser; a tranverse section of the zone in which the two metals in question are in contact is polished metallographically and the concentration of the two constituents is established point by point starting from an arbitrarily selected origin in one of the pure metals. This method has been successfully applied to a number of different systems: U–Zr, U–Mo and U–Ti [Fig. 278 S5(a)]. The equilibrium diagram can be established on the basis of accurate measurements of the concentration. A chart of the distribution of manganese in a Hadfield steel (C = 1·88 %, Mn = 12·5 %) is shown in Fig. 278 S5(b); this chart was obtained by a point-by-point analysis (Philibert and Bizouard, 1959). A rapid study of the surface of the specimen can be made by scanning this surface with an electron probe controlled by a suitable deflection system placed above the objective (Cosslett and Duncumb, 1956); the scanning is synchronized with that of an oscilloscope, the intensity of which is modulated by the counter which collects the X-radiation characteristic of any given element. To increase the resolution of the system, we could alternatively leave the probe stationary —this would allow the spectrograph to be used always at its optimum adjustment—and employ some mechanical means of shifting the specimen.

A thorough analysis of the various possibilities of electron probes is to be found in an article by Castaing (1960).

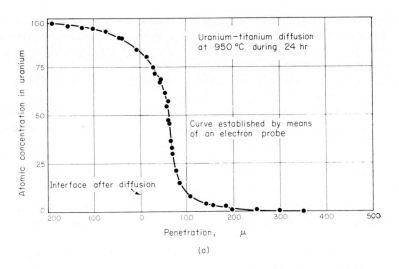

(a)

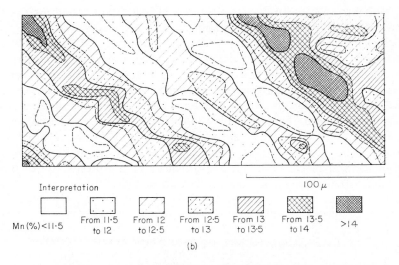

(b)

FIG. 278 S5. Examples of the uses of the micro-probe. (a) The diffusion curve, traced in a direction normal to the zone of contact between uranium and titanium, after heating to 950°C for 24 hours. (b) The distribution of manganese in a special steel (Philibert and Bizouard, 1959).

22.2.8 Other Recent Types of Micro-probe

One of the problems which appear when we employ electron probes is that of determining the exact location of the point of impact. We have already seen how the observation in the instrument described above is effected with the aid of a special mirror objective which is placed inside the last magnetic lens. Alternatively, we may construct a mechanical system to very high precision, with which we can rotate the specimen-holder through 180° about a horizontal axis in such a way as to be able to examine the specimen through a side-window with the aid of a high-resolution optical microscope (Mulvey, 1958, 1959).

The scanning technique which was suggested by Cosslett and Duncumb (1956) and which was inspired by the technique of scanning microscopy (see below) makes it possible to obtain the image of the surface of the specimen rapidly, relative to a given characteristic X-ray line. The probe can then be directed onto one of the details which were visible on the image, by using the afterglow which lasts about 30 sec.

A way of perfecting the method is to collect the secondary electrons which are emitted by the surface in a scintillation counter, and thus obtain the modulation signal; the image is much brighter (Duncumb, 1958).

The scanning is produced by a double system of coils, which are placed above the final reducing lens.

With a recent version due to Nixon (1960), ultra-thin specimens can be analysed and simultaneously examined by transmission microscopy at high magnification. The principle is as follows: the image of the preparation is focused onto the final fluorescent screen in an ordinary microscope; in the centre of this screen is a small hole, which lies on the optic axis of the instrument. Through this hole a second electron beam is directed up the instrument, and forms a micro-probe in the object plane; if the system is perfectly aligned, this probe falls on the region the image of which we are examining. The X-rays which are produced by the probe are collected in a special proportional counter which is placed near the specimen-holder. With this instrument it should be possible to obtain probes one to three hundred Ångströms in diameter with a current of 10^{-11} A.

22.3 THE SCANNING ELECTRON MICROSCOPE

22.3.1 The Principle of the Instrument

We have already mentioned the principle behind this instrument in the preceding section. The electron probe "scans" the surface which is to be examined, propelled by a system of crossed coils which are placed before the final reducing lens, with the same action as the spot of a television camera (Fig. 278 S 6). A receiver which is placed near the surface collects continuously the secondary electrons which are emitted by the surface. The signal which is obtained is amplified, and then fed to the modulation electrode of a television tube, adjusted to scan in synchronism with the probe. In this way, we obtain a highly magnified "image" of the surface; the resolution is in theory equal to the diameter of the probe, provided all parasitic signals have been thoroughly eliminated, and the specimen is perfectly flat. Just as in a microanalyser, the probe diameter must be

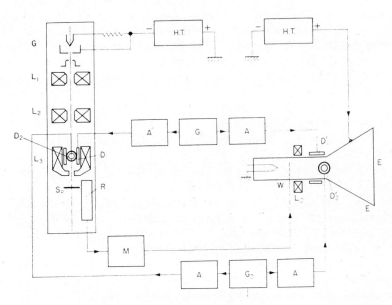

FIG. 278 S6. Diagram of a scanning microscope. On the left—the microscope itself (G: gun; L_1, L_2, L_3: lenses; D_1, D_2: deflexion coils; Sp: specimen; R: receiver). On the right—television tube (W: electrode for modulating the brightness; L_0: concentration; D_1' and D_2': deflexion coils; E: fluorescent screen; G_1 and G_2: generators for the scanning, both the lines and the image, which supply the pair D_1 and D_1' and the pair D_2 and D_2' respectively after amplification in A). The electric signal which is received at R is transmitted to W after amplification in M.

adjusted to be of the same order of magnitude as the penetration of the electrons into the specimen being examined. It is therefore advantageous to employ low energy primary electrons but this in turn necessitates a very careful elimination of the alternating parasitic magnetic fields which would tend to enlarge the spot. Further, the role of the diffraction aberration becomes the more important as λ increases. Finally, at very low voltages, a broadening of the probe is perceptible as a result of space charge. We are thus forced to make a compromise between these three factors: potential, intensity and spot diameter. A satisfactory compromise is obtained with potentials of only a few kV, with which we can obtain a probe intensity of about 10^{-9} A in a probe of a few hundred Ångströms diameter.

22.3.2 The Origins of the Contrast

The secondary electrons are of two kinds:

(a) primary electrons reflected at high energies (between 50 eV and the initial energy of the incident beam);

(b) genuine secondary electrons, emitted with energies of less than 50 eV.

We can collect the electrons of either category to produce the modulation signal but the slow secondaries provide more information about the structure of the surface being studied. The fast electrons are scarcely deflected at all, since the electric fields in the neighbourhood of the surface are weak, and the electrons hence follow almost rectilinear paths. The slow electrons are, on the contrary, strongly influenced by these fields, and they can be collected more easily in a receiver placed near the specimen and slightly polarized to attract the secondaries.

The origins of the contrast which is obtained when these electrons are employed are several:

(a) *The relief*, which affects the rate of emission of secondaries, in consequence of the different orientations of the elementary constituents of the surface of the specimen. Certain regions are hidden from the receiver, and seem darker. This effect of relief is particularly striking if we collect only the fast (reflected) electrons, the trajectories of which are far more acutely influenced by the angle of incidence.

(b) *The chemical nature of the specimen*. Since the emission coefficients of different elements differ widely at low bombardment potentials, we can, for example, pick out the segregations or clusters in an alloy.

(c) *The crystal orientation*, since even for the same metal, the secondary emission varies (if relatively little) with the nature of the crystal faces being bombarded; the presence of this effect has been displayed experimentally.

(d) *The way in which the metal has been worked* (Davoine and Bernard, 1956). If a metal has been drawn, the contrast of certain regions is increased; this effect is perhaps due to a superposition of the two preceding processes during the deformation of the surface, and to the creation of new crystal faces by the impact of the beam.

(e) *The contact potential differences* between neighbouring areas, or "potential contrast"; these are such junctions as the boundary between two monocrystals of the same metal or between two semi-conducting strips doped with different impurities (p–n, p–n–p, junctions . . .). We can completely interchange the brightness of the different strips by biasing the specimen (Everhart, Smith, Wells and Oatley, 1958). The contact potential difference (which may vary between a few tenths of a volt and a few volts) is emphasized by a transition zone, two microns wide, which gives rise to intense electric fields which eliminate the vast majority of the secondary electrons, and this zone has a highly contrasted appearance.

22.3.3 Recent Developments

The most important improvements have been made in the domain of the receivers of secondary electrons, with a view to increasing the useful signal to the uttermost, and eliminating the background noise which the reflected electrons produce. The receiver may be an electrode system similar to that of a photomultiplier. We cannot, however, use the same type of

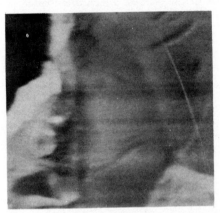 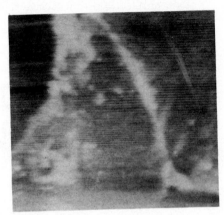

FIG. 278 S7. The appearance of the surface of a zinc sulphide crystal. On the left—the image obtained with the secondary electrons. On the right—the image provided by light radiation (3000–6000 Å) (Photos: Davoine).

layers to perform the multiplication, as the vacuum is relatively poor. We make do with copper–beryllium dynodes (Bernard and Davoine, 1958) which are periodically reactivated. Alternatively the receiver may be a grid, polarized to one or two hundred volts, which attracts the secondary electrons. The latter pass through the grid and are accelerated through a few kV, and finally fall on a fluorescent screen. The light is transported to a photomultiplier with the aid of a perspex light-pipe.

In other attempts to obtain micro-probes the correction of astigmatism and the use of lenses with little spherical aberration have been considered; it should be possible to obtain probes 100 Å in diameter (Smith, 1960).

Descriptions of these devices are to be found in the articles by Davoine (1958), and Smith (1960), together with the characteristics of the associated electronic circuits; recent applications are described in the articles by Thornley (1960), and Davoine, Bernard and Pinard (1960) and also in the articles already quoted.

Figure 278 S 7 shows two aspects of the same zinc sulphide crystal; image (a) was obtained by collecting the secondary electrons, whereas image (b) was formed by collecting the light emitted as fluorescence by the crystal onto a photomultiplier (3000 to 6000 Å).

22.4 THE X-RAY PROJECTION MICROSCOPE

22.4.1 The Principle of the Instrument

Figure 278 S 8 is a schematic cross-section of an X-ray projection micro-scope. The micro-probe S is formed just beyond the final reducing lens L_2, at the surface of a thin target, F. Those of the X-rays, which are created in the zone upon which the electrons fall, which are emitted *through* the

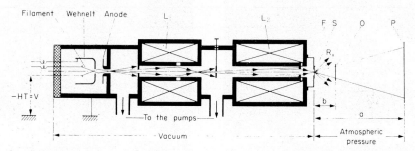

Fig. 278 S 8. Diagram of an X-ray projection microscope (C.N.R.S. Laboratory, Toulouse).

target, pass through the object under examination, O, and produce a pro-jection image of the latter on the screen, P. The linear magnification is equal to the ratio of the distance SO to the distance SP when the object is thin. If the object is thick, however, the magnification varies according to the position on the axis of the point being examined; the zone close to S will be the most magnified.

The more intense the electron current in the probe, the brighter will be the image. As we shall see later, however, the resolution improves as the diameter, d_c, of the spot is reduced and the operating potential V is lowered; it improves, therefore, as i is diminished (i varies with $d^{8/3}$ if we take spherical aberration into account as we have done earlier in this volume, and also with V). Once again, we are forced to have recourse to a compromise, with the result that the operating potentials ordinarily used are of the order of 5 to 20 kV.

22.4.2 Theoretical Resolution and Visibility

The resolution of the instrument is essentially related to the diameter of the source and to the diffraction, always supposing the optical properties of the lenses to be perfectly stable in time.

The image-forming mechanism is the same as in every conical projection system. The non-zero transverse dimensions of the source give rise to

regions of shadow AB and to penumbral zones AC and BD (see Fig. 278 S 9).
If we consider some feature of the object of width d, we find:

$$AB = \frac{ad - d_s(a - b)}{b},$$

$$AC = BD = \frac{d_s(a - b)}{b}.$$

We shall have

$$AB = 0,$$

for $d = d_0$, where

$$d_0 = \frac{a - b}{a} d_s.$$

We regard d_0 as the diameter of the smallest perceptible detail in the object.
Generally speaking, we require high magnification and hence $b \ll a$ so that

$$d_0 \simeq d_s.$$

The smallest perceptible detail is the same size as the source.

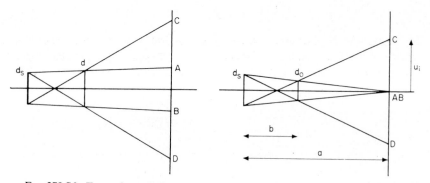

FIG. 278 S 9. Formation of the projection image. d_s: source; d: object; AB,
shadow; AC and BD: penumbrae.

If we now consider two neighbouring points, diameter d_0, the illumination
in the penumbra (u_1) varies linearly with the distance to the centre, and
we shall regard two points as just separated when the total illumination
between the two spots is about 75 per cent of the brightness at the centre
of each spot (Fig. 278 S 10) for the resolving power δ_i in image space;
we thus find:

$$\delta_i = 1 \cdot 25 \, u_1,$$

and transferring back into the plane of the object

$$\delta_0 = 1 \cdot 25 \, u_i;$$

when the linear magnification M is a/b, we obtain:

$$\delta_0 = 1{\cdot}25\,\frac{d_s(a-b)}{b\,M}\,,$$

and hence

$$\delta_0 \simeq 1{\cdot}25\,d_s.$$

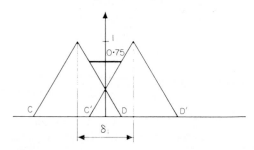

FIG. 278 S 10. Resolution of two neighbouring details.

This justifies the rule of thumb which is normally adopted:

$$\delta \simeq d_s.$$

In reality these penumbral regions are overlaid by Fresnel diffraction fringes. Only the first fringe, which lies at a distance x_1 from the edge of the geometrical shadow (Fig. 278 S 11), is troublesome, as it is very bright.

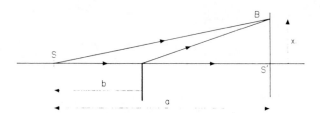

FIG. 278 S 11. Formation of the Fresnel fringes.

The fringes are the result of interference between the direct wave SB and the diffracted wave AB; the maximum of the first fringe lies at a distance x_1 from the line SA, where

$$x_1 = \left[\frac{a(b-a)\lambda}{b}\right]^{1/2}.$$

Transplanting back into the object plane, in the case $b \ll a$, the equivalent distance

$$x_0 = \frac{x_1}{M}$$

is given by
$$x_0 \simeq (b \lambda)^{1/2},$$

in which λ is the wavelength of the radiation employed.

The "perceptibility" will be limited by diffraction except when the condition
$$x_0 \leqq \delta_0 \simeq d_s$$

is satisfied, and this will impose an upper limit on b:
$$b < \frac{\delta_0^2}{\lambda}.$$

If for example we fix the resolution ($\delta_0 = 0 \cdot 1 \mu$, say) and $\lambda = 1$ Å, we find $b < 100 \mu$ which is still feasible; if $\lambda = 10$ Å, however, $b < 10 \mu$ and this is less easy to achieve.

To obtain the resolving power, we must determine the intensity distribution between two neighbouring points of the image. To a first approximation, we arrive once again at the general formula for the resolving power when the limit is provided by diffraction:
$$\delta = \frac{0 \cdot 625 \lambda}{x_0} = 0 \cdot 625 \frac{b \lambda}{d_0}$$

(where α_0 is the angle at which the object element is "seen" from a point on the source: $\alpha_0 = \frac{d_0}{b}$).

We can summarize these results as follows: the limits of perceptibility δ_P and of resolution δ_R will be reduced by diminishing b and λ and are given approximately by
$$\delta_P = (b \lambda)^{1/2},$$
and
$$\delta_R = 1 \cdot 25 (b \lambda)^{1/2} \simeq \delta_P,$$

when the limits are a consequence of diffraction.

If we choose a small enough value for b, we arrive at the simple relation
$$\delta_P \sim \delta_R \sim d_s.$$

Diffraction will thus certainly limit δ_P and δ_R. Since we cannot go on reducing b indefinitely with this method, we can scarcely hope to obtain a resolution better than 100 Å. Usually, however, it is the dimensions of the source which limit the resolution.

22.4.3 The Diameter of the Probe and the Thickness of the Target

As we saw in § 22.1, the maximum intensity I_s which can be attained with a probe of diameter d_s varies in the following way:
$$I_s \simeq \frac{d_s}{C_s^{2/3}} \cdot V \cdot i_0,$$

since the spherical aberration of L_2 imposes a limit on the aperture of the beam which can be usefully employed.

Here, we can use very short focal length lenses, as the potentials are low and the probe need not necessarily be outside the lens, unlike the situation in the micro-analyser where it is the reflected X-rays which we collect. If d_s is reduced, I falls very rapidly, so that when we try to obtain a very high resolution, the brightness is extremely feeble; we shall no longer be able to focus the image directly (focusing consists of bringing the probe onto the target), and it has proved necessary to invent special ways of focusing (see below).

If we are willing to accept a comparatively poor resolution, d_s can be increased, but this in turn cannot be continued indefinitely, as the rate of dissipation of heat in the target is limited. The incident power is almost wholly converted into heat, and increases as d_s^2 if the current density is constant; the rate at which a very thin film can conduct away this heat increases with d_s only, however, since the heat flux is wholly radial. The melting point T_M of the target is quickly reached, and in general, we avoid temperatures higher than $\frac{3}{4} T_M$. For preference, therefore, we use metals which are good conductors, or the refractory metals: tungsten and copper, silver and gold. The thickness of the target is then fixed by the way in which the incident electrons are scattered within it. In fact, the electrons are only completely stopped inside the metal after they have collided enough times to convert all their kinetic energy into thermal energy. It is usually agreed that the large majority of the incident electrons is scattered into a cone of semi-angle 30° to 35°; this leads us to give the X-ray source a diameter d_s which is equal to the thickness h of the target, when the spot diameter is negligible.

When a very fine source is required, we have to reduce both h and d_s. Langner (1956) has calculated the influence of spherical aberration, astigmatism and X-ray scattering in the target on the size of the final "source" d_s, when the geometrical spot diameter is negligible. With gold and a potential of 10 kV, the value of d_s and hence the resolution attainable in practice, is 400 Å when $h = 0 \cdot 1 \mu$.

Reducing h in turn implies a reduction of the X-ray intensity emitted by the target, however. If we require a large resolving power, and at the same time a reasonably bright image, we are obliged to try to increase the intensity of the probe; all things considered, this can only be done, for a given probe diameter, by increasing the specific emission of the cathode (since we are trying at the same time to reduce V, which lessens the brightness), and by correcting the spherical aberration very perfectly.

All these questions are gone into very thoroughly in the book by V. E. Cosslett and W. C. Nixon entitled *X-ray Microscopy* (Cambridge, 1961).

22.4.4 Instrumental Details

The optical system is identical to that of a microanalyser. A system with a single reducing lens could be used, but the two-lens arrangement is more flexible; the second lens is then always operated close to maximum convergence (and thus to the minimum value of C_s) and the diameter of the probe which is finally obtained is adjusted by varying the magnification of the first lens. In such conditions, the axial position of the probe varies very little, and only the slightest of alterations to the excitation of the second lens is necessary to bring it back onto the target.

Once again, magnetic lenses are preferable, as we can obtain at the same time very short focal lengths (with electrons accelerated through a few kilovolts) and values of C_s much smaller than those we should be able to reach with electrostatic lenses. Suppose, for example, $C_s(\text{mag}) = 1$ mm, and $C_s(\text{electr}) = 32$ mm; for the same diameter d_s, we have $\dfrac{I(\text{mag})}{I(\text{electr})} \simeq 10$ or for the same intensity, $d_s(\text{electr}) = 2\cdot4\, d_s(\text{mag})$.

The first instrument of this kind (von Ardenne, 1939) was electrostatic, although the second version—von Ardenne, 1956—is magnetic. An electrostatic instrument of a slightly different kind has been built by Marton *et al.* (1957); the electron source is a very fine point, which works by field emission

Fig. 278 S 12. The X-ray microscope of the Laboratoire d'Optique Électronique du C.N.R.S. at Toulouse (Dupouy and Verdier).

and gives a high brightness; a single reducing lens forms the image of this source on the target. The diameter of the final spot is about $5\,\mu$. With an electrostatic instrument with two lenses, Newberry and Summers (1954) have been able to attain a resolution of $1\,\mu$; the focal length of their lens L_2 was 1 mm and the spherical aberration constant C_s, 2·5 mm. The exposure time was of the order of ten or twenty minutes.

Most projection microscopes are, nevertheless, magnetic, and all are based on the structure of the instrument built by Cosslett and Nixon (1951, 1952 and 1953). The instruments which have been derived from this early form are described in articles by Cosslett and Pearson (1954), Cosslett, Nixon and Pearson (1956), Siegel and Knowlton (1957) and in Verdier's thesis (1960). Figures 278 S 12 and 278 S 13 respectively show the apparatus

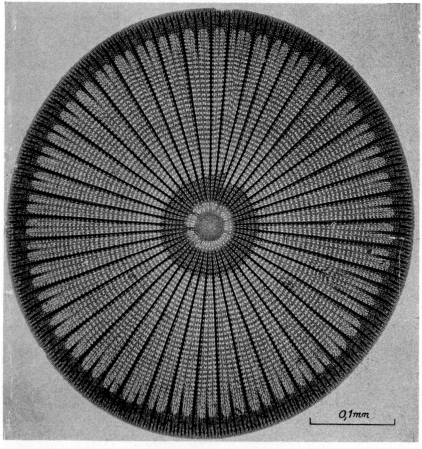

FIG. 278 S 13. Image of a diatom, direct magnification $G = 55$, exposure time 10 minutes, H.T. 7·7 kV.

built by Verdier, and a photograph of a diatom taken with it. With this type of instrument, a resolution of $0\cdot1\,\mu$ has been reached (Nixon, 1955), and this will certainly be improved upon.

Although the final resolution of this type of system is a hundred times poorer than that of a transmission electron microscope, slow variations of potential and mechanical vibrations must be most carefully eliminated, as the exposure times (for high resolution) are extremely long, frequently lasting several minutes.

When the diameter of the X-ray source is less than $0\cdot1\,\mu$ it is often impossible to focus the probe by direct observation of a test grid on the fluorescent screen, as we should normally do if the intensity were adequate. Various methods have been developed which employ the secondary electrons emitted by the target, and with the aid of one of these methods, the focusing can in fact be carried out. In Nixon's solution (1958), the electrons which are emitted by the thin target in the *forward* direction are used (the "forward-scattered electrons"); a large proportion of the primary electrons pass through the target and form the projection image of a very fine grid on the fluorescent screen just as in shadow microscopy. This image is bright enough for the probe to be focused. The X-ray image is formed with the aid of the X-rays which emerge from the source at an angle of about $60°$ with the axis of the instrument [see Fig. 278 S 14(i)]. The resolution which is obtained is of the order of $0\cdot1\,\mu$. Here, however, both target and object are inside the vacuum—this limits the applications of the instrument, but does, on the other hand, allow us to use very soft X-rays.

Ong Sing Poen and Le Poole (1958) have suggested another method in which the electrons which are scattered back towards the face of the target on which the beam is incident, are used. The other face of the target can, therefore, be in air. These back-scattered electrons represent about 1 per cent of the whole beam; they travel back through the lenses in the opposite direction, and by tilting the optic axis very slightly so that the image lies in the same plane as the crossover but a little to one side of it, we can examine the image on a fluorescent screen [Fig. 278 S 14(ii)]. Here again, a resolution of $0\cdot1\,\mu$ has been reached even though the X-ray image is not visible at the final screen.

In conclusion, we should mention that the difficulty can of course be avoided by replacing the fluorescent screen by one of the image intensifiers which are described in Chapter 13.

To end this section, we shall say a few words about the projection technique which has recently been developed by Pattee (1958) and which he calls "Microfluoroscopy". The preparation is spread directly on the fluorescent screen and is illuminated by the beam of X-rays which emerges from a source about $100\,\mu$ away. The screen is about one micron thick, and can thus be examined with the aid of an optical microscope with a high magnification ($\times 1000$) and a very wide objective (and hence a very shallow depth

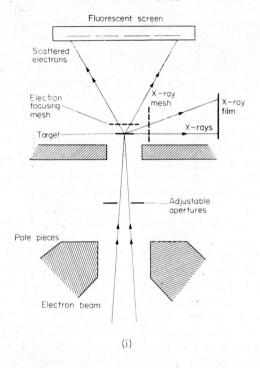

(i)

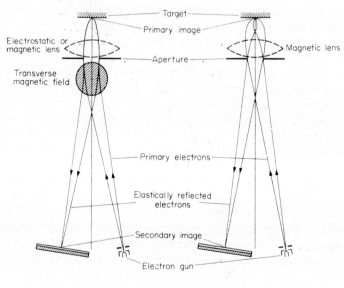

(j)

FIG. 278 S 14. Focusing aids used by (i) Nixon, (j) Ong and Le Poole.

of field). The grain-size of the screen must be smaller than the desired resolution, and it is for that reason that evaporated screens free of structure are employed (Feldman and O'Hara, 1957). A resolution of about $0\cdot25\,\mu$ can be obtained in this way, namely, that of the optical microscope, and we can use a source with a greater diameter, d_s, and hence a high brightness.

22.5 MACHINING BY ELECTRON BOMBARDMENT

22.5.1 The Electron Beam as a Heat Source

When an electron beam of intensity I and potential V falls on a target, the energy $V\,I$ which had been carried by the beam is transferred to the target almost entirely in the form of heat. Even for comparatively high potentials, of 100 to 150 kV, all this energy is distributed into an extremely thin film of matter, a few microns thick. This is far thinner than we could conveniently obtain with high frequency heating, and apart from this, we can localize the zone being heated far more precisely. It is natural therefore, that with the development of guns capable of giving a very high intensity (see Chapter 12, § 12.2.3) and capable of concentrating a power of 10 to 50 kW into a few square millimetres, methods should have been developed with which this energy can be used for melting or for evaporating metals *in vacuo*. (There is the added advantage of the chemical cleanliness of the vacuum.)

Another advantage of heat sources obtained in this way is that it is possible, even when the average power is relatively low, to produce enormous energy densities, which may exceed those which even the most excellent electric arc could provide by a factor of a thousand or ten thousand; this is thanks to the high quality electron optical systems which can now be constructed. If we work in pulsed operation (which enables the current density emitted by the cathode to be increased, and hence the brightness of the source also), it is possible to obtain an energy density of 10^8 or $10^9\,\mathrm{W\,cm^{-2}}$ in a probe a few tens of microns in diameter. This produces an almost instantaneous evaporation of the zone bombarded.

22.5.2 The Heating of a Target Bombarded with Electrons

Several authors have recently calculated in an approximate fashion the temperature distribution and the energy balance in the zone under bombardment (Steigerwald, 1958; Bas, 1960), not to mention earlier work which had been performed in connexion with the heating of the anti-cathodes in X-ray tubes.

We consider first of all the steady state situation. A beam of diameter $2\,r_0$ and uniform density falls on a thick target (Fig. 278 S 15) for which

$h \gg r_0$. We assume that the beam arrives at the bottom of a hole which it has itself made. The incident energy is transformed into heat within a thickness e; this zone is warmed up, melts, and is partially evaporated (the vapour pressure increases exponentially with temperature). A small proportion of the heat is radiated away, but the rest is conducted away into the target. If, in the steady state, we suppose that the lower face of

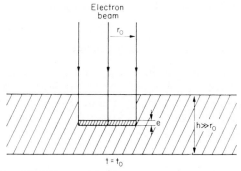

FIG. 278 S15. Punching by electron bombardment.

the target remains at a temperature t_0, an analysis of the energy balance reveals that the conduction outweighs the latent heat, and a *considerable zone must be melted*. The temperature distribution is represented by the curve in Fig. 278 S16a (Bas, 1960), calculated for the case of an alumina

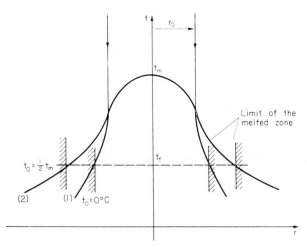

FIG. 278 S16a. The form of the temperature distribution in the neighbourhood of the bombarded zone. t_m: maximum temperature; t_f: melting point; t_0: temperature at the lower edge of the piece being machined (Bas, 1960).

target; the temperature at the centre ($t_m = 2260°C$) is higher than the melting point (2050°C). The radius of the melted zone is of the order of $1.2\, r_0$ if $t_0 = 0°C$, but is of the order of $2\, r_0$ if $t_0 = (1/2)\, t_m$. The metal is evaporated at about $v = 0.1$ mm sec^{-1} at the centre, but at $r = r_0$, this rate of evaporation, v, is practically zero.

The molecules which have been evaporated leave the target with a certain momentum mv, and there is in consequence a reaction which manifests itself as a pressure on the melted zone which hollows out the centre of the latter, spreading the molten matter out towards the cold surrounding region.

We should mention too that there is a very steep radial temperature gradient at the edge of the beam, so that for targets which are poor conductors, there is a danger of cracking the target.

The evaporatation rate is higher, for a given incident energy, when the thermal conductivity λ is lower (and hence t_m higher), and when the metal evaporates more easily. It is possible to calculate the temperature t_1 for which $v_m = 0.1$ mm/sec; this temperature is in most cases higher than the melting point t_f but the difference ($t_1 - t_f$) which varies between very wide limits according to the nature of the target, may even take negative values (for chromium, $t_1 - t_f = -225°C$). **The tendency to evaporate which a metal displays can thus be characterized by the factor $\beta = \lambda\, t_1$, which is expressed in W cm^{-1}**—the smaller β is the more easily can the metal be machined. For everyday substances, Bas (1960) gives values of β between 190 for Al_2O_3 and 7700 for tungsten. The metals which are good thermal conductors—silver, copper, and aluminium—follow immediately behind tungsten. With these figures, an energy density of $\sigma = 7.7 \times 10^6$ W cm^{-2} would be necessary for $\lambda\, t_1 = 7700$ (tungsten) and $r_0 = 10^{-3}$ cm. Experiments on alumina show that the energy density which is necessary is even greater, but nevertheless can be reached.

Provided we have beams of high enough energy density at our disposal, therefore, we can work the most refractory substances in steady state conditions, whether it be a question of boring, welding or melting. The zone which has been worked is, however, surrounded by a region which has melted and resolidified, and by a zone which has been disturbed to a greater or lesser extent from the structural point of view by the sharp temperature increase.

To bore or drill a metal, with machine tolerances of the order of a micron, we avoid this difficulty by using an intense beam in very short bursts (Steigerwald, 1958, 1961). If a beam of 10^9 W cm^{-2} and 100 keV falls on a tungsten surface, the temperature of the layer which stops the electrons is raised to 3650°K in 10^{-8} sec, and a point one micron from the edge of the beam will reach its melting point after 7×10^{-6} sec. If, therefore we apply the energy in short bursts, about a microsecond long and far enough apart, the zone which lies $1\,\mu$ away from the rim of the vaporized region will not reach its melting point. This is the principle underlying the "electron drill", with which extremely narrow machining tolerances can be satisfied by moving the probe about over the target which is to be cut or punched (Schleich, 1958; Opitz, 1960; see Fig. 278 S 16b).

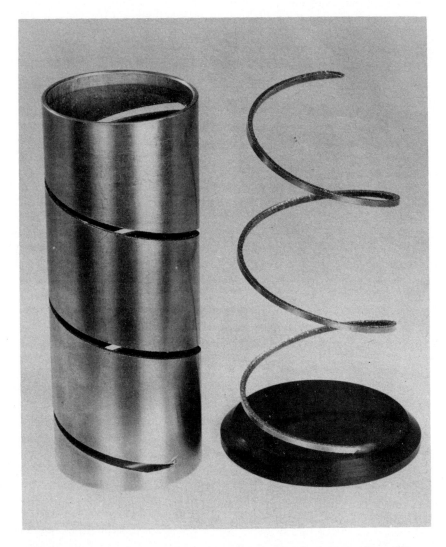

FIG. 278 S 16b. Cutting out with an electron beam. The pipe is made of a special steel; the wall is 2 mm thick, and the cutting speed is 30 cm min⁻¹ (with a Carl Zeiss drill) (Photo: Schleich).

22.5.3 Various Applications of Intense Probes

We shall describe briefly these applications in order of increasing complexity of the optical system.

(i) *Melting and welding*

Processes of this kind were developed in France by Stohr in 1957, to solve the welding problems created by nuclear reactor technology. Here, every effort is made to achieve very intense beams well focused into a crossover a few tenths of a millimetre or a few millimetres across. In general, the "Pierce" optical system is adequate; the gun has a wide tungsten cathode, the back of which is heated by electron bombardment, and is provided with an accelerating electrode and an electrode for concentration. The whole arrangement is planned to give the smallest possible crossover a few centimetres from the anode, and at high intensities we are forced to work with potentials of several tens of kilovolts. The region in which the gun is situated is separated from the working region, so far as the vacuum is concerned, by the opening in the anode, and is provided with a separate pump to ensure that the vacuum is good despite the ejection of gas from the heated metals, and the evaporatation of the latter.

In this way, probes 0·8 mm in diameter can be obtained at 60 kV and a heating power of 5 kW. If we are willing to sacrifice to some extent the fineness of the probe, we can increase the power to 40 or even 100 kW, and thus melt refractory metals in considerable quantities in a crucible to form alloys.

Probes of a few tenths of a millimetre are employed for welding together the edges of metal plates which are very easily oxidized; the performances which can nowadays be obtained with electron welders are better than those of argon arc torches, as regards speed, fineness of the joint and above all, penetration. The melted zone can easily reach two to five centimetres, so that comparatively thick plates can be welded together in this way. A good weld is ensured by scanning the beam laterally, with an amplitude of one or two millimetres, over the line along which the two edges which are to be soldered together are in contact; the plates are meanwhile moved parallel to this line. We can, for example, weld together the edges of two aluminium plates 25 to 50 mm thick in a single operation, at a speed of 30 to 50 cm/min (Stohr, 1961; Fig. 278 S 17). When the plates are made of materials which are good thermal conductors, a zone a few millimetres wide is found on either side of the weld which has been thermally disturbed. When, on the contrary, the substances are poor conductors of heat, stainless steel or the nickel alloys for example, the "penetration-to-width" ratio of the weld can reach 25; with 9·5 mA and 130 kV stainless steel plates 12 mm thick can be welded together at a speed of 35 cm/min. In these substances, a crystallographic analysis shows that the thermally disturbed

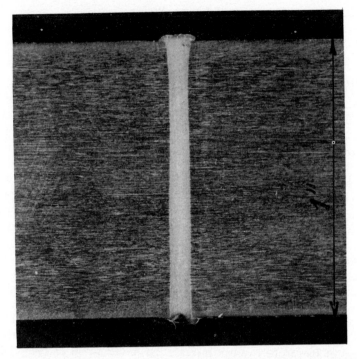

FIG. 278 S17. Welding between two aluminium plates one inch thick; 29 kV electrons, $I = 260$ mA, welding speed 20 inches/min (Photo: Sciacky–C.E.A.).

zone in the neighbourhood of the weld is wholly absent, and that the weld itself is perfectly homogeneous and is composed of micro-crystals.

For refractory substances, such as tungsten, the depth of the melted zone is adequate for plates one or two millimetres thick to be welded, provided the potential is very high: 130 to 150 kV (Hoffmann, 1961).

If the instantaneous power is increased still further, we can cut metal plates just as with an ordinary blow-lamp, but we can obtain far narrower cuts than with the latter.

In this type of application, the radiation is particularly intense, and the working region has to be screened with a layer of lead thick enough to intercept all harmful radiation.

Numerous articles on the results which have been obtained with electron welders are to be found in the *Proceedings* of the Third, Fourth and Fifth Symposia on Electron Beam Processes (Boston, 1961, 1962 and 1963). Figure 278 S18 shows one of the high power welding installations, designed by J. A. Stohr, which is at the Centre d'Études Nucléaires de Saclay (France).

FIG. 278 S 18. Welding machine at the Commissariat à l'Énergie Atomique (C.E.A., Saclay, France).

(ii) *Boring and drilling*

As we have already seen, it is possible even with steady state conditions, to bore relatively fine holes by evaporating away the centre of the melted zone, but if we require the very tightest tolerances, we are forced to use pulsed operation. The power supplied to the target is less than in the preceding case, but the probe is very tiny and the energy density may be considerable.

The first device of this type was described by Schleich (1958). A diagram is given in Fig. 278 S 19. The various parts are as follows:

a "Fernfokus" gun;

a magnetic system for aligning the beam;

a stigmator;

a highly convergent reducing lens;

a deflexion system, which can be supplied with a variable potential, for shifting the probe;

a target-holder, equipped with a high-precision displacement system;

an optical microscope for examining the target.

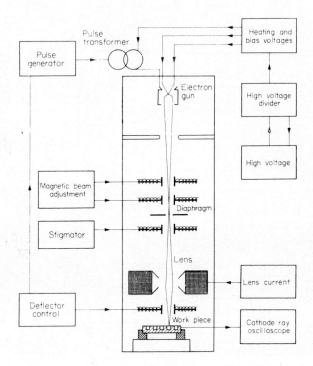

FIG. 278 S 19. Diagram of an electron drill (by courtesy of Carl Zeiss, Oberkochen).

The natural setting of the gun corresponds to a bias beyond cut-off; when the system is in action, the gun is regularly triggered by applying a series of positive pulses to the grid—both the width and the frequency of these pulses can be altered (over ranges of 2 to 70 μsec and 50 to 10^4 c/s respectively). The current into the probe can be varied between zero and eight milliampères, for accelerating potentials between 25 and 150 kV, and probe diameters from 10 to 40 μ.

With this machine, it is possible to bore round holes from 20 to 200 μ in diameter, and slits one millimetre long and 40 μ wide in plates 0·4 mm thick. A "width-to-depth" ratio of 1/50 is attainable; the tolerance is a few microns.

Figure 278 S20 shows several cruciform holes which have been bored through a stainless steel plate, which can serve as a plastic-wire drawing plate or "spinneret". Between each pulse of electrons, the probe is moved to a different point of the shape which is to be cut out, to reduce to the

FIG. 278 S20. Spinneret with various profiles, cut out of a special steel plate 0·5 mm thick (Photo: Schleich).

uttermost any heating of the piece being machined. It is possible to programme beforehand the voltages which are to be applied to the probe deflector, in such a way as to obtain most complicated cross-sections. Figure 278 S21 gives a magnified view of a narrow slit. Such a device as this has a vast range of applications in cutting out the elements of miniature electronic circuits (resistive films deposited on a ceramic material and semi-conductors; see, for example, Opitz, 1960, Steigerwald, 1961).

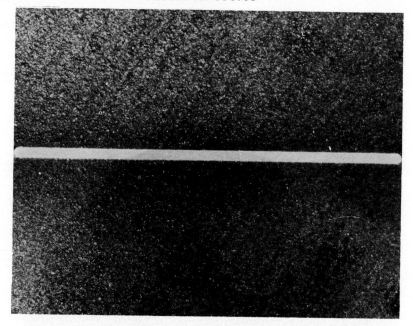

Fig. 278 S 21. Drilling a graphite plate. Slit width: 125 μ, thickness: 0·5 mm.

(iii) *The use of micro-probes for recording*

A micro-probe a few hundred Ångströms across can equally well be used as a heat source with which to form an impression on a structure-free collodion film. A collodion film 500 or 1000 Å thick is decomposed into carbon by the probe, and is at the same time thinned down by the heating. We can therefore machine the plate, and subsequently examine it with a high magnification electron microscope; signs (dots, dashes, letters) which are "written" on the film can be rendered more easily visible by means of the same types of shadowing technique as are used in microscopy (Möllenstedt and Speidel, 1960). The practical arrangement is as follows: a diaphragm a few microns across is placed on the specimen holder of a magnetic microscope, and in this way we can obtain a probe 100 Å in diameter in the focal plane of the objective. The collodion target is placed in this plane, and is sufficiently transparent to electrons for an image of the diaphragm to be visible on the fluorescent screen of the microscope. The probe is moved about by shifting the diaphragm, and can be followed on the final screen. If we compel the final image to follow an arbitrarily chosen figure drawn on the screen, the probe will trace out the same pattern on the collodion film on an extremely reduced scale.

This technique can be used to rule extremely fine lines or groups of parallel lines, which can subsequently serve as micrometers or as test objects for determining the resolving power in electron microscopy (the lines may be 140 Å thick and 1000 Å apart, for example).

Finally, we should mention that this technique may be used to construct very small printed circuits. An insulator is coated with a thin layer of conducting material, the appropriate parts of which are then cut away by the electron beam (Fig. 278 S 22).

Fig. 278 S 22. Cutting away the surface of a resistance with an electron beam. A layer of CrNi has been evaporated onto a ceramic support. The line width is 10 μ. (Drill by Carl Zeiss).

CHAPTER 23

THE ELECTRON (ION) PRISM

23.1 THE HISTORY OF THE MASS SPECTROSCOPE

Goldstein's work on electric discharge phenomena led him in 1886 to the discovery of *electrically positive rays* ("canal" rays). These rays could be deflected with an electric or magnetic field. In 1910, J. J. Thomson working at Cambridge made use of this property to break them down into their several components (by the parabola method). In particular, he found that two components were present in neon. From this observation stemmed the discovery of isotopes, and its importance was clearly displayed by F. Aston who, in 1919, built the first spectrograph (Fig. 279).

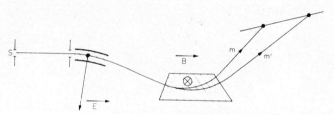

FIG. 279. Aston's spectrograph.

Aston's instrument "focuses energies", a notion which we shall understand more clearly later on; at about the same time, in America, Dempster (1918) was building an instrument which "focused directions", and soon after, he discovered the isotopes of magnesium. During the next few years, Aston established the existence of a large number of isotopes, and improved the accuracy of his spectrograph (1927). He pointed out the differences between the atomic masses and the anticipated integers; the differences, which were also studied by Costa (1925) in France, are now recognized to be a highly significant feature of the energy in nuclear physics.

In 1929, Dempster and Bartky suggested the method which is known as double focusing, by which the quality of these instruments is considerably improved; it was at this point that the high quality spectrographs which are associated with the names of Bainbridge, Jordan and Mattauch, to name but a few, came into being. Meanwhile, instruments with electromechanical methods of recording were being developed—these are the devices known as "spectrometers"—with which quantitative analyses could be made, and which came to be built both in the laboratory and industrially under the stimulus of Nier, Bleakney and Hipple; spectrometers are tending to play a most important role both in ordinary chemistry and in nuclear chemistry.

23.2 THE PRINCIPLE OF MASS SPECTROSCOPY; THE SPECTROGRAPH AND THE SPECTROMETER

23.2.1 The Principle

Mass spectroscopy is a method of analysis based on the fact that ions of the same kinetic energy but different in mass can be separated according to their different masses by a static magnetic field. The features fundamental to the separation process, which of course takes place in a very high vacuum, were all present in essence in Aston's first instrument (Fig. 279). The substance to be studied is first ionized, partially transformed, that is, into ions, in an ion source. This source must be designed in such a way that a narrow beam of positive ions is emitted through the exit slit, S, all with nearly the same energy, or more concisely, in such a way that the ions are *homogeneous*. This energy is, as usual, measured by the accelerating potential V. The beam then passes through an analysing field, the vital part of which is a uniform magnetic field of induction B which is created in the gap of an electromagnet. When it leaves this field, the beam will have been separated into several components which correspond to the various masses, m, m', ... The intensity of each component is recorded by a receiver which may be either a photographic plate or some electrical device linked to a pen-recorder; in this way, we obtain the "spectrum" of the substance, which has then only to be interpreted.

23.2.2 The Two Types of Instrument: Spectrographs and Spectrometers

The analogy with the optical spectroscope which analyses a beam of light is obvious; in the present case, the analysing field replaces the glass prism—its action on the trajectories lies in the domain of the optics of charged particles, although here the particles are not electrons but ions. The ions, each of mass m, should produce as fine as line as possible to ensure that we obtain the optimum separation. The ions arrive, however, in a bundle of rays with different inclinations. Let us suppose to begin with that they are all of the same energy, V; ion optics ensures their angular focusing, that is, ensures that they converge to the line image of S. It is in the neighbourhood of this image, therefore, that we should place the receiver (Fig. 280).

In reality, however, these ions of mass m have various energies, ranging between V and $V + v$ in value; they will not, therefore, all converge to exactly the same line. The image S displays chromatic aberration in the electron optical sense (§ 7.4). This is a defect which can if necessary be eliminated

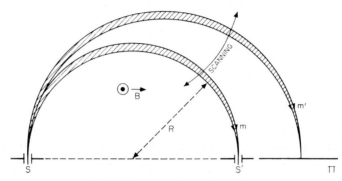

FIG. 280. Dempster's spectrograph.

at the expense of a more complicated optical system; when this is done, however, ions of mass m but of slightly different energies and inclinations all converge to the same line—we say that there is both energy and angular focusing or simply double focusing. This distinction will be an important one in defining the two basic types of instrument which we are about to describe.

(i) *Spectrographs*

The receiver consists of a photographic plate Π, while the field remains fixed; this method is well-adapted for measuring the positions of the lines and hence for determining the values of the masses m, m', ... For extreme accuracy, the lines must be very fine, which implies double focusing; the analysing field therefore becomes more complex as a supplementary electric field E must be added (although Aston's first instrument did contain an electric field, there was in fact only single focusing, which was velocity focusing).

(ii) *Spectrometers*

Here, the "collector" consists of an electrode placed behind a slit S' (Fig. 280); on this slit, we successively bring the ions of each of the various masses present by varying either B or V (magnetic or electric scanning), and we measure the corresponding currents, that is the line intensities.

It is easy to make quantitative analyses with this type of instrument; only angular focusing is required here, and this can be achieved with a magnetic field.

A spectrometer is a simpler instrument than a spectrograph, therefore, and the wide variety and the interest of their chemical applications have led to the construction of industrial models. Spectrometers also exist for separating out isotopes—in essence, the structure of these instruments is the same, but they are considerably more bulky. We quote a few normal orders of magnitude (they will be higher for the separation of isotopes).

Length of the trajectories	50 cm
Size of the ion source and the collector	3 cm
Width of the slits S, S'	< 1 mm
Accelerating potential V	a few kilovolts
Magnetic induction B	a few kilogauss
Ion current carried	$i < 10^{-9}$ A

23.2.3 Preliminary Survey

The analysing field is the characteristic feature of a spectroscope; examining this field, we return into the domain of electron optics, the essential object of this book. In consequence, it is with the study of this field that we shall begin. After summarizing the simplest properties of the field, we shall consider the more general problem of the optics of systems with a symmetry plane, and more precisely, of the sectors with which we are effectively concerned in mass spectroscopy. We shall study their angular focusing action (on ions m of the same energy) and afterwards, the possibility of double focusing. We shall describe some of the more representative instruments, and examine their resolving power, which is the fundamental characteristic of any instrument.

23.3 ELEMENTARY EXAMPLES OF ANALYSING
FIELDS: UNIFORM FIELDS

23.3.1 Equations of Motion

We consider an ion of mass m and charge e which we shall suppose positive, moving with velocity $\boldsymbol{u}$; the equation of motion in an electric field $\boldsymbol{E}$ and a magnetic field of induction $\boldsymbol{B}$ has the form

$$m\frac{\mathrm{d}\boldsymbol{u}}{\mathrm{d}t} = e(\boldsymbol{E} + \boldsymbol{u} \times \boldsymbol{B}). \tag{23.1}$$

In the left-hand side of this equation is implicit the assumption that the motion can be described by classical mechanics, which is easily justifiable since the ion velocity u is always considerably smaller than the velocity of light in mass spectroscopy. We obtain the magnitude of u directly from the conservation of energy relation which is deduced from (23.1), and takes the simple form

$$\tfrac{1}{2}mu^2 = -e\Phi, \tag{23.2}$$

where Φ is the potential from which E is derived, measured relative to the potential of the ion source where the ions are stationary. The minus sign expresses the fact that positive ions can only move through regions of negative potential. The motion of a charged particle is, we should remember, also governed by a principle of least action; the effect of a field is expressed by attributing a certain fictitious optical index n to each point of space. n is proportional to $|\overline{\Phi}|$, and hence independent of m in the case of a purely electrostatic field; in the presence of a magnetic field, on the other hand, n does depend upon m. Given a beam of monoenergetic ions of various masses, therefore, a magnetic field is required to split it according to the different masses. More precisely, the parameter which is characteristic of each ion is m/e (from equation (23.1) above); an ion of mass m doubly charged behaves like an ion of mass $\tfrac{1}{2}m$ singly charged. For this reason, we shall deal henceforward with singly charged ions.

23.3.2 The Uniform Magnetic Field, the Dempster 180° Spectrometer

To begin with, we recapitulate the very simple and well-known case of particles moving through a uniform magnetic induction $B\,(E = 0)$; in what follows, we shall adopt the convention that B acts in the vertical direction, and we shall only be interested in motion in the perpendicular or horizontal directions. This motion has an initial horizontal velocity and can easily be made precise by projecting equation (23.1) on the normal; the trajectories are circles, described at a constant angular velocity:

$$\omega_0 = \frac{e}{m} \cdot B. \tag{23.3}$$

The radius R depends upon u and hence upon the accelerating potential, which is constant in this case, $|\Phi| = V$ so that

$$R = \sqrt{\frac{2m}{e}} \cdot \frac{\sqrt{V}}{B}, \tag{23.4}$$

or if we characterize the (singly charged) ion by its molecular mass M and measure in grams, volts, gauss and centimetres, we find:

$$R = 143 \cdot 6 \sqrt{M} \cdot \frac{\sqrt{V}}{B}. \tag{23.5}$$

If, for example,

$$V = 1600 \text{ V},$$
$$B = 1600 \text{ gauss},$$
$$M = 16,$$

we find that $R = 14\cdot36$ cm.

We idealize the actual situation in a spectroscope by assuming that the field B is bounded laterally by a vertical plane Π; a horizontal ion beam of small aperture enters normally through the slit S which lies in the plane Π (Fig. 280); the ions describe semi-circles inside the field and the masses separate out into well-defined circles of different radii R. A simple geometrical examination shows that in addition, the trajectories which correspond to particles of the same mass converge to virtually the same vertical segment at the point where they leave the field. We say that there is "angular focusing", and it is obviously sensible to place the receiver at the point where this focusing occurs, namely in the plane Π, whether the receiver be a photographic plate or a collector S'. This was the principle of the first spectrometer (Dempster, 1918); the magnitude of the mass M incident at any moment can then be obtained from the relation (23.5). We shall simply point out that M is proportional to B^2 when the scanning is magnetic and to V^{-1} when it is electric. If we are recording photographically, equation (23.5) gives the value of M as a function of the abscissa $X(= 2R)$ of the line: M is proportional to X^2.

23.3.3 B and E Uniform Crossed Fields

We consider the horizontal motion in a region in which a uniform vertical field B and a uniform horizontal field E act.

(i) A particular solution—the Wien velocity filter

One particular solution of the problem consists of uniform rectilinear motion at a velocity u_0 defined by

$$E + u_0 \times B = 0.$$

The vector u_0 forms an orthonormal triad with E and B (Fig. 281a); its magnitude is given by

$$u_0 = \frac{E}{B}. \tag{23.6}$$

If we consider a pencil of ions which emerges from a slit in the initial direction u_0, the only ions which continue in a straight line and hence can be extracted with the aid of slits such as S_2 are those with a velocity v_0, irrespective of their mass. This arrangement is known as Wien's velocity filter (Fig. 281a). If we place such a filter at the exit S_1 of a very heterogeneous ion source, we obtain at the slit S_2 a beam which is homogeneous in velocity for every mass, and hence suitable for spectroscopic analysis (see, for example, Bainbridge, 1932). In reality, of course, this beam is not limited to a single ray $S_1 S_2$, but has a small finite aperture.

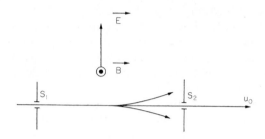

FIG. 281 a. Wien's velocity filter.

If, on the other hand, we place a Wien filter at the exit of a source of ions which are homogeneous in energy, the ions will be separated according to their velocities, and hence to their masses; this we speak of as *velocity spectroscopy* (Oliphant, Shire and Crowther, 1934).

(ii) *Arbitrary trajectories; Bleakney and Hipple's cycloid spectrometer*

We return to equation (23.1), and by writing out the components in the three directions we obtain three linear second order differential equations, the right-hand sides of which result from the existence of an electric field E.

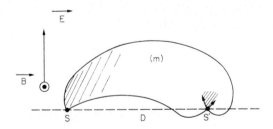

FIG. 281 b. Focusing the cycloidal trajectories.

From the standard theory of differential equations, we know that every possible trajectory is described by a linear combination of two solutions: the general motion in the case $E = 0$ together with a particular value of the trajectory when $E \neq 0$.

Alternatively, we can say the general solutions represent the super-position of uniform circular motion at an angular velocity ω_0 (equation (23.3)) onto a rectilinear uniform motion at a velocity v_0 (equation (23.6)). The trajectories are therefore cycloidal in form with a spatial periodicity

$$D = \frac{2\pi\, v_0}{\omega_0} = 2\pi \frac{m}{e} \frac{E}{B^2}\,.$$

If we consider a beam which emerges from S (Fig. 281 b), all the ions of a given mass will converge into this point a distance D away, irrespective of their initial directions and their energy; we have, therefore, focusing which is theoretically perfect, but it is not always easy to take practical advantage of this fact. Bleakney and Hipple (1938) built a spectrometer on this principle, and more recently the Consolidated Company have made such an instrument available commercially.

23.4 ION OPTICS IN FIELDS WITH A SYMMETRY PLANE

The fields which we have just examined are important special cases of a field-structure whose optics we are about to study. The essential characteristic is a horizontal symmetry plane (the plane of the figures). By this, we mean that at any two points symmetrically placed with respect to this plane, the electrical potential has the same value, and the scalar magnetic potential takes equal and opposite values (with symmetrically placed pole-pieces).

As a result of this symmetry, there are plane trajectories which lie in the plane of the figure, as the electric and magnetic fields in this plane are horizontal and vertical respectively. The complete ion beam will travel in the neighbourhood of the symmetry plane; it begins as a superposition of a number of components of various energies and masses, the geometry of which we are about to study.

23.4.1 The Beam (m, V); the Gaussian Approximation to the Trajectory Equation

A beam which is composed of ions of a given mass m and initial energy V remains a priori in the form of a compact bundle around a "mean" trajectory T_0, which lies in the symmetry plane and which we shall designate the optic axis (Fig. 282). This clustering together has been studied in a masterly fashion by Cotte (1938) in very general conditions; we shall sketch out the method which he used, making the simplifications which the symmetry of the present system allows.

We take a curvilinear coordinate system, in which an arbitrary point N has the coordinates x, y, z, defined in the following way: we erect the plane which passes through N and is normal to the mean trajectory T_0, intersecting the latter in n; z is then the curvilinear abscissa of n, measured in

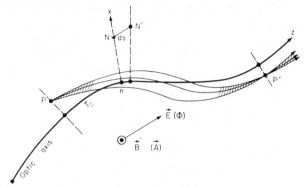

FIG. 282. An optical system with a symmetry plane: a beam of particles of fixed mass m and energy V.

the sense of the motion along T_0. T_0 thus forms a curvilinear axis, Oz. x and y are respectively the horizontal and vertical Cartesian coordinates of N from n in the normal plane, and the relative orientations are such that the triad $x\,y\,z$ is right-handed. To be able to write down the equations of motion, we must know the metric in this coordinate system, the expression for the distance ds, that is, between two infinitely close points N, N'. If $R(z)$ is the radius of curvature of T_0 at z (measured along Ox), and hence negative in the case illustrated in Fig. 282), we obtain the simple expression

$$ds^2 = dx^2 + dy^2 + \left(1 - \frac{x}{R}\right)^2 dz^2.$$

The electric field is defined by the scalar potential $\Phi(x, y, z)$, the origin of which is chosen in such a way that $|\Phi|$ represents the energy of the ions. The magnetic field is defined by the induction $\boldsymbol{B}(x, y, z)$, the vector with components (B_x, B_y, B_z). In the calculations, however, the vector potential $A(x, y, z)$ of the field will be required at a certain point; we represent the covariant components of A by A_x, A_y, A_z, and $\boldsymbol{B}$ is calculated from A by the relation

$$\boldsymbol{B} = \text{Curl}\,A,$$

or in this particular case†

$$B_x = \frac{1}{1 - \frac{x}{R}} \left(\frac{\partial A_z}{\partial y} - \frac{\partial A_y}{\partial z} \right),$$

$$B_y = \frac{1}{1 - \frac{x}{R}} \left(\frac{\partial A_x}{\partial z} - \frac{\partial A_z}{\partial x} \right), \qquad (23.7)$$

$$B_z = \frac{\partial A_y}{\partial x} - \frac{\partial A_x}{\partial y}.$$

To find the trajectories, we start from the principle of least action (§ 6.2.3) which takes the following form when we take into account the fact that it is with positively charged particles that we are concerned (and Φ is negative):

$$\delta \int L \, dz = 0, \qquad (23.8)$$

$$L = \sqrt{|\Phi|} \frac{ds}{dy} + \sqrt{\frac{e}{2m}} A \cdot \frac{ds}{dy}. \qquad (23.9)$$

Writing $\frac{dx}{dy} = x', \ldots$, we have

$$L = L_1 + L_2,$$

$$L_1 = \sqrt{|\Phi|} \sqrt{x'^2 + y'^2 + \left(1 - \frac{x}{R}\right)^2},$$

$$L_2 = \sqrt{\frac{e}{2m}} (A_x x' + A_y y' + A_z).$$

From equation (23.8), the Euler–Lagrange equations can be deduced:

$$\frac{d}{dz} \frac{\partial L}{\partial x'} - \frac{\partial L}{\partial x} = 0,$$

$$\frac{d}{dz} \frac{\partial L}{\partial y'} - \frac{\partial L}{\partial y} = 0, \qquad (23.10)$$

which are differential equations in which time does not appear and from which the required trajectories can be obtained.

Since we are only interested in trajectories close to the axis T_0, we shall assume that the quantities x, y and their derivatives x', y' are sufficiently small for powers higher than unity to be negligible. As a result, the fields can be expressed as expansions of second degree which simplify, by reason

† Cf. Angot: *Compléments de Mathématiques* (2nd ed., page 130).

of the symmetry, to the following forms:

$$\Phi = \varphi + \varphi_1 x + \varphi_2 x^2 + \psi_2 y^2 + \cdots < 0,$$
$$B_x = b_1 y,$$
$$B_y = b + b_1 x + b_2 y^2,$$
$$B_z = b_3 y. \tag{23.11}$$

The coefficients $\varphi, \varphi_1, \ldots, b_1, \ldots$ are functions of z alone; the fact that b_1 appears both in B_x and B_y is a consequence of

$$\text{Curl } \boldsymbol{B} = 0.$$

It is convenient at this point to write down the expressions for the various derivatives which appear in the equations (23.10) in terms of the quantities (23.11), at least to the first order

$$\frac{\partial L_1}{\partial x'} = \frac{\sqrt{|\Phi|}\, x'}{\sqrt{1 + \cdots}} = \sqrt{|\varphi|}\, x' + \cdots,$$

$$\frac{d}{dz}\frac{\partial L_1}{\partial x'} = \left(x'\frac{\partial}{\partial x} + y'\frac{\partial}{\partial y} + \frac{\partial}{\partial z}\right)\frac{\partial L_1}{\partial x'} = \frac{\partial}{\partial z}\frac{\partial L_1}{\partial x'} + \cdots$$

$$= \sqrt{|\varphi|}\, x'' - \frac{\varphi'}{2\sqrt{|\varphi|}}\, x',$$

$$\frac{\partial L_1}{\partial x} = \frac{\partial}{\partial x}\left[\sqrt{|\Phi|}\left(1 - \frac{x}{R} + \cdots\right)\right] = -\frac{\sqrt{|\Phi|}}{R} - \left(1 - \frac{x}{R}\right)\frac{\dfrac{\partial \Phi}{\partial x}}{2\sqrt{|\Phi|}}$$

$$= \frac{2\dfrac{\varphi}{R} - \varphi_1 + \left(3\dfrac{\varphi_1}{R} - 2\varphi_2\right)x}{2\sqrt{|\varphi|} - \dfrac{\varphi_1}{\sqrt{|\varphi|}}\, x},$$

$$\frac{d}{\partial z}\frac{\partial L_2}{\partial x'} - \frac{\partial L_2}{\partial x} = \frac{e}{2m}\left(x'\frac{\partial A_x}{\partial x} + y'\frac{\partial A_x}{\partial y} + \frac{\partial A_x}{\partial z}\right.$$

$$\left. - x'\frac{\partial A_x}{\partial x} - y'\frac{\partial A_y}{\partial x} - \frac{\partial A_z}{\partial x}\right),$$

$$= \frac{e}{2m}\left[-y'B_z + \left(1 - \frac{x}{R}\right)B_y\right] = \frac{e}{2m}\left[b + \left(b_1 - \frac{b}{R}\right)x\right].$$

Substituting these various expressions into (23.10) we obtain a pair of linear second order differential equations for $x(z)$ and $y(z)$ the right-hand sides of which can certainly be put equal to zero, through the very choice of the axis Oz itself, as this latter is a particular solution, $x = 0$, $y = 0$.

These equations are the equations of the required trajectories

$$2\varphi\, x'' + \varphi'\, x' + \left\{3\frac{\varphi_1}{R} - 2\varphi_2\right.$$

$$\left. + \sqrt{\frac{2e}{m}}\left[\frac{\varphi_1}{2\sqrt{|\varphi|}}\, b + |\overline{|\varphi|}\left(\frac{b}{R} - b_1\right)\right]\right\} x = 0\,, \qquad (23.12)$$

$$2q\, y'' + \varphi'\, y' + \left(-2\psi_2 + \sqrt{\frac{2e}{m}}\,|\overline{|\varphi|}\,b_1\right) y = 0\,.$$

23.4.2 Consequences: the Structure of a Beam (m, V) and Angular Focusing

We might say that these equations define the trajectories close to T_0 in the Gaussian approximation, and they do in fact represent a generalization of the ordinary differential equation for electron trajectories close to the optic axis in axially symmetrical electrostatic lenses (§ 5.2.2). Important consequences stem from these equations. First, the two equations (23.12) are independent: they give $x(z)$ and $y(z)$ respectively; we can therefore consider the two projections separately. From the linearity of the equations, we can deduce quite simply (§ 4.3) that for each projection, there are angular focusing properties. The most important concerns $x(z)$; this we shall call *horizontal focusing*, and we shall discuss it in the following way. The trajectories which originate in a point, or more generally, in a vertical "object" focal line P (a slit, for example) all converge onto a vertical "image" focal line P' (which will be a narrow band); the corresponding planes (which are perpendicular to Oz) are conjugate: the Lagrange–Helmholtz relation can be applied, in the form (cf. § 4.3.3)

$$\text{angular magnification} \times \text{linear magnification} = \sqrt{\frac{\varphi_{\text{object}}}{\varphi_{\text{image}}}}\,.$$

The case in which both the objects and their images lie in field-free regions is important, as the useful parts of the trajectories involved are straight lines, and we can perform the same geometrical constructions as in an ordinary centred system.

The focusing of $y(z)$ which we call *vertical focusing* represents a similar correspondence between the horizontal focal lines in the frontal planes. In general, this new pair of conjugates does not coincide with the preceding pair, but in mass spectroscopy, no interest attaches to this vertical focusing in the standard approximation at least, as the effects due to the field boundaries are neglected. In these conditions, the field structure is further simplified, as Φ depends upon x and z only, and B is uniform and vertical, so that the forces which act upon the particles are always horizontal; it is clear that the vertical focusing can only occur with virtual elements; rays which are initially horizontal remain horizontal.

23.5 SECTOR FIELDS; ANGULAR FOCUSING;

THEIR APPLICATION IN SPECTROMETERS

The fields which are used in spectrometry are in general bounded by plane vertical faces L, L', if we neglect edge effects. They can be regarded as occupying a sector of angle Θ, the axis of which we denote by K (Fig. 283)

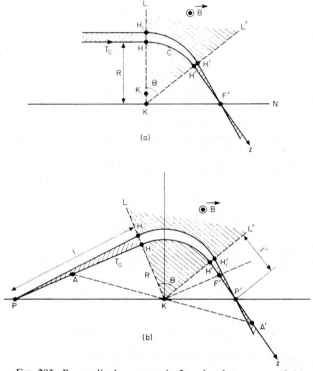

FIG. 283. Perpendicular magnetic focusing by a sector field.

and hence they have become known as sector fields. Their optical properties have been examined by Herzog (1934), Mattauch and Herzog (1934), Cartan (1937) and Mattauch (1940). In general, these properties can be obtained directly without having recourse to the general theory which we have just outlined; the latter is useful, however, for defining *a priori* the (horizontal) angular focusing properties.

23.5.1 The Uniform Magnetic Field (B); Ordinary Spectrometers

The optic axis T_0 is a circular arc C of radius R inside the sector smoothly joined to two straight lines which represent the incident and emergent particles.

(i) *"Perpendicular" focusing: the case in which T_0 at the entry to the sector is perpendicular to T_0 at the exit*

This is clearly the case if T_0 is incident normally upon the face of the sector at a distance R from the point K. We shall first of all establish the *image focus F'*, which is defined by the intersection of T_0 with a neighbouring ray, incident from the same direction (Fig. 283a). The two trajectories $HH'F'$ and $H_1H_1'F'$ can be regarded as two successive positions of the same rigid figure which consists of a circular arc of radius R and a tangent to it. In the displacement involved in changing from one figure to the other, the centre of the circle K moves to K_1, and H' to H_1' so that the instantaneous centre of rotation is situated at the intersection of the corresponding normals KN and $H'z$. As the emergent rays intersect at this point, this is the required focus F'; alternatively, we can say that F' is obtained by drawing a line through K parallel to the direction of incidence.

The optical properties of the system can easily be obtained; if we attach a second sector to the first, similar to it but with a sectoral angle which may be different, and turned back in the opposite sense, we obtain an overall system (Fig. 283b) which clearly gives the image P' of a point object P which lies on the axis. The interrelation is simple: P, P' and K are *collinear*, which we can express as a conjugacy relation between the distances $PH = l$ and $P'H' = l'$. Let us consider in particular the pair of conjugate points A, A' which lie on the external bisector of the sector; the corresponding angular magnification is -1 on account of the symmetry, and hence A and A' are "antiprincipal points" (between which the linear magnification is -1). Once the foci F, F' and the antiprincipal points A, A' of the system are known, the image conjugate to any given object can be constructed in the normal way.

Spectrometers with sector fields

In a spectrometer, the focusing is usually perpendicular and symmetrical, and is produced by a magnetic sector; the two slits S and S' will be situated at the antiprincipal points (Fig. 284). The ion beam falls on the entry face

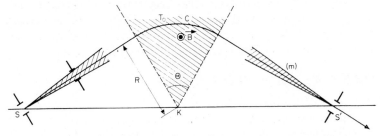

FIG. 284. Symmetrical perpendicular focusing; a sector spectrometer.

at a fixed distance R from K, and the collector receives ions the mass of which satisfies equation (23.4). The sectoral angles which are used are $\Theta = 90°$ (e.g. Hipple, 1942) and $\Theta = 60°$ (Nier, 1940 and 1947). Dempster's instrument (Fig. 280) represents the extreme case for which $\Theta = 180°$.

(ii) *"Oblique" focusing: the case in which the axis T_0 is arbitrary*

Once again, we can use the simple geometrical reasoning due to Cartan (1937). First, we establish the image focus F' in the case in which the optic axis is normal to the entry face alone (Fig. 285). The two neighbouring rays $HH'F'$ and $H_1H_1'F'$ by which this focus is determined are related by an

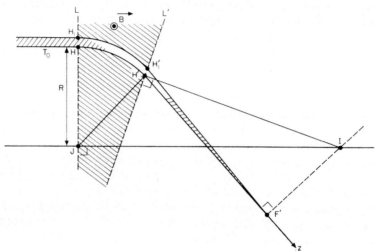

FIG. 285. Oblique magnetic focusing by a sector field: the construction for the image focus.

infinitely small rotation about an instantaneous centre I which lies at the point of intersection of the normals at J (the centre of the arc HH') and H', at the entry and exit faces respectively. The focus F' thus lies at the foot of the perpendicular dropped from I onto $H'z$. To obtain the diagram which gives us the geometrical construction for conjugacy on the axis, we have simply to place a similar sector next to the first, turned back in the opposite sense. With such a construction, we can, for example, obtain the angular focusing of a mass quite different from m which corresponds to the adjustment of the spectrometer shown in Fig. 284.

23.5.2 Fields with an Electric Component

An electric sector field will be, by definition, the field of a cylindrical condenser, restricted to the wedge defined by two meridional planes L and L'. The potential will be equal to the potential outside ($-V$, say) along a circle C of radius R. The field varies as $1/r$, and is described by the potential

$$\Phi = E_0 R \log \frac{r}{R} - V,$$

in which E_0 is the value of the field on R, positive if it is attractive. More generally, the case of a sector field which is at once both electric and magnetic has been considered theoretically by Henneberg (1934) and by Herzog

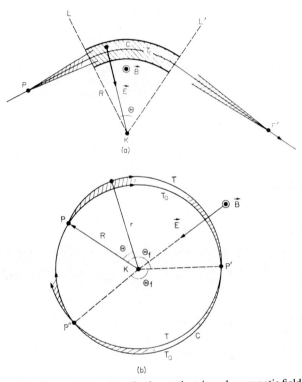

FIG. 286. Perpendicular focusing by an electric and a magnetic field.

(1934); such a field is formed by the superposition within the same sector of a uniform magnetic induction $\boldsymbol{B}$ and a cylindrical electric field $\boldsymbol{E}$ (Fig. 286).

There are ions (m, V) for which the circle C is a particular trajectory inside the sector. They can easily be defined, as the velocity $\boldsymbol{u}$ corresponds to V in such a way that the normal projection of equation (23.1):

$$\frac{m u^2}{R} = e(E_0 + u B),$$

leads to

$$R = \frac{2V}{\sqrt{\dfrac{2eV}{m}} B + E_0}. \tag{23.13}$$

This relation is a more general version of equation (23.4) which concerns purely magnetic fields, but it should be realized that in practice, circular trajectories are the exception. Given ions (m, V) which satisfy this condition, we shall now examine their angular focusing properties in the neighbourhood of the circle C which plays the role of the optic axis T_0, in the case where the field is unbounded.

We shall apply the trajectory theory in curvilinear coordinates; the various parameters have the following values:

$$x = r - R; \quad \varphi = -V; \quad \varphi_1 = E_0; \quad \varphi_2 = -\frac{E_0}{2R}; \quad \text{and} \quad b = B.$$

Equation (23.12) for the focusing takes the simple form

$$x'' + \frac{1 + q^2}{R^2} x = 0,$$

in which we have written

$$q = \frac{E_0 R}{2V}. \tag{23.14}$$

For the trajectories close to C, therefore, we have

$$x = x_0 \cos\left\{\sqrt{1 + q^2}(\theta - \theta_0)\right\}, \tag{23.15}$$

where $\theta = z/R$ is the azimuthal coordinate about K. An object point P therefore gives successive images P', P'' at angular intervals θ_f, $2\theta_f$ where

$$\theta_f = \frac{\pi}{\sqrt{1 + q^2}}.$$

Two important special cases are: the *purely magnetic field* $q = 0$—we have Dempster focusing, $\theta_f = \pi$—and the *purely electric field* $q = 1$ (from equation (23.13)). The angular interval between successive foci is then

$$\theta_f \equiv \frac{\pi}{\sqrt{2}} = 127° 17'.$$

From these results, we could deduce the optical properties of sector fields Φ in the perpendicular focusing case by combining the sectors (Fig. 286a). As an example, we shall calculate the image focus F'; an incident trajectory T which is parallel to the axis T_0 is described by the following expression within the sector, with respect to C:

$$x = x_0 \cos\left(\sqrt{1 + q^2}\theta\right);$$

it emerges at an angle ε given by

$$\varepsilon = -\sqrt{1 + q^2}\frac{x_0}{R}\sin\left(\sqrt{1 + q^2}\,\Theta\right)$$

from which we can calculate the distance F' to the point of intersection with T_0 immediately:

$$H'F' = \frac{R}{\sqrt{1 + q^2}} \cot(\sqrt{1 + q^2}\,\theta).\qquad(23.16)$$

23.6 SECTOR FIELDS; MASS AND ENERGY DISPERSION; DOUBLE FOCUSING; SPECTROGRAPHS

23.6.1 Dispersion of the Beam for $\triangle m, v$

Of the whole beam of ions which emerges from the source, we have examined one component, which consists of the particular family of ions (m, V) and which we assume to be focused perpendicularly. To obtain the structure of the whole ensemble, we shall now consider the adjacent com-

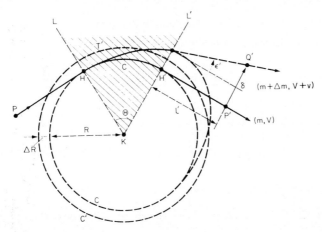

FIG. 287. Mass and energy dispersion in a sectoral field.

ponents, $m + \triangle m$ in mass and $V + v$ in energy. The emergent rays are *a priori* shifted relative to the first case, and we shall simply calculate the extent of this shift at the level of the image P', which is described by the transverse vector $d = P'Q'$ (Fig. 287). The optic axis of the rays (m, V) within the sector was the circle C, radius R; the optic axis of the rays $(m + \triangle m, V + v)$ is another circle C', with radius $R + \triangle R$ which we shall now compute.

In the case of a purely magnetic field, $\triangle R$ can be obtained by differentiating the general expression (23.4) for the radius R:

$$\frac{\triangle R}{R} = \frac{1}{2}\left(\frac{\triangle m}{m} + \frac{v}{V}\right).\qquad(23.17)$$

When the field is purely electric, the problem cannot be solved in this way. C' is the circle along which ions of the same energy as the incident ions, $V + v - E_0 \triangle R$, would travel; the equation of motion along the normal takes the form:

$$\frac{mu^2}{R + \triangle R} = 2\frac{e(V + v - E_0\triangle R)}{R + \triangle R} = e\, E_0 \frac{R}{R + \triangle R}$$

($\triangle m$ does not appear), whence $\left(\text{with } R = \dfrac{2V}{E_0}\right)$,

$$\frac{\triangle R}{R} = \frac{v}{R\, E_0} = \frac{1}{2}\frac{v}{V}. \tag{23.18}$$

As soon as $\triangle R$ is known, the mean trajectory T' of the dispersed ions can easily be established—T' is incident normally along PH, and subsequently is described by the following expression with respect to C':

$$x = -\triangle R \cos(\sqrt{1 + q^2}\theta);$$

T' leaves the sector at an emergent angle

$$\varepsilon' = \sqrt{1 + q^2}\frac{\triangle R}{R} \sin(\sqrt{1 + q^2}\Theta),$$

and the disparity δ at the level of P' which gives the distance l' to L' can easily be evaluated, and is found to be

$$\delta = [1 - \cos(\sqrt{1 + q^2}\,\Theta) + \sqrt{1 + q^2}\,\frac{l'}{R}\sin(\sqrt{1 + q^2}\,\Theta)]\,\triangle R$$

$$= K_m\frac{\triangle m}{m} + K_v\frac{v}{V}. \tag{23.19}$$

Applying (23.17) and (23.18), we obtain the two most important special cases:

(i) B only ($q = 0$)

$$K_m = K_v = \tfrac{1}{2}\{R(1 - \cos\Theta) + l'\sin\Theta\}, \tag{23.20}$$

and if in addition the focusing is symmetrical (Fig. 284), we can easily show that

$$K_m = K_v = R.$$

(ii) Φ alone ($q = 1$)

$$K_m = 0,$$

$$K_v = \tfrac{1}{2}\{R(1 - \cos\sqrt{2}\,\Theta) + l'\sqrt{2}\sin(\sqrt{2}\Theta)\}. \tag{23.21}$$

K_m expresses the dispersion of the beam as a function of the mass, and K_v the chromatic aberration, which is *a priori* harmful.

23.6.2 Double Focusing

An arrangement for which $K_v = 0$ but $K_m \neq 0$ is achromatic, and hence focuses with respect both to angle and to energy; it produces, therefore, double focusing, and the ion beam converges to P' irrespective of variations in the energies of the ions. This is the case in a sector for which $q = -1$, which defines a particular combination of $\boldsymbol{B}$ and Φ—we might have deduced this from the fact that for such a sector, the radius R of C takes a stationary value $(\partial R/\partial V = 0)$.

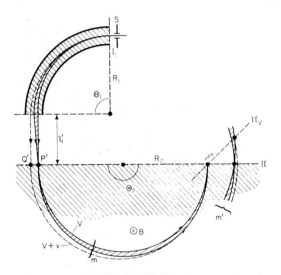

FIG. 288. Dempster's double focusing spectrometer.

We shall now demonstrate that the same property can be attained in a way which is advantageous from the practical point of view; rather than superimposing the magnetic and electric fields, we juxtapose them in the appropriate fashion. We suppose that the ion beam which is emitted at P passes through two successive sector fields, the first of which is electric (denoted by the suffix 1), and the second magnetic (suffix 2) as shown in Fig. 288. The object slit S produces an intermediate image P' and a final image P'' which alone is important. The beam is doubly focused if ions $(m, V + v)$ form rays of the type $SQ'P''$. In these conditions, as we see from Fig. 288, $P'Q'$ is none other than the chromatic aberration of the electric sector, δ_1 say; but $P'Q'$ is also the chromatic aberration, δ_2, of the magnetic sector which would focus P'' into P' in the same geometrical arrangement (reversing the sense of the motion and that of $\boldsymbol{B}$). We see, therefore, that $\delta_1 = \delta_2$, so that

$$R_1(1 - \cos\sqrt{2}\Theta_1) + l_1'\sqrt{2}\sin\sqrt{2}\Theta_1 = R_2(1 - \cos\Theta_2) + l_2\sin\Theta_2. \quad (23.22)$$

This is the double focusing condition when the angles both of incidence and emergence are right angles.

Other arrangements can be found in which the sign of one of the members is changed. An intermediate diaphragm D (Fig. 289) can be used to restrict

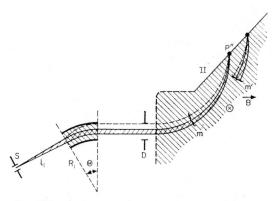

Fig. 289. Mattauch's double focusing spectrometer.

the beam to an energy band which is narrow enough to ensure double focusing although the latter in fact only exists in first order. The condition for a given mass m is not in general valid for other masses m' (except through continuity if the latter are close to m). In other words, given the beam which emerges from S, the position II_v at which "energies" are focused only coincides with the position II of the angular focus for a particular mass m (Fig. 288).

23.6.3 Application to Spectrographs

We shall mention a few spectrographs in which there is double focusing. The general theory of this type of instrument is to be found in Mattauch and Herzog (1934); see also Bainbridge (1953).

(i) *Dempster's instrument* (*1935*)

This is shown schematically in Fig. 288, and is defined by

$$\Theta_1 = \frac{\pi}{2}, \qquad \Theta_2 = \pi, \qquad l_2 = 0,$$

$$R_1 = 8{\cdot}5 \text{ cm}, \qquad R_2 = 10 \text{ cm}.$$

Equation (23.22) leads to l_1', and from conjugacy, we have $l_1 = 1$ cm. The two positions Π and Π_v are markedly separated from one another which is most inconvenient.

(ii) *Bainbridge and Jordan's instrument (1936)*

This instrument is defined by

$$\Theta_1 = \frac{\pi}{\sqrt{2}} = 127° \, 17', \qquad \Theta_2 = \frac{\pi}{3},$$

$$R_1 = R_2, \qquad l_1 = l_1' = 0, \qquad l_2 = l_2' = R_2 \sqrt{3}.$$

(23.22) can be verified straightforwardly and Π and Π_v remain close to one another.

(iii) *Mattauch's instrument (1936); see Fig. 289*

Here, the optical arrangement is such that $l_1' = l_2 = \infty$; S and P'' lie at the object and image foci of the two sectors respectively, and the double focusing condition takes the simple form:

$$\sqrt{2} \cdot \sin \sqrt{2} \, \Theta_1 = \sin \Theta_2 \qquad (23.23)$$

(in which l_1' and l_2 are infinitely large and equal). A particular solution is

$$\Theta_1 = \frac{\pi}{4\sqrt{2}} = 31° \, 50', \qquad \Theta_2 = \frac{\pi}{2},$$

and from conditions (23.16) which provide us with the focus of the first sector we find $l_1 = \dfrac{R_1}{\sqrt{2}}$. It is to be noticed that the relation (23.23) does not depend upon the values of the two radii, and is thus valid for any mass provided the arrangement is such that all the images for $\Theta_2 = \pi/2$ can be received; this can be achieved by halting the field at the "oblique" plane Π (Fig. 289) in which the receiver, a photographic plate, is placed. It is this spectrograph due to Mattauch which has the great advantage of providing double focusing simultaneously for all masses (Π and Π_r are identical). The distribution of the masses M can easily be determined as a function of the abscissa X of the bands. M is proportional to X^2 as in the simplest example (Fig. 280).

(iv) *Double focusing spectrometers*

Spectrometers for measuring masses have also been constructed on the lines described above (Nier and Roberts, 1951; Hannay, 1954). Apart from the advantage of electrical recording, they possess that of needing double focusing only for the particular mass arriving at the collector, and hence for a single point S' alone.

23.7 THE VARIOUS PERTURBING EFFECTS

So far, we have taken for granted the validity of certain approximations and assumed certain ideal conditions, which we must now examine critically. We shall find that some minor adjustments will have to be made to the results, and we shall restrict the discussion to the most important case, namely, the purely magnetic sector.

23.7.1 The Spherical Aberration

The assumption that the rays (m, V) which emerge from P converge onto a point P' (see Fig. 284, for example) is only a "Gaussian approximation": the approximation with the aid of which the equations of motion were

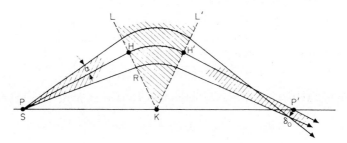

FIG. 290. Spherical aberration in a magnetic sector field (perpendicular focusing).

established. We shall now apply more rigorous optical laws; the emergent rays are tangent only to an element of the caustic surface through P'—their distance at P' remains second order infinitesimal. In other words, if α is the semi-aperture of the beam, the image P' is spread out into a spot by the transverse spherical aberration δ_0 which is proportional to α^2 (Fig. 290). For 180° focusing (Fig. 280), it is no problem to show that

$$\delta_0 = 2R(\cos \alpha - 1)$$
$$\simeq -R\alpha^2. \tag{23.24}$$

The same is true for symmetrical arrangements with normal incidence. For example, with $R = 15$ cm, we have

$$\alpha = 1°, \qquad \delta = 0{\cdot}05 \text{ mm},$$
$$\alpha = 0{\cdot}1°, \qquad \delta = 0{\cdot}5\,\mu.$$

Electric fields would give results of the same order of magnitude. α is made as small as is necessary with the aid of a supplementary slit in a diaphragm, which prevents the aberration from affecting the resolving power.

It is, however, easy to correct the aberration δ_0 of a purely magnetic field in a spectrometer. We have, therefore, the possibility of admitting beams with large apertures, and hence of higher intensity. All that is necessary is that the analysing field should be bounded not by plane faces but by faces L, L' of the appropriate shapes. We select the object point P and the image

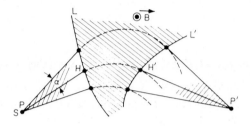

Fig. 291 a. Focusing without spherical aberration.

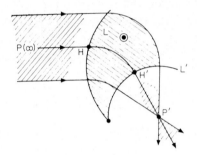

Fig. 291 b. Focusing a parallel beam (SMYTHE *et al.*).

point P' in advance, and hence obtain the entry face L (Fig. 291 a); the circular portions of the trajectories are then defined, and hence so are the exit points H' which must be selected, and which define the shape of the face, L' at which the particles emerge.

The solution to the problem is very arbitrary, therefore, and we extract such solutions as are most convenient. Smythe, Rumbaugh and West (1934) showed how wide parallel beams (P at infinity) could be focused with the aid of a field bounded by two circular cylinders (Fig. 291 b), and subsequently, other solutions have been studied by Bainbridge (1947), Hintenberger (1948), Kerwin and Geoffrion (1949) and Mayne (1952) in particular. Alternatively, plane faces can be retained (with oblique incidence), and the spherical aberration reduced not to zero but to third order infinitesimal.

In conclusion, we should mention that for the beam to be focused into a perfect point, P', it is not necessary that the object P should be a point, and hence S a very narrow slit. P' may even be produced by a caustic which is itself produced by a wide slit source (Cassignol and Nief, 1951); Cassignol (1953) has shown how the spherical aberration of the whole instrument in question can be corrected empirically.

The aberration has also been considered for combinations of magnetic and electric sectors, with a view to perfecting double focusing spectrometers (Johnson and Nier, 1953).

23.7.2 The Influence of Edge Effects of the Analysing Field

The magnetic field B can be considered as being uniform between the ideal faces L and L'; it always extends beyond this region, however, in the form of a fringing field or edge effect $\beta(z)$ (at the edges of the gap), the effects of which will be twofold.

(i) *Displacement of the horizontal focusing (Coggeshall and Muskat, 1944; Ploch and Walcher, 1950)*

We consider one of the rays of the beam which is being focused perpen-

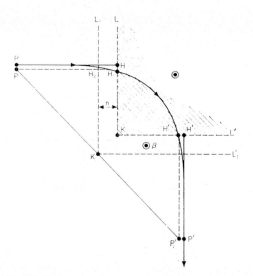

FIG. 292. The influence of the overlap field on the focusing.

dicularly (Fig. 292); the incident portion PH_1 is not perfectly rectilinear, but satisfies the equation

$$\frac{m\,u^2}{\varrho} = -m\,u^2\,x'' = e\,u\,\beta(z),$$

in which z and x are measured with respect to the initial direction PH. At H_1, therefore

$$x' = -\frac{\eta}{R},$$

where we have written

$$B\eta = \int \beta(z)\, dz. \tag{23.25}$$

η is to a certain extent the width of the region of uniform field which integrates to the same value as the fringe field; η is of the order of magnitude of the width of the gap. It is exactly as if the incident ray PH_1 had been replaced by a ray $P_1 H_2 H_1$ emerging from a well-defined object P_1 and the actual field replaced by an ideal field B (without boundary effects) the boundary L_1 of which was situated a distance η from L; the position of P_1 relative to P can equally well be calculated, but the shift is smaller. We proceed in the same way at the exit; the two points P_1, P_1' are collinear with K_1 according to the properties of ideal focusing, and Fig. 292 shows the resulting appearance of the focusing of P at P'. To obtain the image of P at P', we must move the thin edge K with respect to the line PP' through a distance of the order of magnitude of η, and hence of the gap (subject to possible empirical adjustment).

(ii) *The appearance of vertical focusing*

Outside the symmetry plane, the fringe field has a horizontal component. In particular, we consider its projection on the principal normal to the mean ray T_0 of the beam: $\beta_x = b_1 y + \cdots$. If this beam is oblique with respect to the faces L and L', β_x is not zero and the equation of the vertical component of the trajectories is, from (23.12),

$$y'' - \sqrt{\frac{2e}{m} \frac{b_1}{|V}} y = 0. \tag{23.26}$$

The trajectories are slightly curved, therefore, and hence slightly focused vertically. A horizontal incident beam does not remain exactly horizontal if the focusing is oblique. This property, which was underlined by Cotte (1938) in his general theory, is of great interest in β-ray spectroscopy; here, we attempt to accentuate the phenomenon (with mixed fields) so that the beam is focused simultaneously in the horizontal and the vertical directions.

A *point* object P would then give a *point* image P' (Siegbahn and Svartholm, 1946 and 1950). This seems to be of little interest as far as mass spectroscopy is concerned.

23.7.3 Space Charge

We shall summarize only briefly the other possible perturbing influences. The effect of space charge (mutual repulsion between the ions) is to oppose the convergence; the effect is perceptibly larger for an ion beam than it is for an electron beam of the same energy and current density i. In fact, since $i = \varrho u$, we can see that the charge density ϱ is higher for ions, as heavier particles move more slowly for a given energy, and thus the velocity u is smaller.

In mass spectroscopy, the perturbation is always negligible, at least with ordinary values of the current, i, in the region of 10^{-9} A; for beams in which $i = 10^{-6}$ A, however, the effect of space charge is no longer negligible, and is reflected in an increase in the distance between the object and the image (Walcher, 1949; and Robinson, 1949). The intense beams which are used in spectrometers for the separation of isotopes (i of the order of 1 mA) require particularly careful attention. Higher voltages may be used, for example (30 kV or more), or the space charge may be destroyed by admixing slow electrons (Bernas, 1953; Bernas, Kaluszyner and Druaux, 1954).

Finally, we must realize that various types of aberration may result from imperfections in the construction of the instrument: the alignment may be defective, the vacuum inadequate (the ions will be scattered by the residual gas), the conductors may be covered with a film of impurity which is liable to become charged, the source of H.T. or the fields may be unstable and produce fluctuations of the beam. There are no diffraction effects however, as the wavelengths associated with ion beams are so very short (due to the rather large masses).

23.8 THE RESOLVING POWER OF SPECTROMETERS AND SPECTROGRAPHS

23.8.1 The Width of the Line; the Resolving Power of the Field

The problem is to obtain a measure of the ability of the various instruments to discriminate between ions of different masses. If $m + \triangle m$ is the mass closest to m which can be distinguished by the analysing field, we define the resolving power of this field for the mass m to be

$$\left| \frac{m}{\triangle m} \right|.$$

The object is the exit slit S at the ion source, of width s. We shall first of all calculate the width s_1' of the line on which ions of mass m fall in the plane of observation which is conjugate to S. If G is the corresponding optical magnification, the main component of the width is given by the Gaussian approximation, Gs; to this, however, we must add the chromatic aberration which is due to the energy spread v of the beam. If the aperture 2α is sufficiently small and the construction sufficiently careful for there to be no other causes of aberration,

$$s_1' = Gs + K_v \frac{v}{V}. \tag{23.27}$$

The band defined by particles of mass $m + \triangle m$, however, lies a distance $K_m \dfrac{\triangle m}{m}$ away. We usually agree to call the bands separated if this distance is equal at least to the width s_1'; this produces the following expression for the reciprocal of the resolving power of the field:

$$\left| \frac{\triangle m}{m} \right| = \frac{G}{K_m} s + \frac{K_v}{K_m} \frac{v}{V}. \tag{23.28}$$

23.8.2 Spectrographs

The resolving power of spectrographs with double focusing is obtained by putting the second term equal to zero. If, for example, we apply this formula to the spectrograph of Bainbridge and Jordan, where the focusing is magnetic (radius R_2) and symmetrical (antiprincipal points), we have $K_m = R_2$ (from equation 23.20) and $G = 1$, so that

$$\left| \frac{m}{\triangle m} \right| = \frac{R_2}{s}. \tag{23.29}$$

The other arrangements give less straightforward calculations, but the preceding relation gives the order of magnitude. We should remember, however, that only in Mattauch's instrument is the double focusing property valid for all values of the mass. For example, we obtain a theoretical resolving power of about 40,000 if $R_2 = 20$ cm and $s = 5\,\mu$; this demands a semi-aperture α of the order of $0\cdot1°$ or $0\cdot2°$ if the spherical aberration is not to appear (see § 22.7.1).

23.8.3 Spectrometers

For purely magnetic analysing fields, $K_m = K_v$ and from (23.28) we find

$$\left| \frac{m}{\triangle m} \right| < \frac{V}{v}. \tag{23.30}$$

As an examination of types of ion source shows, the relative heterogeneity v/V can hardly be less than 10^{-3}; the resolving power of the field can barely exceed 1000, therefore, a far lower value than the resolution which can be obtained with double focusing in spectrographs. As a result, the width of the object slit s, which always appears as the ratio s/R (or a term of this order) only begins to appear in the expression for the resolving power for widths of about $1/_{10}$ mm, and there is no advantage in choosing it any narrower.

The resolving power of the field (23.28) will not, generally speaking, be that of the apparatus. The spectrometer here receives the useful part of the beam in the slit S' of width s' of the collector; this slit must be at least as wide as the emergent beam, and we may even choose an appreciably wider value, $s' > s'_1$ for reasons which we shall see later. The resolution criterion for the whole spectrograph therefore requires that masses m and $m + \triangle m$ should not fall on S' simultaneously, so that the resolution of the spectrometer itself is less than that of the field alone (23.28). The former is defined by

$$K_m \left| \frac{\triangle m}{m} \right| = s'$$

if $s \ll s'$, but if this condition is not fulfilled, we can easily take account of s; in the usual cases, then,

$$\left| \frac{m}{\triangle m} \right| = \frac{R}{s'}.$$

If, for example, $R = 15$ cm and $s' = 0 \cdot 5$ mm, the resolving power is about 300. This value is for the most part sufficient in the everyday problems of mass spectrometry.

CHAPTER 24

THE INDIVIDUAL PARTS OF A
MASS SPECTROSCOPE

24.1 THE ION SOURCE

We shall first of all study the ion sources which are designed to transform the substance to be analysed into positive ions; so far as is possible, the representatives of each mass m will move at the same velocity (and thus have the same energy V), and the whole bundle of ions will emerge through a slit S in a narrow beam which is the ion object for the analysing field. Every source, therefore, consists of two more or less distinct parts:

(i) The source itself, at which the ionization process is initiated, and which is as nearly equipotential as possible so that the ions are closely homogeneous.

(ii) A mechanism for collimating and accelerating the particles through a potential V, with which an ion beam of the required energy is produced.

Various designs have been proposed, suitably adapted for solids, gases and vapours. We shall only consider ion sources for mass spectroscopy, however.

24.1.1 Sources Produced by Discharge in a Gas

(i) *The ordinary luminous discharge (for gases)*

The earliest of the sources of positive ions is the ordinary luminous discharge in a gas at a pressure p of the order of 10^{-2} mm Hg (the dimensions of the envelope being considerably larger than the mean free path). The cathode is usually perforated by a narrow canal (less than a millimetre in diameter) through which a beam of positive ions which comes from the anode can pass; these are Goldstein's "canal" rays (see Fig. 293a).

This type of source possesses several defects which limit its usefulness. First, the ions are not created in an equipotential region, as their production takes place in the same region as their acceleration; this has the fatal result of producing an ion beam which is extremely heterogeneous in velocity. The energy spread v is an appreciable fraction of the mean energy V which is obtained at the exit; if V is of the order of a few thousand volts, v is about 1000 V. We cannot use a source of this kind in spectroscopy therefore, without either inserting a velocity filter which eliminates the majority of the rays, or using double focusing which has the advantage of retaining a wider band of energies. Further, the ions which are obtained are widely assorted, as the phenomena which occur at the discharge are extremely diverse; single or multiple ionization may occur, with or without

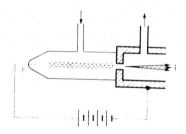

FIG. 293a. A discharge source: the canal rays.

rupture of the molecules, and charge exchange, condensation, and chemical reactions between ions and molecules are all possible. A discharge in oxygen, for example, produces many ions for each isotope: O^+, O^{++}, O_2^+, O_2^{++}, O_3^+ ... If nitrogen too is present, combinations such as NO^+ will also occur. The spectra, therefore, are rich in bands, which can be used as reference positions and for the measurement of masses m, m', ... by the doublet method in spectrography.

On the other hand, such a source is not suitable for quantitative analyses, for spectrometry; the spectra become too complex, and—a more important objection, this—the superposition rule is no longer valid. The spectrum of a mixture is not obtained by superimposing the spectra of its constituents.

The intensity i of the ion beam which emerges from S is much less than 1 mA. Various improvements have been made with a view to increasing i or to working at a lower pressure p (the superposition of an auxiliary magnetic field or use of an HF potential); the sources become more complex, however, and they are not used for spectroscopy. They are used to supply particle accelerators, giving intense beams of atomic ions. Several studies have been devoted to these discharge processes—an analysis of this work is to be found in the book by Ewald and Hintenberger (1953).

(ii) *The gas arc*

An arc is a discharge in which the cathode is rendered incandescent, either spontaneously, or by auxiliary heating which results in a sharp increase in the intensity of the current and reduces the potential across the terminals (1 A, 50 V for example). The current i is larger and the ions are considerably more homogeneous ($v = 10$ V, for example); acceleration is in this case a separate operation. Arcs are too complex in construction and too brutal in action to be used in spectroscopy; the only instruments in which they are used are those for separating isotopes where an ion current of a few milliamps is used (Koch and Bendt-Nielsen, 1944; Bernas, 1953), and as ion sources for accelerators such as cyclotrons and synchrocyclotrons.

24.1.2 High Frequency Sparking (for Solids)

A general method for producing ionization of solids is furnished by high frequency spark discharge which was first applied to mass spectroscopy by Dempster (1935). The spark flies between two tapered electrodes or between an axial electrode A and a plate in which an opening has been cut (Fig. 293b). A is composed of the material to be ionized, or of a metal

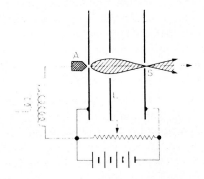

Fig. 293b. A high frequency spark source.

tube in which the material is enclosed in powder form. The pair of electrodes L and S take care of the collimation and the acceleration. The ions which are obtained have a noticeable energy spread v, although less wide than for canal rays—a velocity filter or double focusing is required. Here again, a wide diversity of ions is produced, which gives the method its spectrographic importance, as it complements to a certain extent the luminous discharge—the two phases involved are the solid phase and the gaseous phase respectively.

24.1.3 The Surface Ionization Source (for Solids)

From certain solids, we can instigate evaporation of positive ions by placing the solid over an incandescent anode. This phenomenon occurs for elements which are readily transformed into ions, and which are thus electropositive. We denote the ionization potential by V_i, and the potential barrier at the metal support by V_s; the positive emission varies with $(V_s - V_i)$: the following formula, known as the Langmuir formula (1925), expresses this more precisely:

$$\frac{n'}{n} = \exp\left[\frac{F(V_s - V_i)}{RT}\right] \tag{24.1}$$

(n and n' are the numbers of neutral atoms and positive ions which are evaporated, respectively; $F \simeq 96{,}500$ coulombs $= 1$ faraday).

A tungsten filament, for example ($V_s = 4{\cdot}55$ V), is perfect for obtaining a beam of Cs^+ ions ($V_i = 3{\cdot}86$ for Cs); for alkali metals, alkaline earths and rare earth metals, oxidized tungsten ($V_s = 8$ V) must be used. These sources were first explored by Dempster (1918) and Moon and Oliphant (1932). Alternatively, we can heat not the element which is to be ionized but one of its compounds—the halides (Koch, 1936) or the alumino-silicates (Blewett and Jones, 1936); this technique has been developed extensively by Couchet (1954). Intense sources (1 mA cm⁻²) have been studied by Kunsmann (1927), and employed in particular by Smythe, Rumbaugh and West (1934) in their wide beam instrument designed to separate isotopes (Fig. 291). The theory of the emission from compounds is still virtually unknown, but the practical operation is quite satisfactory.

These surface ionizers have a very distinct character. It is clear, first of all, that they are very selective, and do not give parasitic ions since the ionization phenomenon is strictly localized at the anode. Further, as the latter is to all intents and purposes equipotential, the velocity spread v is very small, as it is equal to the spread of the initial thermal energies, about 0·2 V. The ion beam can be formed in the same way as before (Fig. 293b, electrodes L and S).

Certain negative ions can be evaporated from a hot cathode—in this case, it is with electronegative elements that we are concerned (halogens); these have been studied by Hintenberger (1951) in particular.

24.1.4 Sources Produced by Electron Bombardment (for Gases and Vapours)

The ions emitted by this type of source are produced by electron bombardment and the range of applications is very extensive; sources of this kind were first used by Dempster (1916) and later developed, principally by Bleakney (1932), Tate and Smith (1934) and Nier (1937 and 1947).

(i) *Description and properties of an ordinary source*

The standard model is due to Nier; diagrams of the horizontal xOz projection and the vertical yOz projection are shown in Fig. 294. The substance to be studied is a gas which arrives through the tube G or a vapour which is produced by heating the appropriate solid in a small adjacent compartment (Bernas and Nier, 1948; Louw and de Villiers, 1954; and Cameron, 1954). The gas penetrates at very low pressure p (10^{-4} to 10^{-5} mm Hg) into a small ionization chamber C, where it is ionized by an

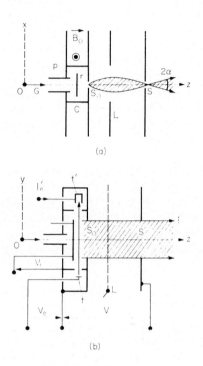

(a)

(b)

FIG. 294. Nier's source.

electron beam which is emitted by a heated tungsten ribbon t. The accelerating potential V_e of the electrons is applied between t and C—it must, of course, exceed the ionization potential of the gas being analysed, and is usually set between 50 and 100 V. The current I_e of the beam is a few hundred microamps. We shall see later that ions which are created in this way have a very low kinetic energy (V_c); a fraction of these ions has then to be extracted from C through the slit S_c, and accelerated and aimed by the field of the collimating and accelerating electrodes L and S. This can be done with the aid of an auxiliary electrode r, for example, which is held at a positive potential V_r of a few volts with respect to the walls of C; r produces a field which has the desired effect upon the ions without sensibly modifying the electron beam. This latter is considerably more perturbed by the divergent lens effect produced by the slit through which the beam enters C. To prevent the beam from spreading, a vertical magnetic field of a few hundred gauss is often employed; the effect of such a field, we recall, is to wind the electron trajectories in helices around the lines of force, and hence to guide them in a direction which is on the average vertical. This operation is controlled with the aid of the current I_s which is incident upon a collector electrode or trap t' which is placed at the exit from C opposite t. Finally, since the chamber C is pierced with several openings, it is important that the areas of these latter should be as small as possible; it must be possible to maintain the internal pressure p without harming the vacuum outside which is at most of the order of 10^{-6} mm of mercury (cf. § 23.4.2).

A source built in this way can be easily and reproducibly adjusted. As the chamber C is almost equipotential, the velocity spread v is only a few volts; this spread is due to V_r and to the initial kinetic energy V_0 of the ions which is always very slight—in general, it is a question of the mean thermal energy of the particles (about 4×10^{-2} V at normal temperatures). On the other hand, the pressure p is low enough here for each ion of the beam to be the result of a single process; unlike the case of a luminous discharge there are no "reactions" in C. The spectra obtained obey the superposition law and are susceptible to quantitative analysis, and therefore to spectrometry. We shall see that the ion currents i which are obtained are of the order of 10^{-3} A at most, which is normally perfectly adequate.

(ii) *High intensity sources*

More intense sources have been produced. The procedure known as *oscillating electrons* represents a large advance, by extending the length of the path of the ionizing beam considerably (Finkelstein, 1940). An application of this procedure is to be found in the source due to Heil (1943), which we can describe by supposing that in Fig. 294, the trap t' is replaced by a second emissive ribbon at the same potential as the first ribbon, t. The reflecting electrodes are distributed on one side and the other in such a way that each ionizing electron traverses the chamber C a large number of times. The increase in intensity can also be obtained by arranging that a higher proportion of ions emerges from the ionization chamber; this can be done by directing the electron beam along the axis and concentrating the resulting ions by means of suitable electrodes at the exit slit (Cassignol, Geller and Moreau, 1953).

Finally, other types of intense source can be obtained by intensifying the ionizing beam I_e and increasing the pressure p; the behaviour then tends towards that of arcs, however, and the spectra no longer obey the superposition law. We might mention as examples the sources of Walcher (1944) and Bernas and Nier (1948) which can be used for isotope separation— the currents i can reach a milliampère.

(iii) *Various ionization processes*

It is useful to introduce the reader at this point to the various ionization phenomena which are produced at the source by the electron bombardment; for further details, the article by Hagstrum and Tate (1941) or the book by Robertson (1954) should be consulted. For the present, we shall examine a simple example, namely, carbon monoxide. This diatomic gas produces the following ions (e = electron):

$$CO + e \rightarrow \begin{cases} CO^+ & + 2e & \text{(i)} \\ CO^{++} & + 3e & \text{(ii)} \\ C^+ & + O & + e & \text{(iii)} \\ C^+ & + O + 2e & \text{(iv)} \\ C & + O^+ + 2e & \text{(v)} \\ C & + O^- & \text{(vi)} \end{cases} \qquad (24.2)$$

(the reaction $CO + e \rightarrow C + O + e$ can obviously not be detected here). The frequency of each of these reactions, that is, the probability with which each occurs, can be calculated; this probability is often defined by the number of times n that the reaction in question is produced by an electron over a distance of 1 cm of its trajectory at a pressure of 1 mm of mercury. At the pressure p of the ionization chamber C, the order of magnitude of n is at most 10; only a small fraction of the electrons is useful, therefore— we shall restate this conclusion more precisely a little later when we are examining the yield of a source.

The determination of n is one of the most important applications of the mass spectrometer; in particular, we study the variation of n as a function of the energy V_e of the ionizing electrons. For the reaction mentioned above, the following results are obtained: (i) and (ii) are ionizations in the proper sense of the word, while (iii) (iv) and (v) are ionizations with simultaneous destruction of the molecule; these five reactions can occur beyond a threshold potential V_i, which is the *ionization potential* or more generally, the potential at which the corresponding ions first appear. For (i) and (ii) the values of this potential are 14 and 41 V respectively. The variation of probability as a function of V_e is of the form labelled a in Fig. 295. Reaction (vi) is a capture accompanied by destruction of the molecule, and its probability curve is different in form; the function has a maximum (or "resonance") for $V_e = 9 \cdot 5$ V (see Fig. 295, curve b). CO thus gives the positive ions which have the following integral masses when V_e is of the order of 100 V, and the various isotopes have been extracted: 12 (C^+), 16 (O^+), 14 (CO^{++}) and 28 (CO^+). If we were to analyse the negative ions in the spectroscope (and for this, we should have simply to change the sign of the potential V), we should obtain only 16 (O^-) which is a considerably less characteristic result. Here we have the first justification for the preference which is accorded to positive ions; another reason is that the reactions which result in negative ions have markedly smaller probabilities than the others.

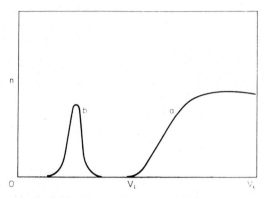

FIG. 295. The two types of ionization curve.

The simple example of CO gives some indication of the complexity of the spectra which will be produced by polyatomic molecules, the hydrocarbons for example. This complexity is sometimes still further increased by the fact that some ions are *metastable,* by which we mean that they are not extant throughout the whole passage through the spectroscope; it is the ions which result from their dissociation which are responsible for the appearance of lines which seem to correspond to non-integral masses in the spectra (Hipple, Fox and Condon, 1946; and Hipple, 1947).

The initial kinetic energy (V_c) of the ions which are produced is necessarily equal to the thermal value for straightforward ionizations, such as (i) and (ii), by virtue of the elementary laws of collision in mechanics. This is no longer true when the ionizations are accompanied by breakdown of the molecules, as the ions which are produced in certain reactions may possess a supplementary kinetic energy; this is at most of the order of one or two volts, however, and hence only slightly disturbs the homogeneity of the complete ion beam.

24.1.5 Focusing and Accelerating the Ions in a Source

As an example, we consider a source due to Nier (Fig. 294), which behaves very much like every other source of fairly homogeneous ions. The problem is, we recall, to produce an ion beam of energy V and small aperture 2α, which emerges through an exit slit S. The latter is earthed, as this helps the insulation of the ionization chamber C which is held at the

positive potential V. With the aid of an electrode with a slit L which is
supplied by a potentiometer, the concentration onto S of the ions which are
produced in C can be improved; simple ion optics shows how this can
be done.

(i) *The existence of lenses*

All the slits are vertical; we shall always be working in conditions where
the optical system possesses a "horizontal" symmetry plane, that of the
analysing field. The optics is here further simplified (Fig. 296); on the one

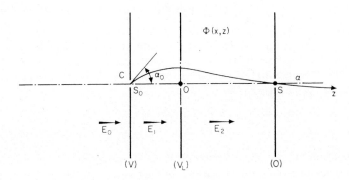

FIG. 296. Focusing within an ion source.

hand, the focusing is only horizontal since the fields depend only upon
x and z so long as the slits are narrow enough; otherwise we should expect
to find vertical focusing (Bertein, 1950). On the other hand, the source
possesses a vertical symmetry plane also, so that the optic axis Oz is rectilinear; the coordinate system $Oxyz$ reverts to the Cartesian form. If we
use the same notation as in the preceding chapter, the focusing equation (23.12)—in the Gaussian approximation—is reduced to:

$$2\varphi\, x'' + \varphi'\, x' + \varphi''\, x = 0. \tag{24.3}$$

This could be demonstrated directly very simply by modelling the analysis
on that of the optics of axially symmetric systems; the Laplace equation
gives

$$\varphi_2 = -\frac{\varphi''}{2}.$$

The slits play the role of a lens, a role which has been analysed to various
approximations (Bertein, 1950, 1951; and Vauthier, 1950, 1954). A particularly simple approximation to the potential distribution on the axis, $\varphi(z)$, is

the distribution created by the juxtaposition of the uniform fields, E_0, E_1, E_2, ... which would be produced by diaphragms without openings; this in fact is Gans' method (§ 4.5.2). In such conditions, the rays are composed of parabolic arcs, joined at known angles of refraction (Davisson and Calbick, 1932) at each diaphragm. We can, however, put less confidence in this kind of analysis for the trajectories within the chamber C; here, we shall simply remark that C forms a kind of immersion objective (§ 8.3 and Chapter 13) with the extractor field E_0, which creates an ionic object at S_0. L has then to image S_0 on S, a requirement which we can analyse by establishing a general conjugacy relation:

$$\frac{E_1}{1 - V'_1} - \frac{E_2}{1 - V'_2} + E_1 - E_2 = 0, \tag{24.4}$$

in which E_1 and E_2 are the fields on one side and the other, V'_1 and V'_2 are the potentials at the object and the image due to these fields; the unit of potential is the potential of L, and the origin is the zero of these energies. In the present application,

$$V'_1 \simeq 0, \qquad V'_2 = \frac{V}{V - V_2}.$$

An examination of the linear magnification γ shows that it is possible to select the position of L in such a way that the passage of the rays through S is assured, when the width of the two slits S_0 and S are comparable and of the usual order of magnitude, namely, a few tenths of a millimetre. With the aid of the Helmholtz–Lagrange relation, however,

$$\gamma \frac{\alpha}{\alpha_0} \sim \sqrt{\frac{v}{V}} \sim \frac{1}{30}, \quad \text{for example.} \tag{24.5}$$

We can foresee that when the concentration is strong ($\gamma < 1$) there is a risk of the angle α being too large for the analysing field to produce good focusing. The sources of Nier (1947)—see Fig. 300—include stops which are placed behind S precisely to limit the aperture of the beam.

(ii) *The influence of the magnetic field of the source (of induction B_0)*

The vertical field exerts a transverse force on the ions, parallel to the axis Ox; the result is simply a slight horizontal displacement of the trajectories and hence of the focusing. Since the slit S is narrow, the ions could miss it; they can, however, always be brought back to S by applying a small potential difference between the two half-plates of which L is composed. We should notice that this phenomenon depends upon the mass m of the ions, so that we shall find discrimination.

24.1.6 The Yield from a Source and Its Perturbations

(i) Calculation of the ion current, i, at the exit, S

We shall always talk in terms of a source of the standard form. We suppose that a given gas enters the ionization chamber at a pressure p. If the electron beam is of intensity I_e and the mean length is l, while n is measured with respect to the total ionization, the ion current created in C is

$$I = n\,I_e\,l\,p. \tag{24.6}$$

For everyday orders of magnitude: $l = 1$ cm, $p = 10^{-4}$ mm Hg, $n = 5$ and $I_e = 100\ \mu$A, we find $I = 5 \times 10^{-8}$ A.

It is, however, clear that the majority of the ions does not reach the slit S_0 so that the current i which emerges through S is only

$$i = K I \quad \text{where} \quad K \ll 1. \tag{24.7}$$

K is difficult to calculate, as the process of extraction of the ions from C is complex. It seems reasonable to take K of the order of $\dfrac{1}{100}$ at most, which would give $i \simeq 5 \times 10^{-10}$ A.

(ii) Discrimination

The formulae (24.6) and (24.7) above can be applied equally well to the total ion current and to the current of any one particular type of ion.

If ions of one mass are emitted by the source more easily than those of another mass—if the ratio $K = i/I$ depends upon the nature of the ions, that is—we say that there is discrimination within the source. This discrimination may have several origins, as K is an overall measure of a complex type of transmission.

In spectrometry where the potential V is scanned, the value of V which corresponds to reception is higher for light ions than for heavy ions; the same is true, therefore, of the values of the field E_0 by which the extraction is for the most part produced. The discrimination, in consequence, favours light ions (Coggeshall, 1944; Washburn and Berry, 1946). We can estimate the importance of this phenomenon by comparing the results of an analysis of CCl_4 obtained by the two methods of scanning (Blears and Mettrick, 1947). The units are arbitrary (see the table on page 826).

When the scanning is magnetic, the magnetic field B_0 of the source is a cause of discrimination, at least in the normal situation where it is held constant; the mass m for which the half-plates have been corrected tends to be favoured with respect to the others. This defect can be overcome by using only relatively low values of B_0, of about 100 or 200 gauss.

THE IONIC INTENSITY OF IONS OF MASS m
AS OBTAINED AT THE RECEIVER

m	B scanning	V scanning
35	100	100
47	133	97
82	118	52
35, 37 (abundant chlorine isotopes)	3·10	3·17

Another type of discrimination affects ions which are formed with an initial kinetic energy, as this energy clearly modifies the trajectories within C, and hence the extraction phenomenon.

In fact, however, we shall see later on that if it is important to reduce discrimination in the measurements of the relative isotope abundance as far as possible, this is no longer true in actual chemical analyses. In the latter case, the important thing is that we should work in constant conditions, and hence we must struggle to overcome the parasitic processes which are more harmful than simple discrimination.

(iii) *The various parasitic effects*

It is the fluctuating nature of these effects which is dangerous. We classify them as hysteresis, "memory" effects, and ageing; we shall mention only the most important.

The substances of which the source is composed tend to be outgassed in the course of operation, because of the relatively high temperature at the ribbon t (which can reach 150°) and the electron bombardment of the surfaces. Parasitic lines appear in the spectra.

The gas to be analysed may be thermally modified (dissociation) in the vicinity of t. This phenomenon varies in time with the ageing of t or its possible replacement.

The gas may affect the emissive power of the ribbon itself (by modifying the work function); the hydrocarbons may increase it by one-third, while oxygenized substances have an even clearer reducing effect. Adjustment of the source electrically considerably attenuates these perturbations.

There is a risk of the metal surfaces becoming charged, by being covered with films of impurity; the trajectories are modified, especially in C.

If we are to operate in the constant conditions which are necessary for analyses, these defects must be reduced as much as possible if fluctuations which clearly exceed 1 per cent are to be avoided. The metal parts should be made of tantalum, stainless steel, or nichrome and the insulators of pyrex or melted silica. The source and the analysing tube must be dehydrated (250 to 300°), especially after exposure to air.

It is recommended that the emissive ribbon should be conditioned and for further details, the reader should consult the book by Barnard.

24.2 COLLECTION OF THE IONS

24.2.1 The Photographic Plate (Spectrographs)

The ion beams which emerge from the analysing field can be made visible with the aid of a fluorescent screen composed of willemite or zinc sulphide, provided their energy (V) is at least of the order of 5 kV. For recording the spectra in the spectrograph, the photographic plate has been in regular use since its introduction by Thomson (1911) in his parabola method. As a result of the very low penetration by the ions, useful emulsions contain very little gelatine, and are of very fine grain (for example, Schumann or Ilford plates). The potentials V should be relatively high, for example 20 kV or more. The exposures last at least a few minutes.

Once the law which describes the blackening of the plate had been determined experimentally, we could in principle measure the intensities i and hence arrive at spectrometry proper—this would be less straightforward than making electrical measurements at the collector, however. We might mention that the darkening is not strictly proportional to the product it, which measures the number of ions received (Lichtblau and Mattauch, 1941).

24.2.2 The Collector (Spectrometers)

By scanning either the magnetic induction B or the potential V, the various beams are brought successively into a small Faraday cage C which is placed behind a slit S'. C thus captures the current i which corresponds to each mass in turn (Fig. 297). The width s' of S' must be chosen to be at least equal to the width of the line s_1' and in fact, we use a decidedly higher value, so that the beam i is wholly received not only for a particular value of B (or of V) but for a small range of neighbouring values; in these conditions, each peak which is recorded has a flat top, which simplifies the interpretation.

The resolving power of the instrument is affected by s'. The slit S' is earthed, and the same is effectively true of C (see below).

The addition of an intermediate stop D' which is held at a negative potential of the order of 20 V prevents the secondary electrons which are produced at C by ion bombardment from leaving, as their departure would falsify the measurements of the current i. Another diaphragm D'', held at an appreciable positive potential (smaller than V, however), improves the spectrum by eliminating the continuous background which is produced by ions which have been scattered by the residual gas. Since the energy of

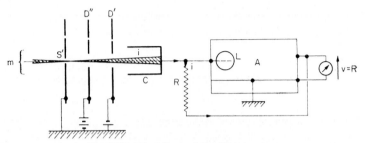

FIG. 297. Reception of the ion current.

these scattered ions is less than V, D'' creates a potential barrier which prevents them from reaching the collector, without disturbing the useful ions at all (Nier, Stevens *et al.*, 1947).

A more drastic means of clarifying the spectra consists in holding the whole collector at the same potential as the source; the most convenient arrangement is to connect both to earth, and the analysing tube will then have to be connected to the high tension $(-V)$. This arrangement has been used to study the ions which are produced by the dissociation of metastable ions which have been formed in the source (§ 24.1.4. See Hipple, Fox and Condon, 1946; and Hipple, 1947). It transpires that all such ions have an energy smaller than V, just as do the scattered ions.

Rather than try to prevent the emission of secondary electrons, we can, on the contrary, take advantage of it as a means of amplifying the value, i, of the current received considerably; it is possible to achieve this by placing a multiplier in front of the Faraday cage (Allen, 1939, 1950; Nier, Roberts and Franklin, 1949; and Barnett, Evans and Stier, 1954). The incident current will, of course, still have to be amplified by a valve circuit, but the value of electron multiplication resides in the fact that the importance of the background noise is reduced, to such an extent that the limiting value of i which can be detected is reduced from 10^{-15} to 10^{-17} A. This is a well-known phenomenon, which was first put to use in television where the particles to be detected are photons. In the present case where ions replace the photons, the theory is the same (Zworykin and Morton, 1955) but to obtain long-term stability seems to be difficult as the flow of ions damages the first plate of the photomultiplier.

24.3 THE ELECTRICAL CIRCUITRY OF A

SPECTROSCOPE

The circuits are fed from an alternating supply, in commercial instruments at least. Arrangements for rectifying and smoothing the potential must be provided therefore, to avoid fluctuations prejudicial both to the resolution and to the sensitivity. These arrangements are not markedly different from the ones we have described in connexion with the microscope (Chapter 18).

24.3.1 Measurement of the Ion Currents i (Spectrometers)

The ion current which falls on the collector is at most of the order of 10^{-10} A, and it is usually measured with the aid of an electrometer, with which currents down to 10^{-15} A can be detected. The current i flows through a very high resistance R, $10^{11} \Omega$ say, which is connected to the collector C (Fig. 297); we have simply to measure the potential across the terminals $v = Ri$. For this, R is placed at the input of an electrometer valve L, which is a vacuum tube in which the control grid g is highly insulated by means of

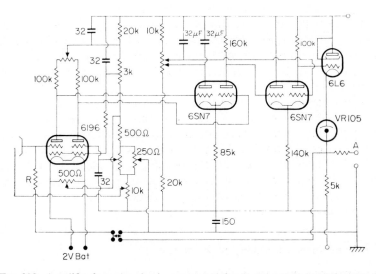

FIG. 298. Amplifier for measuring ion currents (after Bertein, Vastel and Beaussier).

various special features of the construction and operation: the presence of a supplementary grid, an under-heated filament, low potentials to avoid ionization of the residual gas (10 V at most), and a special structure to avoid both the emission of very soft X-rays by the plate and photo-electric emission by the grid in the presence of this weak radiation.

These are serious inconveniences in making a direct reading with a galvanometer, for example, of the measurement obtained at the exit of L. Furthermore, the characteristics of L are very curved, and hence there is a complete lack of proportionality. Finally, parasitic capacities and the capacities of the screening produce a capacity C' shunted across R; the time constant, $C'R$, of this combination may be more than a second which makes readings on the peaks excessively slow. These inconveniences are remedied by placing a D.C. amplifier A after the electrometer valve stage; A has a high gain, of the order of 1000, and R is arranged in such a way as to ensure unity feedback. The principle is shown schematically in Fig. 297; a practical example is illustrated in Fig. 298. In these conditions, a galvanometer connected to the exit indicates the potential $v = R\,i$, and the reading is therefore proportional to i; further, the time constant is now negligible. We can show moreover that the collector C remains virtually at earth potential.

The smallest current which can be detected with an extremely carefully constructed amplifier is of the order of 5×10^{-16} A. The spectrometer is therefore capable of displaying peaks, the relative heights of which are only separated by one part in 10^5 or 10^6; this is the explanation of its extreme sensitivity, and its ability to detect the constituent parts of a mixture even in feeble traces. In this continual striving after sensitivity, it is particularly important to reduce the phenomenon of *drift* which is peculiar to D.C. amplifiers. On this point, the study by Bonne and Margoloff (1951) is to be consulted.

Very low current measurements are tending more and more to use vibratory modulators which allow the steady potentials to be replaced by alternating potentials, and the methods of amplifying alternating currents to be used, as these are considerably better so far as the stability is concerned. We should remember, too, that it is possible to produce the first amplification with the aid of a multiplier placed before the collector; this procedure, which is very sensitive, is advantageous where we are making measurements on very small quantities of isotopes.

24.3.2 Continuous Recording of the Results

The measuring circuitry is completed by a recording device, which must be able to accept spectra in which the various peaks may be of very different heights (Figs. 299 and 300). For this purpose, it is possible to devise a means of recording photographically the spots of a number of galvanometers of different sensitivities, functioning simultaneously. Alternatively, we can use a pen-recorder which is coupled in such a way as to record automatically the sensitivity of each of the peaks (Washburn, Wiley and Rock, 1943; Surdin, 1949; and Bonne and Margoloff, 1951). To obtain the scale on which the masses are measured, which is necessary for interpreting the spectrum, we can use a mass-marker, which is based upon a continuous measurement of the induction B during the magnetic sweep (Hipple, Grove and Hickam, 1945). It usually takes a few minutes to record a spectrum containing masses between 24 and 100.

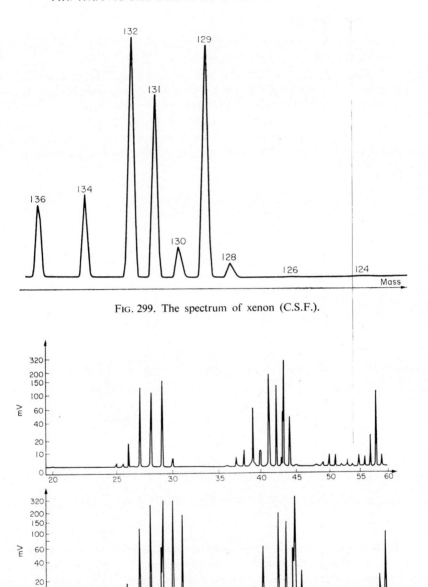

FIG. 299. The spectrum of xenon (C.S.F.).

FIG. 300. Spectra of hydrocarbons (Inst. du Pétrole).

24.3.3 Other Circuits

To obtain the acceleration potential V and the current i_B with which the electromagnet (B) is supplied requires two carefully filtered rectifying circuits; the stability often reaches one part in 10,000 (for fluctuations of 10 per cent at the mains). A still higher stability can be obtained by an original procedure in which an auxiliary spectrometer tube is used (Nier, Roberts and Franklin, 1949). The values of V and B which are produced must be variable; in analytic spectrometers with a high resolving power, typical ranges would be 500 to 3000 V and 1000 to 6000 gauss respectively $(i_B \sim 100$ mA$)$.

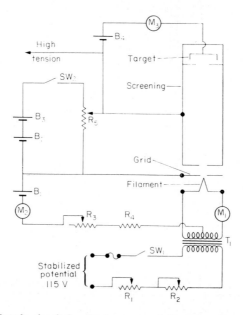

FIG. 301. Supply circuit for the ion source (after Winn and Nier, 1949).

Sources which depend upon electron bombardment require a variable supply for heating the emitter t. Essentially, it is a question of stabilizing the electron beam I_e which produces the ionization. The ways of doing this differ appreciably depending upon whether the feed-back acts upon the filament supply (Ridenour and Lampson, 1937) or upon the electrodes of the supplementary control (Winn and Nier, 1949; Caldecourt, 1951); the latter method has the advantage of not causing thermal variations in t which might influence the spectra (Fig. 301). There are, however, always parasitic fluctuations in a source (§ 23.1.6) which are too complex to be completely removed by straightforward regulation.

24.4 THE VACUUM AND THE GAS CIRCULATION

24.4.1 The Vacuum

Within a spectroscope tube, the vacuum must reach at least 10^{-6} mm Hg. If the residual pressure is higher, some of the ions will be scattered, and both the sensitivity and the resolving power will be reduced; further, the residual gas takes part in the ionization at the ion source, and the result is a number of parasitic lines or peaks. These can be avoided by a careful secondary pumping system: a mechanical pump together with one or more diffusion pumps operating in parallel at various points along the tube, a liquid air trap or a trap consisting of solid carbon dioxide in acetone, the minimum number of demountable joints, and no grease in the secondary vacuum system. Periodic dehydration is recommended (250 to 300°C).

Despite the extreme care which is given to producing the vacuum, the very high sensitivities of mass spectroscopes result in a weak parasitic residual spectrum, which must be taken into account; the spectrum contains in particular H_2O, CO (in certain metallic apparatus), the various mercury isotopes (Hg^+ and Hg^{++}), or traces of the hydrocarbons, depending upon whether the diffusion pump works with oil or mercury.

24.4.2 Circulation of the Gas which Is to Be Analysed

There is also, moreover, an essentially dynamic vacuum as the substance to be analysed has continually to be circulated by the pumping system. As a concrete example, we shall consider the case which occurs the most frequently in which we have a gaseous mixture (Fig. 302). The gas is contained within a flask B_0, which is sealed to the entry duct which has previously been evacuated by an independent pumping unit. By opening the taps, some of this gas is made to flow at low pressure into the reservoir B. A steady flow is released from B, passes through a very narrow opening or leak F, and reaches the ion source; here, the gas moves into the region of the vacuum proper, the tube T and the associated pumps. The spectroscopy can then begin.

The composition of the gaseous mixture is defined by the partial pressures P, P', ... of its various constituents G, G', ... in the flask B. The result of the ionization is that the partial pressures in the chamber C which forms the ion source are now p, p', ...; with the present notation, the overall pressure in C is therefore not p but $p + p' + \cdots$. From the point of view of the analysis, it is of course important to examine the ratios p/P ...; we distinguish between the possible types of flow.

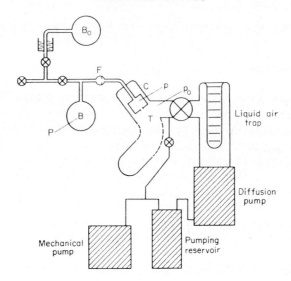

FIG. 302. Circulation of the gas to be analysed, and the pumping system.

(i) *Flow which is everywhere molecular*

This case occurs when the mean free path of the molecules is everywhere appreciably higher than the transverse dimensions of the gas circuit; this simply requires the pressure P to be less than about 10^{-2} mm.

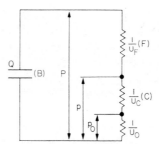

FIG. 303. The electrical circuit equivalent to the circuitry of the circulation and pumping of the gas.

For each constituent $G, \ldots$ the type of flow which is known as molecular is then established independently. This flow is the result of the corresponding pressure p just as the passage of electricity through a conductor is a result of the electric potential difference; we can make use of this analogy to construct the equivalent electrical circuit diagram (Fig. 303). The quantity Q of gas G is measured, for example, by its volume in litres, referred back to N.T.P. Each section of the piping is characterized by its pumping rate U (l. sec^{-1}) which plays the role of conductance. B is thus represented by a condenser with a charge Q at a potential P; this condenser is discharged through a series of resistances which represent the leak F $(1/U_F)$, the slits in the ionization chamber C $(1/U_C)$, and the piping of the actual vacuum, the trap T, and the secondary pump $(1/U_0)$.

We shall summarize the conclusions which can be drawn from an examination of the equivalent circuit diagram; first of all, the pressures which are desirable at the various levels, P, p, p_0—say 10^{-2}, 10^{-5} to 10^{-4}, and 10^{-6} mm Hg—determine the relative orders of magnitude of the pumping rates; these latter can easily be attained, however: $U_F \ll U_c \ll U_0$ (for example, $U_0 = 20$ l. sec^{-1}; $U_c = 1$ l. sec^{-1}; $U_F = 10^{-3}$ l. sec^{-1}). The leak will simply be a circular hole about $1/_{10}$ mm in diameter.

If we compare now the various constituents G, G' ..., the pumping rates in principle vary as $M^{-1/2}$ when we pass from one constituent to another (M is the molecular mass), and ratios such as $\dfrac{U_F}{U_c}$, ... are thus independent of the type of gas. The same is therefore true of the pressure ratios:

$$\frac{p}{P} = \frac{p'}{P'} = \cdots \tag{24.8}$$

In other words, the (relative) composition in C is the same as that in B. The speed U_0 often does not follow an $M^{-1/2}$ law, however; certain gases are condensed at the liquid air trap, which has the effect of sharply increasing the corresponding speed. In fact, the repercussion on p is only slight if $U_c \ll U_0$, and even though the equalities in equation (24.8) are unreliable, the ratios p/P, ... at least are constant provided the trap is constantly topped up, and we shall see later on that this is adequate for the majority of analyses. Since the pressure P in the reservoir B is so low, it is possible for B to become discharged to an appreciable degree in the course of an analysis, and this is one of the inconveniences of the molecular regime. The $M^{-1/2}$ law implies that light gases will be discharged more rapidly than heavy gases—the proportion of heavy gases in the mixture increases steadily, and we say that there is a fractionation.

(ii) *Partially viscous flow*

It is more convenient that the pressure P in B should be of the order of at least a few millimetres or centimetres of mercury. The flow is then viscous (Poiseuille) as far as the leak F; beyond F, the pressure must drop sharply to the value p, and the flow becomes molecular. Here, then, the resistance of the leak F must be considerably higher than in the preceding case. A flow pattern of this kind is very complex, and has been studied by Honig (1945) in particular. It is true, however (Nier, 1948), that if F consists of a long capillary tube (12 cm × 0·15 mm for example), the gas flows out of B as a unity, without fractionating, so that the composition within B remains unaltered. In the other part, however, the light gases flow more rapidly beyond the leak (where the regime is molecular) and instead of (24.8) we shall have

$$\frac{1}{\sqrt{M}} \frac{p}{P} = \frac{1}{\sqrt{M'}} \frac{p'}{P'} = \cdots \tag{24.9}$$

The composition in C is thus no longer the same as that in B, and further, the ratios p/P, ... are no longer constant. Some suitable method for interpreting the results of the analysis is necessary. Adjustable leaks have been adapted for this type of flow (Fig. 304).

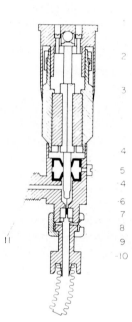

Fig. 304. Adjustable leak (Ewald, 1950)
A steel needle (6), fixed at the end of a brass rod (2), slides in a very narrow channel pierced in a lead plate (7); its displacement is driven by the knob (1), and read accurately on a scale (3). The space between the joints (4) is filled with a mixture of vacuum grease and oil, through the opening (5). A bellows, soldered on the piece (10), is tightened on the flange of the needle valve output by the screw-nut (9).
(11) Gas inlet.

24.5 THE VARIOUS TYPES OF MEASUREMENT
IN MASS SPECTROSCOPY

The measurements fall into two distinct categories; these are concerned either with the positions of the bands and in this case, it is the atomic masses which are determined spectrographically, or with the intensities of the bands, with the object of making quantitative analyses by spectrometry.

In the latter category, we shall also have to consider separately the situations where we are concerned with two isotopes (measurement of relative abundance) and with ordinary chemical analysis.

24.5.1 Determination of the Atomic Masses (Spectrography)

The bands here must be extremely fine, and hence one of the double focusing spectrometers described in Chapter 22 must be used, in particular the Mattauch instrument. Figure 305 shows as an example the spectrum obtained in the vicinity of the mass $M = 20$ together with the corresponding photometric curve. After identifying the ions which correspond to the various lines, their molecular masses M are determined from their positions X

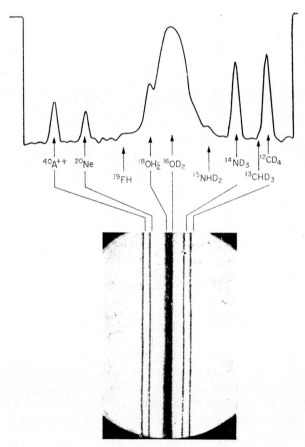

FIG. 305. The multiplet which is obtained in the neighbourhood of mass 20 (after Mattauch, 1938), and its photometric recording.

on the photographic plate Π. In principle, we have simply to apply the relation which is known *a priori* to relate M and X, namely:

$$M = K X^2 \quad \text{(Mattauch's instrument)}. \tag{24.10}$$

The origin of the abscissae X and the theoretical value of K—the situation is complicated by the presence of fringe fields—are not known sufficiently precisely for the rough method above to be satisfactory. It is better to determine the values of M little by little, by making measurements upon small spectral regions, checked section by section.

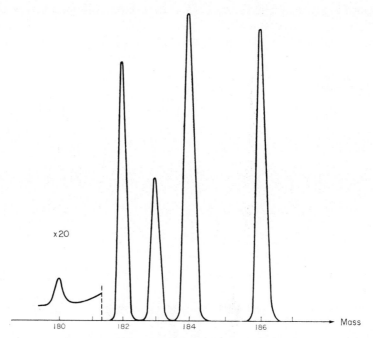

x 20

180   182   184   186   Mass

FIG. 306. Spectrometry: the spectrum of the isotopes of tungsten (after Inghram, 1946).

The most favourable ion sources will be those which emit a wide variety of ions (spark sources, for example).

We shall examine this point more precisely by describing a typical method which is known as the "method of doublets". We recall first of all that with the standard convention according to which the atomic weights are defined with respect to $^{16}O = 16$ (which is known as "the physicists' scale"†) the various masses M are all very nearly whole numbers; the ions which are

---

† In the so-called "chemists' scale" which is usually employed in chemistry, the mass 16 is given to the mixture of isotopes of which atmospheric oxygen is composed.

associated with the same integer thus produce bands which are very close together, and hence form a doublet or, more generally, a multiplet in the spectrum. Close to $M = 20$ for example (see Fig. 305), we find $^{40}A^{++}$, $^{20}Ne^{+}$, $^{18}OH_2^{+}$ and $^{16}OD_2^{+}$. To illustrate the principle of the method of determining a mass, we consider the example of ^{13}C, given the masses of H and ^{12}C, namely

$$H = 1 \cdot 008130, \qquad {}^{12}C = 12 \cdot 003861.$$

It is easy to obtain a spectrum which contains the bands corresponding to ^{12}C, ^{13}C, and ^{12}CH, the last two of which form a doublet. The mutual separations are measured with a microscope, giving

$$({}^{12}C, {}^{12}CH) = \triangle X_1 = 9 \cdot 9624 \text{ mm},$$
$$({}^{12}CH, {}^{13}C) = \triangle X_2 = 0 \cdot 0427 \text{ mm}.$$

If we designate the masses by their actual chemical symbols, we obtain from equation (24.10) the relation

$$\frac{\sqrt{{}^{13}C} - \sqrt{{}^{12}CH}}{\sqrt{{}^{12}C} - \sqrt{{}^{12}CH}} = \frac{\triangle X_2}{\triangle X_1}, \tag{24.11}$$

and hence

$$^{13}C = 13 \cdot 007583.$$

Neither the position of the origin of X nor the value of the constant K has appeared. Measurements of this kind do, of course, require special precautions; on this point we refer the reader to the book by Ewald and Hintenberger. The accuracy is not sufficient to confirm the correctness of all the digits which are normally quoted—only four or five places of decimals, according to the various different cases, can be guaranteed.

24.5.2 Quantitative Analysis; Relative Isotope Abundances (Spectrometry)

Figure 306 shows the spectrum which is obtained for the isotopes of tungsten. In these analyses, the use of an ion source depending upon electron bombardment is recommended, above all for gaseous mixtures. Suppose, for example, we have to determine the relative abundance of the barium isotopes ^{10}B and ^{11}B, that is, to determine the ratio of their concentrations $[^{10}B] : [^{11}B]$ in natural specimens. We analyse spectrometrically the gaseous compound BF_3 which is a mixture of the two constituents $^{10}BF_3$ and $^{11}BF_3$ which we shall designate by G and G'. The relative abundance we require is equal to the ratio of the partial pressures in the analysing flask B, that is

$$\frac{[^{10}B]}{[^{11}B]} = \frac{P}{P'}.$$

It is natural to hope that this can simply be obtained as the ratio of the heights of two homologous peaks, i, i', which are received at the collector;

this would imply that

$$\frac{i}{P} = \frac{i'}{P'},\tag{24.12}$$

This is true provided none of the isotopes is privileged with respect to the remainder, that is, provided there is no discrimination along the chain of processes which leads from P to i, namely:

(a) P (flask) $\rightarrow p$ (ionization chamber): gaseous flow,
(b) $p \rightarrow I$ (ionization chamber): ionization of the gas,
(c) $I \rightarrow i$: passage through the ion source.   (24.13)

If there is to be no discrimination, the process (a) must be a molecular flow; we can, however, use a partially viscous flow provided a correction factor is introduced which takes account of the molecular masses according to (24.9). So far as (b) and (c) are concerned, the ionization probabilities must be equal as must the transmission probabilities in the passage through the ion source [the coefficient K of relation (24.7)]; these requirements are satisfied in practice, as the relative separations between the masses of the isotopes are small, and the magnetic scanning has an averaging effect (see the case of Cl in the table given in § 24.1.6). The same would not be true for a mixture of H and D; in such a case, the standard method of analysis described in the next section would be suitable.

Save for this case, the ratio of the abundances is equal to the ratio of the heights of the peaks; we can obtain this latter ratio to an accuracy of better than 10^{-3} (as far as 0.2×10^{-3}) if a double collector is used so that two isotopes can be received simultaneously and their heights compared by an opposition method (Nier, Ney and Inghram, 1947).

24.5.3 Ordinary Chemical Analysis (Spectrometry)

Here the problem is different; given an arbitrary mixture of gases $G, G', \ldots$ it is no longer a question of obtaining "homologous" peaks which can be assigned to the various different components of which we need only measure the relative heights. We have to measure the individual spectra of each of the constituents beforehand—we suppose them to have been determined at N.T.P. (Fig. 307). These spectra, like the spectrum of the mixture, are defined by the heights of the peaks produced by a certain number of ions with molecular masses $M_1, M_2, \ldots$; the results may be tabulated thus:

Masses		M_1	M_2
Spectra $\Big\{$	G	i_1	i_2
	G'	i_1'	i_2'
	Mixture	j_1	j_2

Here, the question of discrimination does not arise; the results can be interpreted very straightforwardly if the processes which lead from P to i proceed independently for each of the constituents, which implies a molecular flow. In these conditions, the spectrum is formed by superposition;

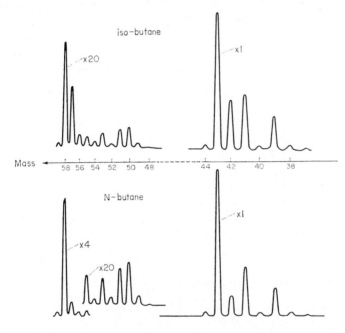

FIG. 307. Spectrometry: the spectrum of normal butane and of isobutane (after Hipple, 1942).

if P and P' are the partial pressures of the constituents, that is, the unknown quantities we are seeking, then

$$j_1 = P\,i_1 + P'\,i_1' + \cdots$$
$$j_2 = P\,i_2 + P'\,i_2' + \cdots \tag{24.14}$$
$$\cdot\;\cdot\;\cdot\;\cdot\;\cdot\;\cdot\;\cdot\;\cdot\;\cdot\;\cdot\;\cdot\;\cdot\;\cdot\;\cdot$$

If there are n constituents, we have simply to write down this system of n linear simultaneous equations for the n lines $M_1, M_2 \ldots$ which have been measured, and to solve for $P, P', \ldots$ When the streaming is partially viscous, and the gases in the mixture flow as a single unit, this method can be used provided the spectra of the constituents are not determined under normal conditions but for equal linear flow velocities. This can be achieved by making measurements in which an internal reference gas or standard is used. A great deal of information about this point is available, primarily in articles dealing with industrial applications (mixtures of hydrocarbons).

Analyses in which six components are present can be performed without difficulty, and if we use an electronic computer to solve equations (24.14), even more constituents can be handled. The accuracy obtained depends upon a great many circumstances; the fundamental limiting factor is the lack of perfect constancy with time. At present the order of magnitude of the error in each constituent is 0·5 per cent.

24.6 AN OVERALL VIEW OF PRESENT-DAY INSTRUMENTS, AND THE FIELDS IN WHICH THEY ARE EMPLOYED

The fundamental part of a spectroscope is its analysing field, and in studying this we have had occasion to review the main kinds of instrument. We shall now point out the essential features of the models at present in use; so far as the detailed description and the details of the construction are concerned, we shall give a few references to typical examples. Detailed bibliographies are to be found in the articles by Roth (1949), Reis *et al.* (1950) and Dibeler and Hipple (1952) and above all, in the article by Reis (1955).

24.6.1 Instruments for Measuring Masses Accurately

The number of spectrographs is still very limited; we recall the particularly advantageous design of Mattauch's instrument. A recent model has been developed by Ewald (1946, 1950 and 1951) which gives a resolving power of 30,000; the analysing tube is built from solid bronze and steel. To achieve the desired properties, the construction must be extremely precise and numerous vacuum-sealed control units are required. We should mention in particular the possibility of employing an ion source using either gas discharge or sparking where the exit slit consists of two small parallel cylinders which can be altered in orientation bodily (Fig. 308).

Under this same heading, we ought also to mention the double focusing spectrometer of Nier and Roberts (1951), which is appreciably more rapid and convenient to use, but with which it does not at present seem possible to obtain the same resolving power.

24.6.2 Analytic Spectrometers

Spectrometers (with single focusing) are considerably more numerous, as a result of the extreme simplicity of their construction and the wider range of their applications; for some years now, they have not only been constructed individually in the laboratory but have also been made commercially.

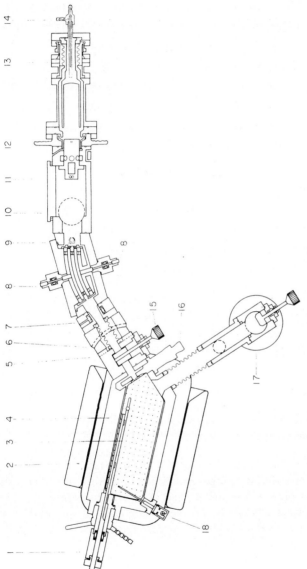

Fig. 308. Mattauch's spectrograph, as built by Ewald. 1: the support for the photographic plate 3. 2: the yoke of the electro-magnet. 4: brass wall of the vacuum chamber. 5, 6, 7, 9: means of adjusting the direction of the beam *in vacuo.* 8: leads for the deflexion prism. 10, 17: pump connexions. 12: insulation between the (earthed) instrument and the source. 13: ion source, to which the gas is admitted through 14. 15: vacuum cock. 16: adjustment for the intermediate diaphragm.

Bleakney's instrument (1932) already displayed a notable evolution in comparison with the first version of Dempster (1918); instruments of truly "modern" conception appeared still later, however: Nier (1940) and Hipple (1942)—see Fig. 309. The sector fields are 60° and 90° respectively, and a large proportion of the instruments used to-day is derived directly

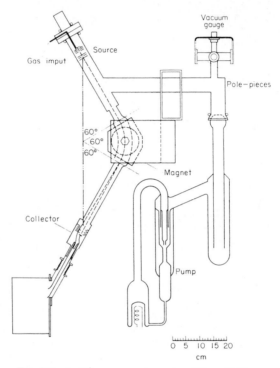

FIG. 309. A 60° mass spectrometer (Nier, 1947).

from these models. There are, however, always partisans of the 180° field, as the arguments over the advantages and disadvantages of the various angles cannot be said to impose any one choice rather than another.

Present instruments tend to be all metal, and easily demountable; they can be separated into several categories according to their various qualities which in turn determine their applications.

(i) *Analytic high resolution spectrometers*

Here, the resolving powers range between about 150 and 400; at present, there are more than 200 of these instruments in the U.S.A., and the resolving power of the most accurate instrument commercially available can reach 400

(Brown, Mitchell and Fowler, 1941; Graham, Harkness and Thode, 1947; Nier, 1947; Bertein and Vastel, 1949). Since these instruments are of commercial importance we shall briefly describe a typical example.

We consider the standard spectrometer due to Nier, the particularly simple structure of which is represented in Fig. 309. The sector is 60° and the radius R of the trajectories is 15 cm. The analyser consists of a copper

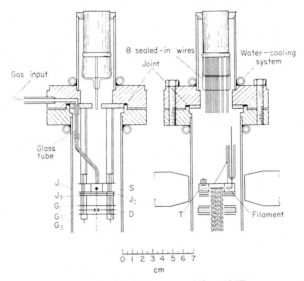

FIG. 310. The ion source (Nier, 1947).

tube 3 cm in diameter which is flattened transversally to 13 mm in such a way as to fit into the gap of the electromagnet. A wider part made of inconel carries the ion source, and is coupled to the pump circuit by a spherical joint. The diffusion pump is not made of metal, here, but of pyrex; the pumping rate is about seven litres per second at the spherical joint. The vacuum is measured with a Penning gauge (a current of 1 μA for a vacuum of 10^{-6} cm Hg).

The ion source (Fig. 310) can easily be removed, as it is fixed to a cover made leakproof by a Teflon joint; a thorough baking is possible, since the fixing clamps are water-cooled. The electrodes of the source are made of nichrome, and its magnetic field is 150 gauss. A double collector is employed for collection of the ions as this permits two neighbouring peaks to be received simultaneously and compared by a null method. The focusing is adjusted by shifting the analysing tube with respect to the electromagnet which is held fixed.

(ii) *Spectrometers with a medium resolving power* (*50 to 80*)

These spectrometers are particularly suitable in industry for controlling chemical reactions or separations (as "monitors"); they are much lighter, more robust, and less expensive. They may be fitted with a permanent magnet instead of an electromagnet, and with an oscilloscope on which the spectra can be displayed directly. This class of instruments is at present passing through a period of particularly rapid development (Siri, 1947; Walcher and Gifford, 1954).

(iii) *Simplified spectrometers, for detecting leaks*

It is these which are at present the most numerous; they are used to detect the presence of gases such as helium in tubing or in some given space. They are, of course, equally convenient for analysing sufficiently light gases (Nier, Stevens, Hustrulid and Abbott, 1947; Cassignol, Geller and Moreau, 1953).

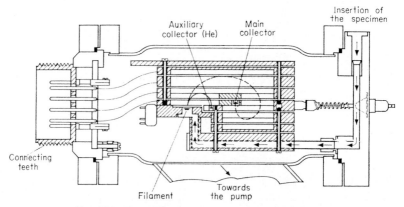

FIG. 311. The "Diatron" spectrometer (Consolidated).

It is interesting to notice that the focusing principle, which is theoretically perfect, in which crossed fields are used (cycloidal trajectories—Bleakney and Hipple, 1938), has recently been taken up again under the name of the "diatron"; the corresponding resolving power reaches 150 despite the extremely small dimensions, of the order of centimetres (Fig. 311).

24.6.3 Spectrometers for the Separation of Isotopes

The possibility of separating isotopes in considerable quantities with a spectrometer is related to the existence of very intense ion sources which were not available to the early workers who first envisaged isotope separation (Aston, 1927; Morand, 1927). Progress has been made in this direction, and there are now several separators which furnish the isotopes which are

required for the increasing number of applications of radioactive isotopes; the instruments at Oak Ridge in the U.S.A. ("calutrons") produce
more than 200 different isotopes. In France, we mention the separator of
Bernas (1953; see Fig. 312), with which masses of 10 to 100 mg can be

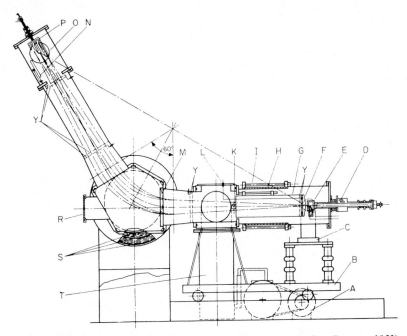

Fig. 312. A spectrometer for the separation of isotopes (after Bernas, 1953).
E = ion source; C = electro-magnet of the source; D = air-lock; F = accelerating
electrode supported by the copper cylinder G; H = pyrex tube; I = isolator rods;
K = cam for adjusting the inter-electrode distance EF; L = high tension plug;
M: demountable part of the pole-pieces; P = collector; O = pyrex window; N = baffle of the collector chamber; R = window for studying the undeflected beam;
S = cooling circuit of the main-magnet coils; T, A = vacuum pumps; B = car-
riage; Y = polarized grids for neutralizing space-charge.

obtained in 24 hours, using ion beams of several mA; the instrument
consists of a spectrometer with a sector field of 60°, and larger dimensions
than the analysing instruments: the radius of curvature of the trajectories,
$R = 50$ cm; the width of the gap is 7 cm; $B_{max} = 12,000$ gauss; $V_{max} = 50$ kV.
An instrument of the same type has been constructed in Amsterdam by
Kistemaker.

24.6.4 Present-day Applications

In conclusion, we shall list the various fields in which the mass spectrometer is employed:

(a) To make accurate measurements of the atomic masses—to obtain the mass defects involved in balancing the energy in nuclear reactions.

(b) To measure relative abundances—to study the evolution of reactions by the indicator technique.

(c) To perform analyses by isotopic dilution (in biology) and to determine the geological time-scale.

(d) To make chemical analyses, as industrial monitors; in particular, to study the chemistry of the hydrocarbons.

(e) To study molecules: determination of the ionization potentials or ion threshold potentials, and hence calculation of binding energies; to study dynamic chemical processes involving metastable ions and free radicals.

(f) To study nuclear reactions.

(g) To detect small traces of matter—leaks in vacuum systems, impurities in metals.

(h) To separate isotopes.

The diversity of these applications explains the considerable importance of mass spectroscopes both in ordinary chemistry and in the most modern kinds of research.

24.7 MASS SPECTROSCOPES WITH A HIGH FREQUENCY FIELD

We shall conclude this study by outlining briefly the principle of a class of instruments which also separate ions according to their masses, but in a manner very different from that of the normal spectrometer. In these instruments, a high frequency electric field is employed—for a given supply frequency, ions of a certain mass will be given an oscillatory motion, the frequency of which matches that of the supply, and will pass through the instrument. Only these ions fall on the collector. The remaining ions cannot resonate, and the instrument is so designed that they are rejected either towards the walls or towards the source.

The omegatron (Goudsmit 1948, Hipple and Thomas 1949) is a small cyclotron. The ions are produced in the gap of an electromagnet by an ionizing electron beam, as in an ordinary source. If the frequency of the supply to the "Dees" is f, the only ions which can resonate are those with a mass M given by

$$M = \frac{e\,B}{2\pi f};$$

e is the ionic charge and B is the magnetic induction. These ions trace out a spiral and fall on the collector which is situated on the periphery. Ions with masses other than M remain near the centre. This instrument has been used to determine the masses of the light elements with very high precision. An instrument which is comparable but which is based upon the principle of the linear accelerator has been built by Bennett (1950). This is a linear omegatron, which does not require a magnetic field. The high frequency electric potential is applied to a series of equidistant grids, and the only ions which can pass through the instrument are those for which the time of transit between the grids is equal to half the period.

The omegatron (Wagener and Marth, 1957) is extremely sensitive and still gives useful signals at pressures of 10^{-10} mm; this is the instrument which is best suited to the study of the composition of the residual atmosphere in evacuated apparatus and sealed-off tubes. Recently, ordinary spectrometers have equalled this performance in this application, but at the cost of special construction. The metal spectrometer tube is welded to the glass entry and exit tubes; both the taps and the joints are metallic (either lead or aluminium). We can therefore bake the whole apparatus at more than 450°C, and draw off the residual gases by means of an absorption effect; this latter can either be produced by means of titanium vapour pumps or by actual absorption in an ionization gauge. The research performed recently by Bayard and Alpert (1950) has led to the construction of ionization gauges which are capable of simultaneously measuring pressures of 10^{-11} mm Hg accurately, and providing an efficient "pumping" action as soon as the pressure falls below 10^{-6} mm Hg.

Another invention which also allows these very high vacuums to be attained in practice is the "ion pump", in which the residual atmosphere is ionized and drawn off by electrical forces, and this whole class of instruments opens up a new and interesting field of research; the phenomena associated with absorption by surfaces can in particular be measured with certainty.

Recently Paul and his collaborators (1953 and 1958) have proposed a type of instrument depending upon the principle of strong focusing which we described in Chapter 10. These lenses are convergent in the plane xOz and divergent in the plane yOz. With a chain of such lenses we can produce a mean overall convergence in all directions—a result of the mutual compensation—and hence direct a beam of particles along the axial zone (§ 10.5). Paul suggested that the electrodes of a single but very long lens should be fed with an alternating potential. It is clear that the particles will then be subjected to a focusing field for one half-period and a defocusing field during the subsequent half-period. So far as the particles are concerned, the situation is exactly the same as if they were passing through a continuous chain of strong convergence lenses supplied with fixed potentials.

First of all, we set out the equations of motion for a particle (mass m, charge e and velocity $v = \dot{z} = \sqrt{\dfrac{2e\Phi_0}{m}}$) in a field with quadrupole symmetry, independent of z but varying sinusoidally with time at a frequency f. We assume that the length of the trajectory which the particle covers in each period is short in comparison with the length L of the lens.

We have

$$\Phi(x, y) = \frac{\Phi_1}{a^2}(y^2 - x^2)\cos\omega t,$$

in which $\pm \Phi_1$ represents the peak potential applied to the electrodes; the latter are all tangent to a circle of radius a (see Chapter 10). ω is the angular frequency, $\omega = 2\pi f$.

The equations of motion take the form:

$$m\ddot{x} + \frac{2e\Phi_1}{a^2}\cos\omega t \cdot x = 0,$$

$$m\ddot{y} - \frac{2e\Phi_1}{a^2}\cos\omega t \cdot y = 0,$$

$$m\ddot{z} = 0,$$

in which we have assumed that $\dot{z} = v = $ constant. These equations can be written in the form:

$$\frac{d^2x}{d\tau^2} - (a_x - 2q\cos 2\tau)\,x = 0,$$

(for the x component); with $a_x = 0$,

$$q = \frac{4e\Phi_1}{m\,a^2\,\omega^2}, \quad \text{and} \quad \tau = \frac{\omega t}{2}.$$

The solution is of the form:

$$x = A e^{\mu\tau}\sum_{n=-\infty}^{\infty} C n\, e^{in\tau} + B e^{-\mu\tau}\sum_{n=-\infty}^{\infty} C n e^{-in\tau}.$$

If μ is real or complex, the amplitude increases indefinitely; if, on the contrary, μ is purely imaginary, the solution will be finite for all values of τ. The trajectory will be stable and can pass through the lens. The first ranges of values of q which correspond to stability are

$$0 < q < 0.92,$$

and

$$7.5 < q < 7.52,$$

but q is *independent of the velocity of the incident ions*. The only parameter of interest is the mass m; only those particles which have a mass greater than m_0 (m_0 is the solution of the equation $q = 0.92$), can pass through the lens, and the first zone thus constitutes a high-pass mass filter. The second domain represents a band-pass filter. For values of the mass which correspond to instability, the amplitude of the oscillations of the trajectories will grow exponentially; the resolving power of the combination is better in proportion to this growth and hence improves as the number of oscillations is increased. If the electrode length L is fixed, therefore, we should use slow ions ($v/f \ll L$).

If we employ the second zone, the values of Φ_1 become inconveniently high. We can alternatively construct a band-pass filter with the aid of the first zone only, by superimposing a steady quadrupole voltage $\pm \Phi_G$ on the high-frequency voltage. This gives:

$$\ddot{x} + \frac{2e}{m\,a^2}(\Phi_G + \Phi_1 \cos \omega t)\,x = 0,$$

$$\ddot{y} - \frac{2e}{m\,a^2}(\Phi_G + \Phi_1 \cos \omega t)\,y = 0,$$

and now we require that

$$a_x = -a_y = \frac{8e\,\Phi_G}{a^2\,m\,\omega^2}\,q = \frac{4e\,\Phi_1}{a^2\,m\,\omega^2},$$

the ratio $u = a/q = \pm 2\Phi_G/\Phi_1$ depends only upon the ratio of the applied voltages.

In the first domain, u has an upper limit, $u_{max} = \dfrac{2\Phi_G}{\Phi_1} = 0.333$; the domain of stable operation is infinitely narrow at this point, and corresponds to

$$q_1 = 0.706.$$

To each mass, therefore, corresponds a ratio u or a frequency f, for which the value of q lies within the selected band in the neighbourhood of u_{max}.

For a constant field we can calculate the mass resolving power:

$$\frac{m}{\triangle m} = \frac{0.75}{1 - (u/u_{max})}.$$

If we vary Φ_1 (or the frequency f) and it is with a source of slow ions of various masses that we are concerned, the different masses in question appear successively at the detector—the system behaves as a mass spectrograph.

When the beams are intense, the role of space charge is important and the equations of motion are in consequence far more complex.

The accuracy of the simple theory can be demonstrated with rubidium ions (Paul and Raether, 1953) when the current is weak. For $f = 2.55$ Mc/s, $\Phi_1 = 1000$ V, $a = 1$ cm and $L = 50$ cm, the positive rubidium ions of masses 85 and 87 are separated with the best resolution for $u = 0.328$. Very good resolution has been obtained with light ions (lithium, oxygen, nitrogen) by Gaudaire (1960), with a relatively short instrument (50 cm).

Increasing L (to several metres), we can reach a resolving power $m/\triangle m$ of several thousand. Paul and his fellow workers have also shown how the same system can be transformed into an isotope separator giving a current of several milliampères (Paul, Reinhard, and von Zahn, 1958). The electrical supply becomes considerably more complex.

β-RAY SPECTROGRAPHS WITH ELECTRON LENSES

25.1 INTRODUCTION

β-rays are emitted directly by radioactive substances, and are created indirectly when a secondary target is bombarded with γ-rays. The energy spectrum of the β-radiation is a rich source of information about the nuclear structure since the position of the lines in the spectrum is related to the separation of the nuclear energy levels, and the intensity of the lines is a function of the transition probability between the various levels (See Siegbahn, 1955; and Deutsch and Kofoed-Hansen, 1959). For this reason, there is an immediate interest in constructing spectrographs which are both bright and dispersive, if by these terms we understand the electron optical counterparts of their ordinary optical meanings.

Very many kinds of spectrograph have been successfully tried; we shall not be concerned with those in which there is only a plane of symmetry, which resemble glass prisms from the point of view of symmetry; we shall consider only those instruments in which actual lenses are employed, and in which there is axial symmetry.

There would, in principle, be no obstacle to prevent us from developing a theory which was also valid for instruments containing only a plane of symmetry; we have in fact already described the main properties of such devices when we were dealing with mass spectrographs, and there is very little we can add to this treatment, without invoking the considerably more complicated theory such as is to be found in the reviews by Siegbahn (1955, Chapter III: *Beta-ray spectrometer theory and design*) and by Bainbridge (1953, Chapter V: *Charged particle dynamics and optics*). Here, we shall only remark that there are objections to using iron to shape the magnetic field when high accuracy is sought: hysteresis in the yoke introduces uncertainty into the relation between current and field intensity. This small defect is less harmful in mass spectrographs because the mass numbers are integers; this explains why iron-free prisms are found in the domain of β-ray spectroscopy, although they are unknown elsewhere (Moussa, 1952). For lenses, on the other hand, the task is far simpler since Glaser (1941) and Glaser and Lammel (1943) have developed a whole battery of formulae which describe very adequately the properties of magnetic lenses.

In the β-ray spectrograph, we make use of the extreme dispersion of electron lenses, that is, the fact that the image position is highly dependent upon the emission velocity of the particles; this velocity is inversely proportional to the de Broglie wavelength, which explains why the phenomenon is called dispersion. The energies in question range between very wide limits, from a few tens of kilo-electron volts up to 25 MeV. Electrostatic lenses are out of the question, for their convergence would be too small in the megavolt region. We use magnetic lenses, therefore, which remain suitable throughout the whole energy range.

The following highly simplified outline explains the principle behind the operation which was first suggested by Kapitza (see Tricker, 1924); it recalls the optical method of focal separation used by Wood (see Tricker, 1924) to study infra-red spectra.

25.2 THE GENERAL PRINCIPLE

The principle upon which the instrument depends has been discussed earlier, and can be understood by examining Fig. 313. We shall first of all consider the theoretical behaviour of the instrument when it is provided with a point source; this is not the usual case, but it can be reached with intense conversion lines.

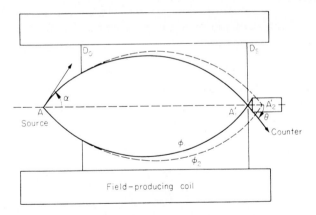

Fig. 313. The principle of the spectrograph.

A point source at A emits β-rays in all directions. The aperture diaphragm, D_0, restricts the angular aperture of the beam which reaches the lens, L, to 2α; rays which are homogeneous in energy, φ_1, are converged into a point image A_1', by the action of L. Rays corresponding to a different energy φ_2 (greater than φ_1) will, however, meet again in a different point A_2'

(beyond A_1') as the convergence of the lens is less for faster particles. If a diaphragm D_s, in which a small hole has been pierced, is placed at A_1', it will separate the rays of energy φ_1 from all the others provided that it is protected from direct radiation by a lead screen, which is generally placed at the centre of the lens.

25.3 ENERGY SELECTION BY AN APERTURE

25.3.1 Optical Theory

If the image A_1' were a point, the selection would be rigorous; the lens L, however, is not perfect, and the image of the point A is in reality a small circular spot, the radius ϱ_s of which is given by the following formula:

$$\varrho_s = G\,C_s\alpha^3, \tag{25.1}$$

with the usual notation, namely:

G is the magnification of the lens for the pair of points A, A_1' (and is usually equal to unity).

C_s is the spherical aberration coefficient referred back to the object.

The rays with energy φ_2 will be well separated from those with energy φ_1 only if they are spread out over D_s, and cover a circular area of which the radius, ϱ_c, is much larger than ϱ_s. If we denote $\varphi_2 - \varphi$ by $\delta\varphi$ and write $\dfrac{\varphi_1 = \varphi}{A_1'A_2'}$, the optics of the situation gives us the following expression for $\overline{A_1'A_2'}$:

$$\overline{A_1'A_2'} = G\,C_c\,\frac{\delta\varphi}{\varphi}. \tag{25.2}$$

If there were no spherical aberration at A_2', the value of ϱ_c would be

$$\varrho_c = \overline{A_1'A_2'}\,\theta = G\,C_c\,\theta\,\frac{\delta\varphi}{\varphi}, \tag{25.3}$$

in which 2θ is the angular aperture of the emergent beam at A_2' which is given to a first approximation by

$$\theta = \frac{\alpha}{G}, \tag{25.4}$$

with the result that

$$\varrho_c = C_c\,\alpha\,\frac{\delta\varphi}{\varphi}. \tag{25.5}$$

The ratio ϱ_c/ϱ_s will give us an idea of the purity of the spectrum which is obtained:

$$\frac{\varrho_c}{\varrho_s} = \frac{1}{G}\,\frac{C_c}{C_s}\,\frac{1}{\alpha^2}\,\frac{\delta\varphi}{\varphi}. \tag{25.6}$$

In reality, equation (25.5) gives too low a value for ϱ_c, and we should substitute

$$\varrho_c' = \varrho_c + \varrho_s,$$

in order to take into account in the simplest way possible the spherical aberration which is present at A_2' just as it is at A_1'. This leads to

$$\frac{\varrho_c'}{\varrho_s} = 1 + \frac{1}{G}\frac{C_c}{C_s}\frac{1}{\alpha^2}\frac{\delta q}{q}. \tag{25.7}$$

25.3.2 Resolving Power

We can see, therefore, that two monochromatic lines of the same intensity and with wavelengths λ_1 and λ_2 will be separated if $\dfrac{\varrho_c'}{\varrho_s} = 2$ since the transmitted intensity, which is inversely proportional to $\left(\dfrac{\varrho_c'}{\varrho_s}\right)^2$, will fall in relative value from unity, when the spectrometer is adjusted exactly onto one or other of the lines λ_1 or λ_2, to 0·71 when it is on the average wavelength $\dfrac{\lambda_1 + \lambda_2}{2}$ (see Fig. 314).

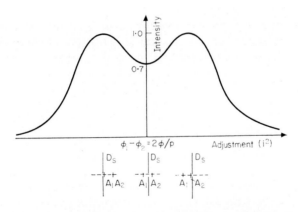

FIG. 314. The intensity distribution between two lines at the limit of resolution.

This calculation is admittedly schematic, but it is adequate to justify our selecting the value of the quantity $\varphi/\delta\varphi$, or better (Grivet, 1950a, 1951; Persico, 1949), of

$$P = \frac{\lambda}{\delta\lambda} = \frac{2\varphi}{\delta\varphi},$$

which corresponds to ϱ_c/ϱ_s equal to unity, to characterize the resolving power of the spectrograph. (We define P in this way in order to bring the notation into conformity with standard optical terminology.) For a point source, we find

$$P = \frac{\lambda}{\delta\lambda} = \left(\frac{2\,q}{\delta q}\right)_{\varrho_c = \varrho_s} = \frac{2\,C_c}{G\,C_s\,\alpha^2}. \tag{25.8}$$

25.3.3 A Typical Example: the Long-coil Spectrograph

The uniform field spectrograph is of very widespread application. Its high quality makes it a very representative type. We shall use it here, therefore, as a standard for comparison. We shall first evaluate its resolving power by means of formula (25.8); formula (9.3) shows that

$$C_s = C_c,$$

for a uniform field, and since the magnification G is equal to unity, we obtain simply

$$P = \frac{2}{\alpha^2}, \tag{25.9}$$

with α expressed in radians.

It is usual, in β-ray spectrography, to relate P to the solid angle Ω_s of the cone of useful rays. Ω_s is expressed as a fraction of the whole sphere. The real solid angle is equal to $2\pi(1 - \cos\alpha)$, so that

$$\Omega_s = \frac{2\pi(1 - \cos\alpha)}{4\pi} \simeq \frac{\alpha^2}{4}, \tag{25.10}$$

and hence

$$P = \frac{1}{2\Omega_s}. \tag{25.11}$$

For a point source, the resolution of the long-coil spectrograph is inversely proportional to twice the brightness coefficient Ω_s.

The case of a short circular coil, of mean radius R_m, is slightly less favourable, as the ratio C_c/C_s is then smaller than unity, and varies between 0·66 and 0·18 when the length, L, of the instrument satisfies the relation

$$1\cdot5\,R_m < L < 5\,R_m.$$

It is worth noticing that here, the resolution of an instrument appears to be independent of its size; this feature is connected with the hypothesis that a point source is being used, and the length L will be found to be shortened as soon as the radius of the source is introduced in a more realistic fashion.

25.4 CLARITY, THE SIZE OF THE INSTRUMENT, AND THE RADIUS OF THE SOURCE

25.4.1 The Two Factors Governing the Intensity: s and α

Sources of β-rays are often very weak, and the brightness is an essential quality in a spectrograph. The intensity which is received is proportional not only to the solid angle of the useful beam, but also to the surface area of the source, S, and is thus proportional to the product $S \Omega_s$. The source is usually in the form of a small disc of radius s, which lies normal to the axis at A, and is of surface area $S = \pi s^2$. The intensity, I, which falls upon the counter is therefore proportional to the product $s^2 \Omega_s$, or again, to $s^2 \alpha^2$:

$$I = \text{const. } s^2 \Omega_s = \text{const. } s^2 \alpha^2. \tag{25.12}$$

It is therefore advantageous to choose s as large as possible, while retaining the same value for α; the resolving power is not affected provided formula (25.8) remains valid, that is, provided s is small. On the contrary, the intensity, by which we mean the number of useful pulses recorded at the counter, strongly increases, without the number of parasitic pulses—and it is these which now constitute background noise—being increased.

25.4.2 The Incompatibility of Resolution and Brightness

When the radius of the source becomes appreciable in comparison with the radius of the spherical aberration disc, however, the theory of § 25.2 has to be modified; equation (25.8), which was obtained for a point source, ceases to be valid, and the effective resolution drops markedly. Only rarely can the spectroscopist choose between high resolution and high brightness; a specialized instrument has to be adopted, in which one of these qualities is sacrificed to the advantage of the other. More often, he will seek a "universal" instrument, which should achieve a reasonable compromise between these mutually opposed requirements.

To attain this result, there are two parameters at the designer's disposal: the length, L, and the angular aperture, α, and thus also the diameter of the instrument. The values taken by these dimensions determine not only the quality of the spectrograph, but also its bulk, and in consequence, its price and the electrical power supply which will be necessary.

25.4.3 The Advantages of Large Spectrographs

Increasing the length offers an effective, if costly, means of resolving the intensity versus resolution dilemma. The diameter of the spherical aberration disc, $C_s \alpha^3$, is proportional to the length L, like C_s itself, so that by lengthening L, we can always arrive at a situation in which the source diameter is negligible. Large spectrographs are the best. For economic reasons, however, L is never greater than two metres in modern versions, and is usually situated between 0·75 and 1·5 metres. It is possible, however, that designers may be led to exceed these limits in the near future, in order to unravel the complex spectra of the heavy nuclei.

After protracted experimental study, therefore, the length L which has come to be adopted is of the order of a metre. We can thus certainly use a source whose diameter lies between one and a few millimetres, and this provides a suitable intensity.

These, then, are the average values of the principal data necessary for the construction of a good commercial spectrograph.

25.4.4 Source-width and Angular Aperture—the Best Compromise

One parameter is still free—the angle α. Its choice remains at the disposal of the operator, and need not be fixed by the designer. The spectroscopist can easily change the aperture diaphragm, and choose the value of α as straightforwardly as that of s. His choice can then be made in the following way. First, the final intensity which will be necessary is fixed at an exact value and this determines the value of the product $s\alpha$. An optimum value of the ratio of the source radius s and the angular aperture α then exists, for which the resolving power is a maximum. A simple calculation demonstrates this.

First, we calculate the broadening of the aberration disc which is the result of the finite size of the source. If we denote the radius of the source by s, then the Gaussian image, B', of a point B which lies on the edge of the source, will lie in the frontal plane containing A', the Gaussian image of A, the centre of the source. The distance between A' and B' is given by

$$\overline{A'\,B'} = Gs.$$

The rays which emerge from B are very close to those which emerge from A, and they produce a disc in the focal plane which is very similar to that due to A. We shall neglect the distortions of this figure, and only take account of the displacement which occurs on transferring from A to B. If we combine all the elementary discs which correspond to different points around the source (Fig. 315), we obtain an overall enlargement of the figure in the focal plane which is equal to twice $A'B'$. (This is because the circles which define the opposite extremities of the resultant ring correspond to points on the edge of the source taken in pairs, at the ends of the same diameter.) The total diameter of the image spot is therefore given by

$$\varrho'_s = \varrho_s + 2Gs = C_s \alpha^3 + 2Gs, \tag{25.13}$$

so that to obtain the best resolving power with a constant brightness, we must look for the minimum value of (25.13), subject to the following condition:

$$s^2\alpha^2 = \text{const.} = K^2. \tag{25.14}$$

This ensures that the intensity at the receiver is constant when the radius of the source is varied, the brightness of the latter remaining unchanged

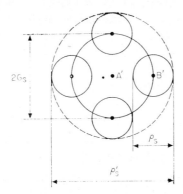

FIG. 315. The effect of the finite size of the source on the aberration spot.

(this means, from the practical standpoint, that the thickness of the radio-active deposit must remain unaltered).

The derivative $\dfrac{\partial \varrho_s'}{\partial \alpha}$ is zero when the value of α is a root of the equation:

$$3C_s\alpha^2 - \frac{2GK}{\alpha^2} = 0, \tag{25.15}$$

and this gives

$$Gs = (3/2)\varrho_s \tag{25.16}$$

as the optimum condition. We then find

$$\varrho_s' = 4\varrho_s, \tag{25.17}$$

and the resolving power, P_s falls to one-quarter of the ideal value, P, which would correspond to a point source:

$$P_s = (1/4)P; \tag{25.18}$$

α and s are connected by the relation

$$s = \frac{3}{2}\frac{C_s}{G}\alpha^3.$$

In the ordinary case for which $G = 1$, equation (25.19) simply expresses the fact that the optimum diameter of the source is equal to one-and-a-half times that of the spherical aberration disc.

The relations (25.9), (25.18) and (25.19) determine the dimensions of the various parts of the spectrograph completely, once the length, L, and the resolving power have been settled.

Numerical calculation based upon these formulae brings out once again the difficulty of reconciling resolving power and brightness which was found in the older instruments where the simple system which we have just analysed was used—where, that is, the selective diaphragm was simply a hole. The value of Witcher's discovery, which is described in the following sections and which avoids this problem, is obvious.

Let us consider a numerical example, in which the field is uniform, the length $L = 1$ m, and the resolving power with an extended source, $P_s = 100$.

Then

$$P = 400,$$

$$\alpha = 1/14 \text{ radian} = 4° 5',$$

$$C_s = 0.5 \text{ m},$$

and from (25.19)

$$s = 0.55 \text{ mm}.$$

The brightness would then be very low since $\Omega_s = 0.0013$. It was for this reason that the spectrographs which were in use up to 1944 were designed to give a very modest resolution, of the order of $P_s = 20$, for this leads to

$$P = 80, \quad \alpha \simeq 9°, \quad s = 6 \text{ mm} \quad \text{and} \quad \Omega_s = 0.006.$$

25.5 THE ANNULAR SELECTIVE DIAPHRAGM (WITCHER)

25.5.1 The Witcher Ring: Point Source

In all modern spectrographs, whatever type of lens be chosen, the brightness of the instrument is considerably increased by means of a procedure introduced by Witcher (1941) and developed by Frankel (1948). These two authors pointed out that the solid angle Ω_s would be considerably increased if a hollow conical beam containing all the rays emitted at angles between $2(\alpha - \varepsilon)$ and $2(\alpha + \varepsilon)$ were used, rather than a solid cone of rays of angle 2α (see Fig. 316). In the plane at which they leave the lens, these rays all touch the caustic surface S_1 which has its apex at A_1'; the mean rays, emitted at an angle α, touch the caustic along a circle Γ which lies in the frontal plane Π; the whole beam corresponds to emission angles between $\alpha - \varepsilon$ and $\alpha + \varepsilon$, and thus describes in the plane Π a narrow ring, the thickness $\delta_s R$ of which is second order in ε, and the outer contour of which is the circle Γ; the radius of the latter we shall call R. We can thus place the diaphragm which performs the energy selection in the plane Π: we have simply to cut out an annular opening of radius R and thickness $\delta_s R$, which will allow only the monochromatic beam of energy φ_1 to pass. So far, the theory of this system has only been worked out for the case of a uniform focusing field, but it is easy to make a qualitative estimate of the order of magnitude

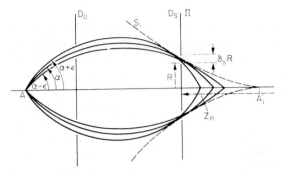

FIG. 316. The geometry of a hollow conical beam; the Witcher ring.

of the improvement which is possible. With the new value of the solid angle, we find

$$\Omega_s = \frac{\Omega}{4\pi} = \frac{4\pi\,\varepsilon\,\sin\alpha}{4\pi},$$

or

$$\Omega_s = \varepsilon\,\sin\alpha. \tag{25.20}$$

The brightness now depends upon the product $\varepsilon\sin\alpha$ and as the fineness of the focusing ring is now virtually independent of α, and dependent only upon ε, we can make α as large as 45°. Further, the values given to ε are of the same order of magnitude as those given to α in the simpler arrangement, so that by using a Witcher ring, we can hope to increase the brightness by a factor of about $\dfrac{\sin\alpha}{\varepsilon}$; about seven times, for example, when the final brightness is 10°/$_\circ$ ($\varepsilon = 18°$, $\alpha = 45°$).

Experiment confirms these predictions in the uniform field case, and shows that this device is equally effective in all the other kinds of spectrograph.

25.5.2 Position and Breadth of the Focusing Ring

(i) Methods

If the brightness is strongly increased, the energy selection mechanism remains substantially unchanged, and the arguments set out for the spectrograph with a circular opening remain valid, provided the diameter of the hole is replaced by the width of the annular slit. The calculations become far more complicated when the position of the slit makes simple Gaussian optics inadequate. The ring is defined by three parameters, its position on the axis, its radius and its thickness, and the determination of these requires an exact or approximate knowledge of the form of the caustic; this in turn implies that the calculation will be at least third order. Two theories have been developed.

One theory is rigorous, but is restricted to the uniform homogeneous field; it is due to Dumond (1949) and Persico (1949). The other theory is general, but is approximate and only correct up to the third order (Grivet, 1950a, 1951). This gives good accuracy provided that the rays are not inclined to the axis more steeply than 30°; beyond this value, in the range which is often employed nowadays, for which the angles lie between 30° and 45°, the orders of magnitude which this theory yields are still useful, and above all, it gives a clear indication of the relative importance of the various parameters. The laws remain dimensionally correct, and the coefficients can be obtained by measurement, so that the numerical shortcomings of the theory can be corrected.

(ii) *The theory of the uniform field; Persico (1950)*

We call $B\varrho$ the (magnetic) "rigidity" of the particle $\left(B\varrho = \dfrac{m\,v}{e} = \dfrac{p}{e} \right)$, where ϱ is the radius of the projection of the trajectory on a frontal plane. The equation of motion which gives the fictitious trajectory in a moving meridian plane is

$$r = 2\varrho \sin\alpha \sin\left(\frac{z}{2\varrho \cos\alpha} \right), \tag{25.21}$$

in which α is the angle between the axis and the direction of the velocity with which the particle leaves the source. Since the ring is, geometrically, the place at which the trajectories touch their envelope (the caustic), the trace of the ring in the moving meridian plane will be the point of contact of the family of trajectories (25.21) with its envelope when α varies. The second equation for the position will therefore be obtained by setting the derivative of (25.21) with respect to α equal to zero:

$$2\varrho \cos\alpha \sin\left(\frac{z}{\varrho \cos\alpha} \right) + \frac{z \sin^2\alpha}{\cos^2\alpha} \cos\left(\frac{z}{2\varrho \cos\alpha} \right) = 0. \tag{25.22}$$

It is simpler to replace the pair of equations (25.21) and (25.22) by the set

$$r = 2\varrho \sin\alpha \sin\Theta, \tag{25.23}$$

$$\frac{\tan\Theta}{\Theta} = -\tan^2\alpha, \tag{25.24}$$

$$\Theta = \frac{z}{2\varrho \cos\alpha}, \tag{25.25}$$

in which we have introduced the parameter Θ which has a simple physical meaning: it is the angle between the moving meridian plane and the direction of the initial velocity. The function $\Theta(\alpha)$, as defined by (25.24), is plotted in Fig. 317, after Dumond (1949). On substituting for Θ in equations (25.23) and (25.25), we obtain equations for the possible positions at which

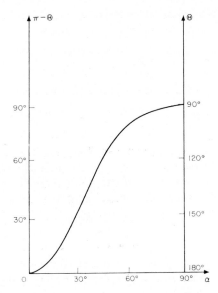

Fig. 317. The form of the function, $\Theta(\alpha)$.

the ring may be placed (Fig. 318; see Persico and Geoffrion, 1950). To each value of α, there corresponds a value of z $[z = z_s(\alpha)]$ which defines the abscissa of the ring, and a value of r $[r = R]$ which defines its radius. This correspondence is plotted in Fig. 318.

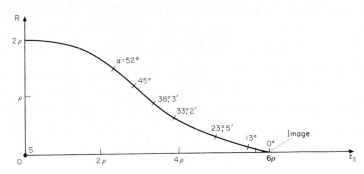

Fig. 318. The position (z_s) and radius (R) of the annular diaphragm, for different values of the angle α.

The width which the annulus should be given, δR, is obtained by differentiating (25.23) with respect to α, and by taking as the value of $\delta\alpha$ the angle 2ε, which measures the angular aperture of the useful conical region of space occupied by the beam, and is defined by

$$\alpha - \varepsilon < \alpha < \alpha + \varepsilon.$$

Let us expand (25.23) as a Taylor series; for $z = z_0$, we have $\dfrac{\partial r}{\partial \alpha} = 0$. If we neglect powers of ε higher than 2, we obtain

$$\delta R_s = \left(\frac{\partial^2 r}{\partial \alpha^2}\right) \frac{\varepsilon^2}{2}. \tag{25.26}$$

By differentiating (25.23), and taking (25.24) into account, we obtain

$$\delta R_s = \varrho \left(\frac{\sin\Theta}{\sin\alpha}\right) (3 + \cos^2\alpha \tan^2\Theta)\, \varepsilon^2. \tag{25.26'}$$

The expressions which give r and z [namely, (25.23) and (25.25)], depend linearly upon ϱ. A variation, $B\,d\varrho$, of the momentum increases all the dimensions homocentrically about the point S in the ratio $(1 + d\varrho/\varrho)$.

In particular, the chromatic dilatation of the radius, δR_c, is given by

$$\frac{\delta R_c}{R} = \frac{d\varrho}{\varrho}. \tag{25.27}$$

Two lines will be completely separated if the difference δR_c is equal to δR_s. The value of the resolving power,

$$P = \frac{\varrho}{\delta\varrho},$$

is therefore obtained by simply replacing δR_c by δR_s, according to (25.26'). This results in

$$P = \frac{2}{3\varepsilon^2}, \tag{25.28}$$

or, if we use (25.20) to express this as a function of the useful solid angle Ω_s,

$$P = \left(\frac{2}{3} \frac{\sin\alpha}{\varepsilon}\right) \frac{1}{\Omega_s}. \tag{25.29}$$

Comparing this with (25.11), we see that the relation between P and Ω_s is far more favourable when the ring is used; the coefficient $\frac{1}{2}$ is replaced by $\dfrac{2\sin\alpha}{3\,\varepsilon}$ which is of the order of 3 to 5.

The progress which we have made by introducing the ring is thus expressed by K in

$$P = \frac{K}{\Omega_s};$$

there is a marked advantage, and K may increase from 0·5 to 5.

(iii) *The arbitrary lens. The straight ray approximation*

(A) *The method.* The same advantage is obtained whatever type of magnetic lens be employed, as experiment has shown in every recent design.
Theory, however, can no longer give the position of the ring, its breadth,
or its chromatic displacement, with the excellent accuracy which characterized the formulae which have been obtained for the uniform field. Now,
we have to be satisfied with a third order theory, the predictions of which
become only qualitative for $\alpha > 30°$; and in practice α is often given a
value greater than 30° in order to improve the clarity. It is for this reason
that only the principle underlying the calculations will be given.

We need to find an expression describing the ends of the trajectories,
which is valid at least to the third order approximation; we obtain such an
expression by representing the stretch of trajectory between the plane in
which the focusing ring lies and the plane of the Gaussian image A' of the
source by a straight line, and by expressing the equation for this straight
line as a series expansion in α (the initial emission angle) up to α^3 inclusive.
It is legitimate to represent the emergent ray by a straight line in the case
of the short-coil spectrograph since the magnetic field is completely negligible
in the region between the ring and the Gaussian image A'. In the other
models, this supposition is only approximate, and can only be justified
by considering the form of the trajectories, which usually pass through a
point of inflexion at A'; this is true, for example, of the sinusoidal path in
the uniform field spectrograph.

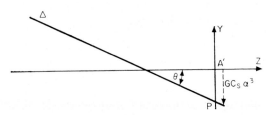

FIG. 319. The intersection of the straight emergent ray with the Gaussian image
plane.

As the emergent ray Δ is assumed to be straight, it will be completely
defined (Grivet and Paquien, 1950) by the point P at which it intersects
the Gaussian image plane A' and by its gradient, $\tan \theta$, at this point (see
Fig. 319); the distance $A'P$, of P from the axis is given by the spherical
aberration coefficient

$$A'P = G\, C_s\, \alpha^3, \tag{25.30}$$

and the slope is given by a third order expansion of the following kind:

$$\tan \theta = \frac{\alpha}{G} - \mu\, \alpha^3. \tag{25.31}$$

With the aid of the theory of third order aberrations (Grivet, 1950e), it is possible to calculate μ as a function of the coma coefficient F, the spherical aberration coefficient C_s, the magnification G and the object focal length, $f_0 = \overline{FH_0}$ by means of the general expression for μ:

$$\mu = \frac{1}{G}\left(\frac{1}{6} - F\right) + \frac{C_s}{g\,M_0}. \tag{25.32}$$

This expression contains both the algebraic distance M_0 between the source and the aperture diaphragm, and the magnification g relative to this plane, but only two cases are of practical interest:

(a) spectrographs with unit magnification G, for which μ vanishes from the results;

(b) spectrographs with a high magnification G; the role of the constant μ is then an important one, but in instruments of this kind, the source is always placed at, or very close to, the focus of the lens (large G) and equation (25.32) can then be simplified. We find

$$\mu = \frac{1}{G}\left(\frac{1}{6} - F\right) + \frac{C_s}{f_0}. \tag{25.33}$$

Details of the proofs of these formulae are to be found in an article by Grivet (1951). To lighten the labour involved in writing down the full expression, we shall continue to use "μ" in the equations, and only substitute the value given by equation (25.33) when we arrive at the final results.

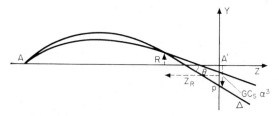

Fɪɢ. 320. Trajectories in a rotating meridian plane.

For simplicity, we shall always study the trajectory which lies in' the rotating meridional plane which contains at every instant the electron which is describing the ray in question; Oz will be the axis of revolution, and the axis normal to Oz at A' we call Oy. The origin of coordinates is thus at A', the Gaussian image of A (see Fig. 320).

In this system, the equation of the emergent ray is of the form

$$y = G\,C_s\,\alpha^3 + z\tan\theta, \tag{25.34}$$

in which the algebraic values of G, C_s, α and θ are defined according to the usual optical conventions. The angle θ is related to the principal parameter, α, by relation (25.31).

(B) *The focusing ring*

(a) *The position and radius of the ring.* The point of contact of the straight line and the meridian of the caustic surface is obtained from equations (25.31) and (25.34) together with the derivative of (25.34) with respect to α, namely:

$$0 = 3G\,C_s\alpha^2 + z\,\frac{d(\tan\theta)}{d\alpha}. \tag{25.35}$$

The ordinate of the point of contact is the radius R of the focusing ring Γ, and its abscissa z_R fixes the position of the circle A with respect to the origin, that is, to the Gaussian image A'. Equations (25.31), (25.34) and (25.35) lead to the following representation of the various quantities with α as parameter:

$$z_R = -\frac{3G^2\,C_s\,\alpha^2}{1 - 3\mu\,G\,\alpha^2}, \tag{25.36}$$

$$R = \frac{2G\,C_s\,\alpha^3}{1 - 3\mu\,G\,\alpha^2}. \tag{25.37}$$

(b) *The width of the ring.* We shall obtain the width of the focusing ring, $\delta_s R$, when the emission angle varies about a mean value α_0 between the limits $\alpha_0 + \varepsilon$ and $\alpha_0 - \varepsilon$ (see Fig. 321). The equation for the emergent

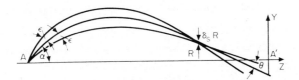

Fig. 321. The breadth of the ring.

ray which corresponds to the value $\alpha = \alpha_0 + \varepsilon$ is obtained by differentiating equation (25.34) with respect to α. We write $\delta\alpha = \varepsilon$, and for δY, which, for geometrical reasons, is second order in ε (it is the distance of a point on a curve to the tangent at a neighbouring point), we obtain the following expression:

$$\delta Y = G\,C_s[3\alpha_0^2\,\varepsilon + 3\alpha_0\,\varepsilon^2 + z\,\delta(\tan\theta)], \tag{25.38}$$

$$\delta(\tan\theta) = \frac{\varepsilon}{G} - 3\mu\,\alpha_0\,\varepsilon^2. \tag{25.39}$$

We write α simply, instead of α_0, and we consider the plane $z = z_R$, and finally obtain:

$$\delta_s R = \frac{3G\,C_s\,\alpha}{1 - 3\mu\,G\,\alpha^2}, \tag{25.40}$$

for the width which the annulus in the diaphragm ought to be given.

(c) Some special ring positions

(1) *The normal position—the unit magnification spectrograph.* In the majority of spectrographs which have been used so far, which have either a uniform field, a short coil, or a Siegbahn field (Siegbahn, 1944, 1946; Siegbahn and Svartholm, 1946), the magnification G is unity ($G = -1$), the angle α

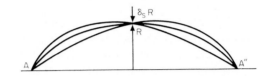

FIG. 322. The Siegbahn and Slätis ring position.

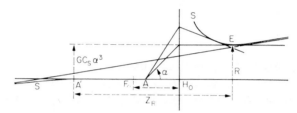

FIG. 323. The caustic in the Siegbahn and Slätis arrangement.

is about twenty degrees, and the ring remains close to the Gaussian image A'. These are cases in which we may neglect the terms $\mu\,G\,\alpha^2 = \mu\,\alpha^2$ in comparison with unity, and the formulae reduce to:

$$z_R = 3\,C_s\,\alpha^2, \tag{25.41}$$

$$R = 2\,C_s\,\alpha^3, \tag{25.42}$$

$$\delta_s R = 3\,C_s\,\alpha\,\varepsilon^2. \tag{25.43}$$

(2) *The Siegbahn and Slätis position: the high magnification spectrograph.* The arrangement which Siegbahn and Slätis suggested (1949) is characterized by the fact that the mean ray of the beam emerges from the first lens parallel to the axis of the instrument (see Fig. 322). It is therefore defined by $\theta = 0$, or, using equation (25.31), by

$$\frac{1}{G} - \mu\,\alpha^2 = 0, \tag{25.44}$$

so that G is very large and remains considerably greater than one even when α is itself large. In practice, we go up to $\alpha = 45°$. The lens characteristics then give values of G of the order of 5, as is shown in the references quoted above. In this extreme case, the theory can no longer hope to be accurate, but it still provides a useful model which takes the peculiarities of the arrangement satisfactorily into account. In every case, the term in α in the denominators of formulae (25.36), (25.37) and (25.40) becomes first order; if we take equation (25.25) into account and leave G and α as parameters (linked as they are to the value of μ which is characteristic of the lens), we find:

$$z_R = \frac{3}{2} G^2 C_s \alpha^2, \tag{25.45}$$

$$R = G C_s \alpha^3, \tag{25.46}$$

$$\delta_s R = \frac{3}{2} G C_s \alpha \, \varepsilon^2. \tag{25.47}$$

Equation (25.45) is not very suggestive here, as the Gaussian image A' is very distant on the left-hand side, but it is possible to show that the ring remains at a reasonable distance from the lens, and that this type of spectrograph will, in consequence, be of about the same length as the others.

The reader will find numerous references in the Bibliography which complement those already cited; for an overall survey of all types of β-ray spectrograph, the reader should refer to the article by Persico and Geoffrion (1950), which we have already mentioned, and to a long article by Gerholm (1956) which includes a very extensive bibliography.

25.6 MODERN TRENDS

25.6.1 Choice of Diaphragms, and Line-shape

The resolving power which we have calculated for various situations in the preceding paragraphs gives a fair estimate of the quality of the device for general comparison, but nowadays, the spectroscopist requires a more refined measure of the performance of his spectrograph. A very satisfactory answer to this demand is provided by a detailed study of the line-shape produced by the instrument for a hypothetical monochromatic source. This way of characterizing the "quality" was introduced in 1949 for the solenoid spectrograph by Dumond, who later considerably developed it (1952); Hubert (1953) went into it still more deeply, and showed that the large pedestal of the line which is obtained with a simple stop could be largely suppressed by using instead a set of diaphragms, placed at rather widely separated points along the axis. With this arrangement, the ordinary stop is still used, but its radius (or its width, in the case of a ring focus) must be adjusted, and is, in fact, a new parameter; the width which was calculated earlier and found to be equal to the spherical aberration spot is no longer the best choice. Hubert's theory has been substantiated by all recent theoretical work (cf. Beard, 1957) and by experiment: his baffles are incorporated in all modern instruments (Dumond, 1957; Jungermann et al.,

1962). The action of this system of diaphragms is very different from that of the ordinary "helical baffle" commonly used in nearly all magnetic devices, which cuts down the parasitic background by distinguishing between the useful rays—helical in shape—and the parasites, which are more randomly curved.

The calculations of Dumond and Hubert apply only to the uniform field, but it is also possible to develop a line-shape theory for other kinds of lens, starting from Grivet's theory; this development was successfully undertaken by Schneider (1957), who was able to calculate the optimum choice of ring diaphragm in terms of the line-width in a Siegbahn–Slätis spectrograph.

25.6.2 Correction of the Spherical Aberration

It was early discovered by Kervin (1949) and Persico (1950) that spherical aberration could be corrected in a simple way for the hollow conical bundle in the ring spectrograph. In principle, this may be achieved by placing a magnetic dipole of suitable strength at the centre of the spectrograph, inside the basic stop. This dipole can consist of a fairly small coil. In fact, the early attempts to correct a particular spectrograph by this means met with very little success (see Grivet, Hubert and Seiden, 1952), probably because the correction requires a far higher degree of mechanical and magnetic accuracy than does the standard apparatus. Recently, however, a Russian team has constructed an instrument which is appreciably improved by this method, after a careful sequence of preliminary investigations (Dolmatova and Kel'man, 1957, 1959).

TABLE 1

SOME USEFUL CONSTANTS

(in R.M.K.S. units)

Dielectric constant of free space:
$$\epsilon_0 = \frac{10^{-9}}{36\pi} 8 \cdot 855 \times 10^{-12} \text{ F m}^{-1}$$

Permeability of free space:
$$\mu_0 = \frac{4\pi}{10^7} = 1 \cdot 257 \times 10^{-6} \text{ henry m}^{-1}$$

Velocity of light in free space:
$$c = 2 \cdot 998 \times 10^8 \simeq 3 \times 10^8 \text{ m sec}^{-1}$$

Charge of the electron or proton:
$$e = 1 \cdot 602 \times 10^{-19} \text{ coulomb (C)}$$

Rest mass of the electron:
$$m_0 = 9 \cdot 107 \times 10^{-31} \text{ kg}$$

Rest mass of the proton:
$$M_0 = 1 \cdot 672 \times 10^{-27} \text{ kg}$$

Ratio e/m_0 (electron):
$$\frac{e}{m_0} = 1 \cdot 759 \times 10^{11} \text{ C kg}^{-1}$$

Ratio e/M_0 (proton):
$$\frac{e}{M_0} = 0 \cdot 958 \times 10^8 \text{ C kg}^{-1}$$

Ratio $\dfrac{m_0}{M_0}$:
$$\frac{m_0}{M_0} = 1836$$

Relativistic correction factors;
electron:
$$\frac{e}{m_0 c^2} = 1 \cdot 957 \times 10^{-6} \text{ V}^{-1}$$

proton:
$$\frac{e}{M_0 c^2} = 1 \cdot 066 \times 10^{-9} \text{ V}^{-1}$$

Boltzmann's constant:
$$k = 1 \cdot 380 \times 10^{-23} \text{ J deg.}^{-1}$$

Reciprocal $1/k$:
$$\frac{1}{k} = 11{,}600 \text{ deg. eV}^{-1}$$

Planck's constant:
$$h = 6 \cdot 624 \times 10^{-34} \text{ J sec}^{-1}$$

Electron volt:
$$1 \text{ eV} = 1 \cdot 602 \times 10^{-19} \text{ J}$$

Velocity of a 1 eV electron:
$$v_e = \sqrt{\frac{2e}{m_0}} = 5 \cdot 932 \times 10^5 \text{ m sec}^{-1}$$

Velocity of a 1 eV proton:
$$v_H = 1 \cdot 384 \times 10^4 \text{ m sec}^{-1}$$

TABLE 2

SOME USEFUL FORMULAE

(The numerical values are given in R.M.K.S. units for electrons)

---

(1) *Low velocities* $(v \ll c)$

$$v = \sqrt{\frac{2e}{m}} \cdot \sqrt{V} = 0.5932 \times 10^6 \sqrt{V}$$

$$p = m v \qquad = 5.425 \times 10^{-25} \sqrt{V}$$

$$B \varrho = \frac{m v}{e} \qquad = 5.685 \times 10^{-12} v$$

$$B \varrho = \sqrt{\frac{2 m V}{e}} \quad = 3.372 \times 10^{-6} \sqrt{V}$$

$$v = \frac{e}{m} B \varrho \qquad = 1.759 \times 10^{11} (B \varrho)$$

$$\lambda = h/p \qquad = \frac{12.26}{\sqrt{V}} \times 10^{-10}$$

or

$$\lambda = \frac{12.26}{\sqrt{V}} \text{ Å}$$

v in m sec^{-1}; B in teslas; ρ in metres; V in volts; m in kg; e in coulombs.

(2) *Relativistic velocities*

$$\varepsilon = \frac{e}{2 m_0 c^2} \qquad = 0.978 \times 10^{-6}$$

$$V^* = V(1 + \varepsilon V)$$

$$\frac{m}{m_0} = \frac{1}{\sqrt{1 - \frac{v^2}{c^2}}} = 1 + 2 e V$$

$$p = m v \qquad = 2 m_0 c \sqrt{e V(1 + \varepsilon V)}$$

$$\qquad = \sqrt{2 m_0 e V(1 + \varepsilon V)}$$

$$\beta = v/c \qquad = 2 \frac{\sqrt{\varepsilon V(1 + \varepsilon V)}}{1 + 2 \varepsilon V}$$

$$B \varrho = \frac{p}{e} \qquad = 2 \frac{m_0 c}{e} \sqrt{\varepsilon(1 + \varepsilon V)}$$

$$\lambda = \frac{h}{p} \qquad = \frac{h}{2 m_0 c} \frac{1}{\sqrt{\varepsilon(1 + \varepsilon V)}}$$

TABLE 3

FUNDAMENTAL CHARACTERISTICS OF FREE ELECTRONS IN MOTION

In the following table, the following quantities are tabulated to four significant figures:

V – The potential necessary to raise the kinetic energy of an electron initially at rest to a value E. V is expressed in ·volts.

V^* – The potential, relativistically corrected in such a way that the relativistic value of the momentum is obtained by simply replacing V by V^* in the non-relativistic expression. V and V^* are related by the equation:

$$V^* = V + \frac{e}{2\,m_0\,c^2}\,V^2,$$

and the relativistic momentum p has the form:

$$p = m_0\,c\,\sqrt{\frac{2E}{m_0\,c^2} + \left(\frac{E}{m_0\,c^2}\right)^2} = (2\,e\,m_0\,V^*)^{1/2}.$$

$B\rho$ – The product of the magnetic induction B, measured in teslas (1 tesla $= 10^4$ gauss) and the radius of curvature of the electron trajectory ρ, measured in meters, in a plane perpendicular to the field (1 tesla $\cdot$ m $= 10^6$ gauss $\cdot$ cm).

λ – The de Broglie wavelength of the electron in the absence of a magnetic field; λ is expressed in metres (1 m $= 10^{10}$ Å).

(a) Thermal and photoelectric velocities
(0·1 to 10 eV)

V volts	V^* volts	$B\rho$ tesla · m	λ m
0·1	0·1000		
0·2	0·2000		
0·3	0·3000	$1·846 \times 10^{-6}$	$2·240 \times 10^{-9}$
0·4	0·4000	2·132	1·940
0·5	0·5000	2·385	1·734
0·6	0·6000	2·616	1·583
0·7	0·7000	2·821	1·469
0·8	0·8000	3·024	1·371
0·9	0·9000	3·184	1·293
1	1·0000	3·372	1·226
2	2·0000	4·769	$8·671 \times 10^{-10}$
3	3·0000	5·840	7·080
4	4·0000	6·744	6·132
5	5·0000	7·540	5·484
6	6·0000	8·260	5·006
7	7·0000	8·922	4·635
8	8·0000	9·538	4·336
9	9·0000	10·12	4·088
10	10·00	10·66	3·878

(b) Vacuum tube region (10 to 10^4 eV)

V volts	V^* volts	$B\rho$ tesla · m	λ m
10	10·00	10·66	$3·878 \times 10^{-10}$
20	20·00	15·08	2·742
30	30·00	18·47	2·239
40	40·00	21·32	1·939
50	50·00	23·84	1·734
100	100·0	33·72	1·226
150	150·0	41·30	1·001
200	200·0	47·69	$8·670 \times 10^{-11}$
250	250·0	53·32	7·755
300	300·0	58·41	7·079
500	500·2	75·42	5·483
1000	1001	106·7	3·876
2500	2506	168·8	2·449
5000	5024	239·0	1·730
7500	7555	293·1	1·410
10000	10097	338·8	1·220

(c) X-ray tubes, microscopes, diffraction cameras
(10⁴ to 2 · 10⁵ eV)

V volts	V^* volts	$B\rho$ tesla · m	λ m
10000	10097	338·8	$1·220 \times 10^{-11}$
20000	20391	481·5	$8·588 \times 10^{-12}$
30000	30880	592·5	6·979
40000	41560	687·5	6·015
50000	52440	772·2	5·355
60000	63520	849·9	4·866
70000	74790	922·2	4·484
80000	86260	990·4	4·175
90000	97920	1055	3·918
100×10^3	$109·8 \times 10^3$	1117	3·701
120	134·0	1235	3·349
140	159·1	1345	3·073
160	185·0	1450	2·850
180	211·7	1551	2·665
200	239·1	1649	2·507

(d) β-ray devices and accelerators
(2×10^5 to 20×10^6 eV)

V volts	V^* volts	$B\rho$ tesla · m	λ m
$2 \quad \times 10^5$	$2·391 \times 10^5$	1649	$2·507 \times 10^{-12}$
3	3·880	2100	1·968
4	5·565	2515	1·644
5	7·446	2909	1·421
6	9·522	3290	1·257
$8 \quad \times 10^5$	$1·426 \times 10^6$	4027	1·026
$1 \quad \times 10^6$	1·978	4743	$8·718 \times 10^{-13}$
1·5	3·701	6487	6·374
2	5·914	8200	5·042
10	107·8	35020	1·180
20	411·4	68390	$6·046 \times 10^{-14}$

TABLE 4

THE LEGENDRE FUNCTIONS OF FRACTIONAL ORDER

$$P_n(\cos\theta) \quad 0.1 \leqq n \leqq 2.0$$

θ	n = 0.1	0.2	0.3	0.4	0.5	0.6	0.7	0.8	0.9	1.0
5°	0.365201	−0.254581	−0.813813	−1.272544	−1.599553	−1.774742	−1.791031	−1.652930	−1.378654	−0.996444
10°	0.501717	0.005894	−0.453932	−0.847492	−1.150000	−1.343918	−1.420160	−1.378607	−1.227945	−0.984808
20°	0.638358	0.268268	−0.088558	−0.411669	−0.683193	−0.888957	−1.019355	−1.069887	−1.041354	−0.939693
30°	0.718190	0.423320	0.130467	−0.145688	−0.391682	−0.596024	−0.749809	−0.847187	−0.885632	−0.866025
40°	0.774511	0.534092	0.289416	0.051166	−0.170483	−0.366364	−0.528746	−0.651698	−0.734616	−0.766044
50°	0.817704	0.620144	0.414869	0.209624	0.012012	−0.170817	−0.332468	−0.467557	−0.574395	−0.642788
60°	0.852374	0.690081	0.518406	0.342882	0.169084	−0.002434	−0.151995	−0.289652	−0.406673	−0.500000
70°	0.880955	0.748422	0.606031	0.456324	0.307261	0.157754	0.016269	−0.116849	−0.237227	−0.342020
80°	0.904886	0.797813	0.681210	0.557670	0.430035	0.301038	0.173516	0.050203	−0.066312	−0.113648
90°	0.925086	0.839927	0.746089	0.645288	0.539353	0.430189	0.319752	0.209982	0.102787	0.000000
100°	0.942171	0.875872	0.802069	0.721889	0.636309	0.546730	0.454374	0.360536	0.266528	0.173648
110°	0.956571	0.906416	0.850092	0.788227	0.721505	0.650659	0.576469	0.499745	0.421314	0.342020
120°	0.968597	0.932102	0.890814	0.845072	0.795249	0.741748	0.685007	0.625480	0.563646	0.500000
130°	0.978471	0.953322	0.924694	0.892752	0.857676	0.819665	0.778935	0.735715	0.692387	0.642788
140°	0.986362	0.970362	0.952059	0.931521	0.908821	0.884042	0.857275	0.828616	0.798169	0.766044
150°	0.992387	0.983428	0.973140	0.961544	0.948665	0.934528	0.919163	0.902601	0.884877	0.866025
160°	0.996635	0.992665	0.988095	0.982927	0.977168	0.970822	0.963895	0.956393	0.948322	0.939693
170°	0.999161	0.998171	0.997028	0.995735	0.994289	0.992699	0.990947	0.989050	0.987003	0.984808
180°	1.000000	1.000000	1.000000	1.000000	1.000000	1.000000	1.000000	1.000000	1.000000	1.000000

Table 4 (continued)

θ	n = 1·1	1·2	1·3	1·4	1·5	1·6	1·7	1·8	1·9	2·0
5°	− 0·542651	− 0·058480	+ 0·413527	+ 0·833237	1·165932	1·385435	1·476543	1·433916	1·265476	0·989351
10°	− 0·672324	− 0·318188	+ 0·047601	+ 0·395130	0·696918	0·930007	1·077683	1·130627	1·087486	0·954769
20°	− 0·775450	− 0·562976	− 0·319350	− 0·063159	+ 0·186880	0·412909	0·599730	0·735204	0·811308	0·824533
30°	− 0·792525	− 0·672251	− 0·514788	− 0·331564	− 0·135126	+ 0·061648	+ 0·246321	0·407651	0·536224	0·625000
40°	− 0·756377	− 0·705113	− 0·617004	− 0·498494	− 0·357352	− 0·202235	− 0·042220	+ 0·113682	+ 0·260689	0·380236
50°	− 0·679240	− 0·680298	− 0·648801	− 0·587376	− 0·500235	− 0·392577	− 0·270327	− 0·139860	− 0·007413	+ 0·119764
60°	− 0·567487	− 0·607957	− 0·621232	− 0·608117	− 0·570350	− 0·510520	− 0·431952	− 0·338567	− 0·234713	− 0·125000
70°	− 0·428902	− 0·496128	− 0·542575	− 0·567190	− 0·571540	− 0·555254	− 0·520788	− 0·468913	− 0·402560	− 0·324533
80°	− 0·269688	− 0·352628	− 0·421019	− 0·473792	− 0·510309	− 0·530315	− 0·533992	− 0·521924	− 0·495078	− 0·454769
90°	− 0·096662	− 0·185629	− 0·265506	− 0·335101	− 0·393447	− 0·439820	− 0·473745	− 0·495011	− 0·503662	− 0·500000
100°	+ 0·083160	− 0·003726	− 0·085870	− 0·162209	− 0·231828	− 0·293895	− 0·347694	− 0·392684	− 0·428461	− 0·454769
110°	0·262707	+ 0·184212	+ 0·107352	+ 0·032911	− 0·038356	− 0·105747	− 0·168608	− 0·226363	− 0·278485	− 0·324533
120°	0·435048	0·369302	0·303277	0·237487	+ 0·172439	+ 0·108625	+ 0·046528	− 0·013395	− 0·070707	− 0·125000
130°	0·593260	0·542960	0·491146	0·438457	0·385144	0·331534	0·277906	+ 0·224546	+ 0·171992	+ 0·119764
140°	0·732358	0·697230	0·660787	0·623160	0·584483	0·544892	0·504529	0·463536	0·422057	0·380236
150°	0·846085	0·825099	0·803106	0·780153	0·756288	0·731556	0·706009	0·679698	0·652678	0·625000
160°	0·930510	0·920782	0·910521	0·899731	0·888427	0·876616	0·864311	0·851520	0·838256	0·824533
170°	0·982463	0·979971	0·977330	0·974544	0·971609	0·968530	0·965306	0·961937	0·958424	0·954769
180°	1·000000	1·000000	1·000000	1·000000	1·000000	1·000000	1·000000	1·000000	1·000000	1·000000

This table has been kindly provided by M. Schelkunoff.

BIBLIOGRAPHY

Books of a more general nature are to be found listed in the Bibliography for Part 1. More specialized works, devoted to a single electron optical instrument, are mentioned in the appropriate section of Part 2.

BIBLIOGRAPHY FOR PART 2†

I. CATHODE RAY TUBES

(Chapter 12)

(A) Books

DELABY, H. (1951) *Bases techniques de la télévision*, Eyrolles, Paris.
GRIVET, P., and P. HERRENG (1953) *La télévison*, P.U.F. (Collection *Que sais-je?* No. 30), Paris.
LEVERENTZ, H. W. (1945) *Luminescence of Solids*, Chapman & Hall, London.
MALOFF, I. G., and W. D. EPSTEIN (1938) *Electron Optics in Television*, McGraw-Hill, New York.
PUCKLE, O. S. (1951) *Time-base*, Chapman & Hall. London.
RIDER, J. F., and S. D. USLAN (1950) *Encyclopedia of Cathode Ray Tubes*, Rider Publications, New York.
SAY, M. G. *et al.* (1954) *Cathode Ray Tubes*, Newnes, London.
SOLLER, T., M. A. STARR and G. E. VALLEY (1948) *Cathode Ray Tubes*, McGraw-Hill (*MIT. Rad. Lab.* No. 22), New York.
WILSON, W. (1953) *The Cathode Ray Oscillograph in Industry*, Chapman & Hall, London.
ZWORYKIN, W. K., and G. A. MORTON (1954) *Television*, Wiley, New York, 2nd ed.

(B) Articles

BEAM, W. R. (1955) *RCA. Rev.* **16**, 242.
BLOOM, L. R., and H. M. VON FORSTER (1954) *Rev. Sci. Instrum.* **25**, 649.
BONVALET, P. A. (1952) *Techniques de l'Ingénieur "Electronique"* IIE, 2170.
BROADWAY, L. F., and J. R. PIERCE (1939) *Proc. Phys. Soc.* **51**, 335.
CAZALAS, A. (1952) *Techniques de l'Ingénieur "Electronique"* IE, 770.
COOPER, H. G. (1961) *Bell Syst. Techn. J.* **40**, 723.
DARBYSHIRE, J. A. (1955) *Electr. Eng.* **27**, 523.
DAWSON, H. G. (1898) *Proc. Lond. Math. Soc.* **29**, 521.
DELBORD, Y. L. (1951) *Techniques de l'Ingénieur "Electronique"* IIE, 3140.
DESERNO, P. (1953) *Arch. Elektrotechn.* **29**, 139.
DOSSE, J. (1940) *Z. Physik* **115**, 530.
DUFOUR, C., and G. WENDT (1954) *Le Vide* **54**, 249.
EPSTEIN, D., and L. PENSAK (1946) *RCA Rev.* **7**, 55.
FIELD, L. M. (1946) *Rev. Mod. Phys.* **18**, 353.
GLASER, W. (1949) *Ann. der Phys.* **4**, 389.

† A number of the references in the bibliography are to communications to scientific congresses. The text in each case is to be found in the Proceedings, which used always to be published with an appreciable delay. A list of these congresses, with their full titles, is to be found in Section IVB. In the text, only the date and the plane where the congress was held, and the number or page number of the communication are given.

HARDY, D. R., B. JACKSON and R. FEINBERG (1956) *Electr. Eng.* **28**, 8.
HOADLEY, H. (1951) *Rev. Sci. Instrum.* **22**, 706.
HOPKINSON, R. G. (1946) *J. Inst. Electr. Eng.* **93**, part IIIa, 779, 795, 808.
HUTTER, R. G. E. (1947) *J. Appl. Phys.* **18**, 740, 797; (1948) *Adv. in Electronics* **2**, 167.
IVEY, H. F. (1954) *Adv. in Electronics* **6**, 133.
JACOB, L. (1939) *Phil. Mag.* **28**, 81; (1952) *Proc. Phys. Soc.* B**65**, 421.
LAMBERT, I. E., and R. FELD (1946) *Proc. Inst. Radio Engrs.* **34**, 432.
LANGMUIR, D. B. (1937) *Proc. Inst. Radio Engrs.* **25**, 977.
LAW, R. R. (1937) *Proc. Inst. Radio Engrs.* **25**, 954.
LEE, S. B. (1946) *Proc. Inst. Radio Engrs.* **34**, 121.
MOAK, C. D. (1960) *Nucl. Instr. and Meth.* **8**, 19.
MORTON, G. A. (1946) *Rev. Mod. Phys.* **18**, 362.
MOSS, H. (1945) *Wireless Eng.* **22**, 316; (1946) *J. Television Soc.* **4**, 206; (1946a) *J. Brit. Inst. Rad. Eng.* **6**, 99; (1950) *Adv. in Electronics* **2**, 2.
NGUYEN, TIEN CHI (1952) *Techniques de l'Ingénieur "Electronique"* 1.
OWAKI, K., S. TERAHATA, T. HADA and T. NAKAMUBA (1950) *Proc. Inst. Rad. Eng.* **38**, 1173.
PICHT, J., and J. HIMPAN (1941) *Ann. der Phys.* **39**, 409, 436 and 478.
PIERCE, J. R. (1941) *Proc. Inst. Radio Eng.* **29**, 28.
RECKNAGEL, A. (1938) *Z. Physik* **111**, 61.
ROGOWSKI, W., and H. THIELEN (1939) *Arch. Elektrotechn.* **33**, 411.
SANGSTER, M. (1955) *Appl. Sci. Res.* B**4**, 4.
SCHAEFER, H., and W. WALCHER (1943) *Z. Physik* **121**, 679.
SCHLESINGER, K. (1961) *Trans. Inst. Radio Eng.* **8**, 224.
SCHWARTZ, E. (1938) *Ferseh A.G.* **1**, 19.
SMITH, S. I., R. V. TALBOT and E. K. SMITH (1952) *Proc. Inst. Radio Eng.* **40**, 297.
SPONSLER, G. S. (1955) *J. Appl. Phys.* **26**, 676.
TERRILL, H. M., and L. SWEENY (1944a) *J. Frank Inst.* **237**, 495; (1944b) **238**, 220.
WALLRAFF, A. (1935) *Arch. Electrotechn.* **29**, 351.
WATSON, E. E. (1927) *Phil. Mag.* **3**, 849.
WENDT, G. (1954) *Ann. Radioél.* **9**, 286.
WORONCOW, A. (1946) *Proc. Inst. Elec. Eng.* Part III A, **93**, 1564.
ZWORYKIN, V. K. (1933) *J. Frank Inst.* **215**, 535.

II. THE EMISSION MICROSCOPE
(Chapter 13)

ARNAL, R. (1958) *C. R. Acad. Sci. Paris* **247**, 2110.
ASHWORTH, F. (1948) Thesis, Bristol; (1951) *Adv. in Electronics* **3**, 1.
BARKER, A. N., H. O. W. RICHARDSON and N. FEATHER (1950) *Research Correspondence*, suppl. 3–9, 431.
BAS, E. B. (1960) *Le Vide* **16**, 303.
BAYH, W. (1958) *Z. Physik* **150**,10.
BEHNE, R. (1936) *Ann. der Phys.* **26**, 372 and 385.
BENJAMIN, M., and R. O. JENKINS (1940) *Proc. Roy. Soc.*, A.**176**, 262; (1942) A.**180**, 225.
BERNARD, R., and R. GOUTTE, (1958) Berlin, 215.
BETHGE, H., H. EGGERT and K. HERBOLD (1958) Berlin, 1, 217.
BOERSCH, H. (1942) *Naturwiss.* **30**, 120.
BRÜCHE, E. (1933) *Zeit. Phys.* **86**,448.
BRÜCHE, E., and H. JOHANNSON (1932) *Naturwiss.* **20**, 353; (1934) *Ann. der Phys.* **20**, 161.
BRÜCHE, E., and W. KNECHT (1934) *Z. Techn. Phys.* **15**, 461; (1935), **16**, 95.
BRÜCHE, E., and H. MAHL (1935) *Z. Techn. Phys.* **16**, 623; (1936) **17**, 81 and 262.

BURGERS, W. G., and J. J. A. PLOOS VAN AMSTEL (1937a) *Physica* **4**, 5; (1937b) **4**, 15; (1938a) **5**, 305; (1938b) **5**, 313.

CASTAING, R., B. JOUFFREY and G. SLODZIAN (1960) *C. R. Acad. Sci. Paris* **251**, 1010.

CASTAING, R., and G. SLODZIAN (1960) Delft, 169; (1962) *J. de Microscopie*, **1**, 395,

COUCHET, G. (1954) *Congrès Soc. Savants*, Alger.

COUCHET, G., M. GAUZIT and A. SEPTIER (1951) *C.R. Acad. Sci. Paris* **233**, 1087; (1952) *Bull. Mic. Appl.* **2**, 85.

DÜKER, H. (1960) *Zeit. Metallkunde* **51**, 314 and 377

DÜKER, H. and A. ILLENBERGER (1962) Philadelphia I. D-5

DYKE, W. P., and W. W. DOLAN (1956) *Adv. in Electronics* **8**, 90.

FERT, C., F. PRADAL, F. SAPORTE and R. SIMON (1958) Berlin I, 197.

FERT, C., and R. SIMON (1956) *C. R. Acad. Sci. Paris* **243**, 1309; (1957) **244**, 1177.

GAUZIT, M., and A. SEPTIER (1951) *Microscopie*, (2) **1**, 109.

GROSS, H., and G. SEITZ (1937) *Z. Physik* **105**, 734.

HEIDENREICH, R. D. (1955) *J. Appl. Phys.* **26**, 757, 879.

HILL, A., W. BUCHNER, J. CLARK and J. FISK (1939) *Phys. Rev.* **55**, 463.

HOLLAND, L., and L. LAURENSON (1959) *Le Vide* **81**, 141.

HUBER, H., A. M. SCHROFF and M. WARNECKE (1959) *Le Vide* **82**, 214.

HUGUENIN, E. L. (1954) *C. R. Acad. Sci. Paris* **239**, 404; (1955) **241**, 307; (1956) Thèse Ing. Doct. Paris.

HUGUENIN, E. L., and M. GAUZIT (1955) Toulouse, 219; (1956) *C. R. Acad. Sci. Paris* **241**, 379; (1956a) **242**, 1311.

JENKINS, R. O. (1943) *Rep. Prog. Phys.* **9**, 177.

JEPSEN, R. L. (1959) *Le Vide* **80**, 80.

JOHANNSON, H. (1933) *Ann. der Phys.* **18**, 385.

JOHNSON, R. P. (1938) *J. Appl. Phys.* **9**, 508.

JOHNSON, R. P., and W. SHOCKLEY (1936) *Phys. Rev.* **49**, 436.

KINDER, E. (1944) *Z. Physik* **122**, 192.

KOCH, W. (1958) *Z. Physik* **152**. 1.

MAHL, H. (1936) *Z. Techn. Phys.* **17**, 653; (1938) *Ann. der Phys.* **31**, 425; (1942) *Z. Techn. Phys.* **117**, 23.

MAHL, H., and J. POHL (1935) *Z. Techn. Phys.* **16**, 219.

MARTIN, S. T. (1939) *Phys. Rev.* **56**, 947.

MARTON, L., and P. H. ABELSON (1947) *Science* **106**, 70.

MECKLENBURG, W. (1942) *Z. Physik* **120**, 21.

MESCHTER, E. (1938) *Rev. Sci. Instrum.* **9**, 12.

MÖLLENSTEDT, G., and H. DÜKER (1953) *Optik* **10**, 192.

MÖLLENSTEDT, G., and W. HUBIG (1954) *Optik* **11**, 528.

MÖLLENSTEDT, G. and F. LENZ (1963) *Adv. in Electronics* **18**, 251.

MÜLLER, E. W. (1937) *Z. Physik* **106**, 541; (1939) *Naturwiss.* **27**, 820; (1943) *Z. Physik* **120**, 261 and 270; (1950) *Naturwiss.* **37**, 333; (1951) *Z. Physik* **131**, 136; (1955) *J. Appl. Phys.* **26**, 732; (1960) *Adv. in Electronics* **13**, 83.

POHL, J. (1934) *Z. Techn. Phys.* **15**, 579.

POPP, G., and W. WALCHER (1957) *Ann. der Phys.* **20**, 293.

RATHENAU, G. W., and G. BAAS (1951) *Physica* **17**, 117.

RECKNAGEL, A. (1943) *Z. Physik* **120**, 331.

SCHENK, D. (1935) *Ann. der Phys.* **23**, 240.

SEPTIER, A. (1953a) *C. R. Acad. Sci. Paris* **237**, 231; (1953b) Thèse, Paris; (1954) *Ann. Radioél.* **9**, 374; (1955) *C. R. Acad. Sci. Paris* **240**, 1200.

SEPTIER, A., and M. GAUZIT, (1950), Paris, 246.

SEPTIER, A., M. GAUZIT and P. BARUCH (1952) *C. R. Acad. Sci. Paris* **234**, 105.

SIMON, R. (1959) Thèse, Toulouse.

SMOLUCHOWSKI, R. (1941) *Phys. Rev.* **60**, 661.

STRANSKI, I. N., and R. SUHRMANN (1947) *Ann. der Phys.* **1**, 153 and 169; **61**, 153.

Stühmer, S. (1960) *Nukleonik* **2**, 1.
Willax, H. A. (1960) *Nukleonik* **2**, 7
Zworykin, V. K. (1933) *J. Frank. Inst.* **215**, 535.

III. IMAGE-CONVERTERS AND INTENSIFIERS

(Chapter 14)

(A) Books and Surveys on Photo-electronic Image Devices

Eckart, F. (1956) *Elektronenoptische Bildwandler und Röntgenbildverstärker*, Barth, Leipzig.
Advances in Electronics (1960, 1962), Academic Press, New York; Volumes 12 and 16.

(B) Articles

Anderson, A. E. (1960) *Trans. IRE-NS* **7**, 133.
Beurle, R. L., and N. A. Slark (1960) *Adv. in Electronics* **12**, 247.
Burns, J., and M. J. Neumann (1960a) *Adv. in Electronics* **12**, 97; (1960b) *Trans. Inst. Radio. Eng.* N **57**, 142.
Castaing, R., B. Jouffrey and G. Slodzian (1960) *C. R. Acad. Sci. Paris* **251**, 1010.
Coeterier, F., and M. C. Teves (1937) *Physica* **4**, 33.
Duchesne, M. (1953a) Thèse, Paris; (1953b) *Bull. Astr.* **17**, 88; (1953c) **17**, 107; (1961) *C. R. Acad. Sci. Paris* **253**, 417.
Eckart, F. (1954) *Ann. der Phys.* **6**, 1.
Essig, S. F. (1960) *Adv. in Electronics* **12**, 73.
Farnsworth, P. T. (1934) *J. Frank. Inst.* **218**, 411.
Ferrell, R. A. (1956) *Phys. Rev.* **101**, 554.
Gibbons, D. J. (1960) *Adv. in Electronics* **12**, 203.
Haine, M. E., A. E. Ennos and P. A. Einstein (1960) *Adv. in Electronics* **12**, 317.
Hall, J. S., W. K. Ford and W. A. Baum (1960) *Adv. in Electronics* **12**, 21.
Heimann, W. (1960) *Adv. in Electronics* **12**, 235.
Hiltner, W. A., and P. Pesch (1960) *Adv. in Electronics* **12**, 17.
Holst, G., J. H. de Boer and C. F. Veenemans (1934) *Physica* **1**, 297.
Iams, H., G. A. Morton and V. K. Zworykin (1939) *Proc. Inst. Radio Eng.* **27**, 541.
Lallemand, A. (1936) *C. R. Acad. Sci. Paris* **203**, 243 and 990; (1955) Toulouse, 227.
Lallemand, A., and M. Duchesne (1951) *C. R. Acad. Sci. Paris* **233**, 305; (1952) **235**, 503; (1954) **238**, 335; (1955a) **240**, 1329; (1955b) **241**, 360.
Lallemand, A., M. Duchesne, J. Duflo and J. Banaigs (1959) *C. R. Acad. Sci. Paris* **248**, 2191.
Lallemand, A., M. Duchesne and G. Wlerick (1960) *Adv. in Electronics* **12**, 5.
Le Poole, J. P. (1954) Thesis, Delft.
Linden, D. R., and P. A. Snell (1957) *Proc. Inst. Radio Eng.* **45**, 513.
McGee, J. D., E. A. Flinn and M. D. Evans (1960) *Adv. in Electronics* **12**, 87.
McGee, J. D., and G. Lubzinsky (1939) *J. Inst. Electr. Eng.* **84**, 468.
Mandel, L. (1955) *J. Sci. Instrum.* **32**, 405.
Marton, L., L. B. Leder and H. Mendlowitz (1955) *Adv. in Electronics* **7**, 183.
Marton, L., J. A. Simpson, H. A. Fowler and N. Swanson (1962) *Phys. Rev.* **126**, 182.
Möllenstedt, G., Innsbruck, 1953.
Morton, G. A., and L. E. Flory (1946) *RCA. Rev.* **7**, 385.
Morton, G. A., and E. G. Ramberg (1936) *Physics* **7**, 451.
Perl, M. L., and L. W. Jones (1960) *Adv. in Electronics* **12**, 153.
Schaffernicht, W. (1948) *Fiat. Res. German Science, Electronics* **1**, 78.
Sommer, A. H. (1955) *Rev. Sci. Instrum.* **26**, 725.
Sternglass, E. J. (1954) *Research Memo.* **60.8.10.43.** MI.

STOUDENHEIMER, R. G. (1960) *Adv. in Electronics* **12**, 41.
STOUDENHEIMER, R. G., J. C. MOOR and L. PALMER (1960) *Trans. Inst. Radio Eng.* N **57**, 136.
STUDER, F. J., D. A. CUSANO and A. H. YOUNG (1951) *J. Opt. Soc. Amer.* **41**, 559.
TEVES, M. G., and T. TOL (1952) *Philips techn. Rev.* **14**, 1.
TIMOFEEV, P. V., and V. V. SOROKINA (1960) *Radio Eng. and Electronics* **5**, 201.
WACHTEL, M. M., D. D. DOUGHTY and A. E. ANDERSON (1960) *Adv. in Electronics* **12**, 59.
WENDT, G. (1955) *Ann. Radioél.* **10**, 74.
WILCOCK, W. L., D. L. EMBERSON and B. WEEKLEY (1960) *Trans. Inst. Radio Eng.* N**57**, 126.
ZACHAROV, B., and S. DOWDEN (1960) *Adv. in Electronics* **12**, 31.
ZWORYKIN, V. K., and G. A. MORTON (1936) *J. Opt. Soc. Amer.* **26**, 181.
ZWORYKIN, V. K., and E. G. RAMBERG (1936) *Physics* **7**, 451.

IV ELECTRON MICROSCOPY
(Chapters 15 to 19)

(A) Books

ARDENNE, M. VON (1940) *Elektronen-Übermikroskopie*, Springer, Berlin.
BORRIES, B. VON (1949) *Die Übermikroskopie*, Werner Langer, Berlin.
BROGLIE, L. DE (1950) *Optique électronique et corpusculaire*, Hermann, Paris.
GABOR, D. (1945) *The Electron Microscope*, Hutton, London.
HAINE, M. E. (1954) *Advances in Electronics* **6**, 298, Academic Press, New York.
HAINE. M. E. and V. E. COSSLET **(1961)** *The Electron Microscope*, E. & F. N. Spon Ltd., London.
LEBEDEV, A. A. (1954) *Electron Microscopy* (in Russian), Moscow.
LEISEGANG, S. (1956) *Elektronenmikroskope* in *Encyclopaedia of Physics* **33**, 396, Springer, Berlin.
MAGNAN, C. (1961) *Traité de Microscopie électronique* (2 vols), Hermann. Paris.
MAHL, H., and E. GOLZ (1951) *Elektronenmikroskopie*, V.E.B. Bibl. Inst., Leipzig.
MARTON, L. (1946) *Electron Microscopy, Rep. Progr. Phys.* **10**, 204.
PICHT, J., and R. GAIN (1955) *Das Elektronenmikroskop*, Fachbuch, Leipzig.
ZWORYKIN, V. K., G. A. MORTON, E. G. RAMBERG, J. HILLIER and A. W. VANCE (1949) *Electron Optics and the Electron Microscope*, Wiley, New York.

(B) Conferences on Electron Microscopy (*some of the communications to these conferences are included in the bibliography*)

1949 (DELFT), *Proceedings*, published in 1952 at Delft.
1950 (PARIS), *Comptes rendus*, published in 1952, Éditions de la Revue d'Optique, Paris.
1951 (WASHINGTON), *Proceedings*, published in 1954 under the auspices of the N.B.S., with the title: *Electron Physics, circular 527.*
1953 (INNSBRUCK), Verhandlungen, unpublished.
1954 (GHENT), *Comptes rendus*, published in 1956 under the auspices of Ghent University.
1954 (LONDON), *Proceedings*, published in 1957 by The Physical Society.
1955 (TOULOUSE), *Comptes rendus*, published in 1956, Éditions du C.N.R.S. with the title: *Les techniques récentes en microscopie électronique et corpusculaire.*
1956 (STOCKHOLM), *Proceedings*, published in 1957 by Alunqvist & Wiksell, Stockholm.
1956 (CAMBRIDGE), *Proceedings* of a symposium entitled "*X-ray Microscopy and Micro-radiography*", published in 1957 by Academic Press, New York.
1958 (BERLIN), *Verhandlungen*, published in 1960 by Springer, Berlin.
1960 (DELFT), *Proceedings*, published by the North Holland Publishing Company, Amsterdam.

1962 (PHILADELPHIA), *Abstracts*, published by Academic Press, New York.
1964 (PRAGUE) *Abstracts* of the Third European Conference, published by the Czechoslovak Academy of Sciences.
1966 (KYOTO) *Abstracts* of the Sixth International Conference, published by the Maruzen Co. of Tokyo.
1968 (ROME) *Abstracts* of the Fourth European Conference, published by Tipografia Poliglotta Vaticana of Rome.
1970 (GRENOBLE) *Abstracts* of the Seventh International Conference.

(C) Articles

ADLER, H., R. MINKNER, G. REINHOLD and J. SEITZ (1960) Delft, 222.
AGAR, A. W., R. S. S. M. REWELL and R. A. SCOTT (1949) Delft, 52.
AMERICAN ELECTRON MICROSCOPY SOCIETY (1946) *J. Appl. Phys.* **17**, 989.
ANDERSON, H. W. (1935) *Trans Amer. Inst. Electr. Eng.* **54**, 1315; (1935a) *Rev. Sci. Instrum.* **6**, 309.
ARDENNE, M. VON (1938) *Z. Phys.* **108**, 338; (1939) *Z. Techn. Phys.* **20**, 235; (1940a) *Z. Physik* **116**, 736; (1940b) *Kolloid Z.* **93**, 158; (1940c) *Naturwiss.* **28**, 248; (1941) *Phys. Z.* **42**, 72; (1941a) *Z. Physik* **117**, 657.
ARNAL, R. (1955) *C. R. Acad. Sci. Paris* **240**, 610; (1956) *Ann. Phys. Paris* **11**, 232; (1958) *C. R. Acad. Sci. Paris* **247**, 2110.
ATTA, T. VAN, and R. J. VAN DE GRAAFF (1933) *Phys. Rev.* **43**, 158.
BACKUS, R. C., and R. C. WILLIAM (1949) *J. Appl. Phys.* **20**, 224.
BAKER, R. F., E. G. RAMBERG and J. HILLIER (1942) *J. Appl. Phys.* **13**, 450; (1943) **14**, 39.
BALDINGER, E. (1959) *Encyclopaedia of Physics* **44**, 1.
BAS, E. B. (1960) Delft, 126.
BASSETT, G. A. (1960) Delft, 270.
BASSETT, G. A., and J. W. MENTER (1957) *Phil. Mag.* **2**, 1482.
BASSETT, G. A., J. W. MENTER and D. W. PASHLEY (1958a) Berlin, 353; (1958b) *Proc. Roy. Soc.* A **246**, 345.
BAUER, S. H. (1943) *Rev. Sci. Instrum.* **14**, 30.
BEAUFILS, R. (1959) *C. R. Acad. Sci. Paris* **248**, 3145.
BERTEIN, F. (1946) *C. R. Acad. Sci. Paris* **222**, 64; (1947a) **224**, 106; (1947b) **224**, 560; (1947c) **225**, 801; (1947d) **225**, 865; (1948) *Ann. Radioél.* **3**, 49; (1949) *C. R. Acad. Sci. Paris* **229**, 291.
BERTEIN, F., and P. GRIVET (1951) Washington, 355.
BERTEIN, F., and E. REGENSTREIF (1949) *C. R. Acad. Sci. Paris* **228**, 1854.
BEYERSDORFER, K. (1950) *Optik* **7**, 192.
BISHOP, F. W. (1950) *Electronics* **23**, 110.
BLACKSTOCK, A. W., R. D. BIRKHOFF and J. C. SLATER (1955) *Rev. Sci. Instrum.* **26**, 274.
BOERSCH, H. (1936) *Ann. der Phys.* **26**, 631 and **27**, 75; (1939) *Z. Techn. Phys.* **20**, 346; (1940) *Jahrb. AEG Forsch. Sonderh. Übermikroskop* **7**, 34; (1942) *Naturwiss.* **30**, 73; *Z. Naturforsch.* **121**, 746; (1944) *Kolloid Z.* **106**, 169; (1947) *Experientia* **4**, 1; (1947a) (1943) *Z. Physik* **2**a, 615; (1949) *Optik* **5**, 436; (1953) *Z. Physik* **134**, 156; (1954) **139**, 115.
BOERSCH. H.. O. BOSTANJOGLO and K. GROHMANN (1966) *Z. angew. Phys.* **20**, 193.
BOERSCH. H.. O. BOSTANJOGLO and B. LISCHKE (1966/67) *Optik* **24**, 460.
BOHM, D., and O. PINES (1953) *Phys. Rev.* **72**, 609.
BOLLMANN, W. (1960) Delft, 330.
BOROVIK, E. S., and B. P. BATRAKOV (1958) *Sov. Phys. Techn. Phys.* **3**, 1811.
BORRIES, B. VON (1942) *Phys. Zeit.* **43**, 190; (1943) *Z. angew. Photogr.* **4**, 43; (1944) *Z. Physik* **122**, 539; (1948) *Optik* **3**, 321 and 389.
BORRIES, B. VON, and E. RUSKA (1940a) *Naturwiss.* **28**, 366; (1940b) *Z. Physik* **116**, 249.

BOTHE, W. (1933) *Encyclopaedia of Physics* **22**, a.

BRAGG, W. L. (1950) *Nature* **166**, 399.

BRAGG, W. L., and G. L. ROGERS (1951) *Nature* **167**, 190.

BRAUCKS, F. W. (1958) *Optik* **15**, 242.

BREMMER, H., (1950) Paris, 40.

BRUCK, H. (1947a) *C. R. Acad. Sci. Paris* **224**, 1553; (1947b) **224**, 1628; (1947c) **224**, 1818; (1953) *Rapport CEA* No. 189.

BRUCK, H., and M. BRICKA (1948) *Ann. Radioél.* **3**, 339.

BRUCK, H., and P. GRIVET (1947) *C. R. Acad. Sci. Paris* **224**, 1768; (1950) *Rev. Opt.* **29**, 164.

BRUCK, H., and L. ROMANI (1944) *Cahiers Phys.* **24**, 1.

BUHL, R. (1960) *Z. Physik* **158**, 35.

CASTAING, R. (1950a) *C. R. Acad. Sci. Paris* **231**, 835; (1950b) **231**, 994; (1954a) London, No. 5, 58 and 87; (1954b) *C.R. Acad. Sci. Paris* **238**, 1885; (1955) Toulouse, 117.

CASTAING, R., and J. DESCAMPS (1954) *C. R. Acad. Sci. Paris* **238**, 1506 and 2132.

CASTAING, R. and L. HENRY (1962) *C. R. Acad. Sci. Paris* **255**, 76.

CHALLICE, C. E. (1950) *Proc. Phys. Soc.* B **63**, 59.

CHANSON, P., and C. MAGNAN (1951) *C. R. Acad. Sci. Paris* **233**, 1436.

CHARLES, D. (1947) *Ann. Radioél.* **2**, 75.

CHOUMOFF, S., and J. LAPLUME (1957) *Rev. Techn. CFTH* **25**, 115.

CLIFFORD, D., and R. FORTESCUE (1952) *Nature* **170**, 503.

CONRADY, A. E. (1919) *Mon. Not. Roy. Astr. Soc.* **79**, 575.

COSSLETT, V. E. (1954) London, No. 71.

COSSLETT, V. E., and W. C. NIXON (1952) *Nature* **170**, 436; (1953) *J. Appl. Phys.* **24**, 616.

COUCHET, G. (1951) *C. R. Acad. Sci. Paris* **233**, 1013; (1954) *Ann. Phys. Paris* **9**, 731.

COUCHET, G., M. GAUZIT and A. SEPTIER (1951a) *C. R. Acad. Sci. Paris* **233**, 1087; (1951b) *Bull. Micr. Appl.* (2), **2**, 85.

CRANBERG, L. (1952) *J. Appl. Phys.* **23**, 518.

CUCKOW, F. X (1946) *J. Appl. Phys.* **17**, 66.

DANIELSON, W. E. *et al.* (1952) *J. Appl. Phys.* **23**, 860.

DEUBNER, A., and H. HIEBER (1954) *Z. angew. Phys.* **6**, 112.

DIETRICH, W. (1958a) *Z. Physik* **151**, 519; (1958b) **152**, 87; (1958c) **152**, 306.

DLUGOSZ, S. (1950) *J. Sci. Instrum.* **27**, 77.

DOLBY, R. M., and D. W. SWIFT (1960) Delft, 124.

DORSTEN, A. C. VAN (1948) *Philips techn. Rev.* **10**, 135; (1950) **12**, 33.

DORSTEN, A. C. VAN, H. NIEUWDORP and A. VERHOEFF (1950) *Philips techn. Rev.* **12**, 33.

DORSTEN, A. C. VAN, W. J. OOSTERKAMP and J. B. LE POOLE (1947) *Philips techn. Rev.* **9**, 193.

DORSTEN, A. C. VAN, and H. F. PREMSELA, (1960) Delft, 101.

DOSSE, J. (1941) *Z. Phys.* **117**, 316, 437 and 722.

DOWELL. W. C. T. (1963) *Optik* **20**, 535.

DOWELL, W. C. T., J. L. FARRANT and A. L. G. REES (1956) *Proceedings of the Tokyo Conference on Electron Microscopy* 320; (1958) Berlin, 367.

DUGAS, J., P. DURANDEAU and C. FERT (1961) *Rev. Opt.* **40**, 277.

DUPOUY. G.. and F. PERRIER (1960) Delft, 25; (1960a), 110; (1964a) *J. Micr.* **3**, 233; (1964b) *C.R. Acad. Sci. Paris* **258**, 4213; (1966) *J. Micr.* **5**, 369.

DUPOUY, G., F. PERRIER and L. DURRIEU (1960b) *C. R. Acad. Sci. Paris* **251**, 2836.

DUPOUY, G., F. PERRIER and R. FABRE (1961) *C. R. Acad. Sci. Paris* **252**, 627.

DUPOUY. G.. F. PERRIER and J. TRINQUIER (1964) *J. Micr.* **3**, 115.

DURANDEAU, P. (1955) *J. Phys. Radium* **17**, 18A; (1956) **17**, 33.

DURANDEAU, P., and C. FERT (1957) *Rev. Optique* **36**, 205.

DURANDEAU, P., C. FERT and J. TARDIEU (1958) *C. R. Acad. Sci. Paris* **246**, 79.

DYSON, J. (1950) Parjs, 127.

EAVES, G. (1952) *Nature* **169**, 975.

EISENHANDLER, C. B. and B. M. SIEGEL (1966a) *J. Appl. Phys.* **37**, 1613; (1966b) *Appl. Phys. Letters* **8**, 258.

EITEL, W., and E. GOTTHARDT (1940) *Naturwiss.* **28**, 367; (1942) *Z. Physik* **118**, 714.

ELLIS, S. G. (1951) *J. Appl. Phys.* **23**, 728.

ELMORE, W., and M. SANDS (1950) *Electronics: Experimental techniques*, McGraw Hill, New York.

ENGEL. A.. G. KOPPEN and O. WOLFF (1962) Philadelphia **1**, E.13.

ENNOS, J. (1953) *Brit. J. Appl. Phys.* **4**, 101; (1954) **5**, 27.

EVERITT, W. F., and K. J. HANSSEN (1956) *Optik* **13**, 385.

FAGET, J., M. FAGOT and C. FERT (1960) Delft, 18.

FAGET, J., and C. FERT (1956) *C. R. Acad. Sci. Paris* **243**, 2028; (1957) *Cahiers Phys.* **83**, 285; (1958) Berlin, 234.

FAGOT, M., J. FERRÉ and C. FERT (1958) *C. R. Acad. Sci. Paris* **246**, 1404.

FAGOT, M., J. FERRÉ and C. FERT (1961) *C. R. Acad. Sci. Paris* **252**, 3766.

FAGOT, M., and C. FERT (1960) *C. R. Acad. Sci. Paris* **250**, 94.

FARRANT, J. L., and A. J. HODGE (1948) *J. appl. Phys.* **19**, 840.

FELDMAN, C., and M. O'HARA (1957) *J. Opt. Soc. Amer.* **47**, 300.

FELICI, M. (1957) *Onde Electr.* **37**, 7.

FERNÁNDEZ-MORÁN, H. (1965) *Proc. Nat. Acad. Sci.* **53**, 445; (1966a) *Kyoto* **1**, 147; (1966b) *Proc. Nat. Acad. Sci.* **56**, 801.

FERRELL, R. A. (1956) *Phys. Rev.* **101**, 554.

FERT, C. (1954) *C. R. Acad. Sci. Paris* **238**, 333; (1954) London, 161.

FERT, C., and J. FAGET (1958) Berlin, 234.

FERT, C., and A. LAFFITTE (1961) *C. R. Acad. Sci, Paris* **252**, 3213.

FERT, C., and P. SELME (1957) *Bull. Micr. Appl.* **7**, 12.

FORST, G. (1958) *Z. angew. Phys.* **10**, 546.

FREI, E. H., and F. L. HIRSFELD (1951) *Rev. Sci. Instrum.* **22**, 231.

FRÖHLICH, H., and H. PELZER (1955) *Proc. Phys. Soc.* A **68**, 525.

GABOR, D. (1948) *Nature* **161**, 777; (1949) *Proc. Roy. Soc.* A **197**, 454; (1951) *Research* **4**, 107.

GARDEZ, J. (1959) *C. R. Acad. Sci. Paris* **249**, 2034.

GAUTHÉ, B. (1955) *C. R. Acad. Sci. Paris* **241**, 188; (1957) Thèse, Paris.

GAUZIT, M. (1951) *C. R. Acad. Sci. Paris* **233**, 1586; (1953) *Bull. Micr. appl.* (2) **3**, 58; (1954a) *Ann. Phys. Paris* **9**, 683; (1954b) London, No. 66.

GEROULD, C. H. (1950) *J. Appl. Phys.* **21**, 183.

GILROY, J., R. HART and C. LAVERICK (1956) *Scientific Research* **1**, No. 9, 31.

GLASER, W. (1940) *Z. Physik* **116**, 19 and 96; (1943) **121**, 647; (1949) *Acta Phys. Austriaca* **3**, 38; (1950) Paris, No. 9.

GLEICHAUF, P. (1951) *J. Appl. Phys.* **22**, 535; (1951a) **22**, 766.

GOLTZ, J. (1940) *Jahrb. AEG Forsch. Sonderh.* **7**, 57.

GOTTARD, E. (1942) *Z. Physik* **118**, 714.

GREEN, M. (1962) Thesis, Cambridge.

GRIBI, M., M. THURKAUF, W. VILLIGER and L. WEGMANN (1959) *Optik* **16**, 65.

GRIGSON. C. W. B.. W. C. NIXON and F. TOTHILL (1966) *Kyoto* **1**, 157.

GRIVET, P., (1948) Cambridge, No. 22.

GRIVET, P., F. BERTEIN and H. BRUCK (1947) *Ann. Radioél.* **2**, 249.

GRIVET, P., and G. REGENSTREIF, (1950) Paris, 230.

GRUBE, W. L. (1949) *J. Appl. Phys.* **20**. 125.

HADDEN, R. J. B. (1951) *Report A.E.R.E.* GM. 92.

HAILER, K. (1938) *Wiss. Veröff. Siemens* **17**, 115.

HAINE, M. E. (1947) *J. Sci. Instrum.* **24**, 61; (1950) Paris, No. 31; (1954) London, No. 18; (1957) *J. Brit. I.R.E.* **17**, 211.

HAINE, M. E., and J. DYSON (1950) *Nature* **166**, 315.

HAINE, M. E., and P. A. EINSTEIN (1952) *Brit. J. Appl. Phys.* **3**, 49.

HAINE, M. E., P. A. EINSTEIN and P. H. BORCHERDS (1958) *Brit. J. Appl. Phys.* **9**, **482**.

HAINE, M. E., and P. HIRST (1953) *Brit. J. Appl. Phys.* **4**, 239.

HAINE, M. E., and T. MULVEY (1952) *J. Opt. Soc. Amer.* **42**, 746; (1954) *J. Sci. Instrum.* **31**, 325.

HAINE, M. E., R. S. PAGE and R. G. GARFITT (1950) *J. Appl. Phys.* **21**, 173.

HALL, C. G. (1947) *J. Appl. Phys.* **18**, 588; (1948) **19**, 198 and 271; (1949) **20**, 631; (1951) **22**, 655.

HAMM, F. A. (1951) *Rev. Sci. Instrum.* **22**, 895.

HANAK. J. J. (1964) *RCA Rev.* **25**, 551.

HANSZEN. K. J. (1958) *Z. Naturforsch.* **13a**, 409.

HANSZEN. K. J. and B. MORGENSTERN (1965) *Z. angew. Phys.* **19**, 215.

HANSZEN. K. J.. B. MORGENSTERN and K. J. ROSENBRUCH (1964) *Z. angew. Phys.* **16**, 477.

HARTL, W., and H. RAETHER (1961) *Z. Physik* **161**, 238.

HARTMAN, R. E. and R. S. HARTMAN (1966) *Kyoto* **1**, 159.

HASHIMOTO, H., T. NAIKI and M. MANNAMI (1958) Berlin, 331.

HEARD, H. G., *et al.* (1952) *Amer. UCRL Report*, 1622; (1953) 2252; (1953a) 2321.

HEIDENREICH. R. D. and R. W. HAMMING (1965) *Bell System Techn. J.* **44**, 207.

HEIDENREICH, R. D., and L. A. MATHESON (1944) *J. Appl. Phys.* **15**, 423.

HEIDENREICH, R. D., and L. STURKEY (1945) *J. Appl. Phys.* **16**, 97.

HEISE, F. (1949) *Optik* **5**, 479.

HELMCKE J. G. (1954) *Optik* **11**, 201.

HERCHENBACH, W., and H. DUCKER (1956) *Optik* **13**, 375.

HERRENG, P. (1943) Thèse, Paris.

HIBI, T., (1954) London, 636.

HILLIER, J. (1945) *J. Appl. Phys.* **16**, 469; (1946) **17**, 411; (1950) **21**, 785.

HILLIER, J., and R. F. BAKER (1944) *J. Appl. Phys.* **15**, 663.

HILLIER, J., and E. G. RAMBERG (1947) *J. Appl. Phys.* **18**, 48.

HINDERER, H. (1942) *Z. Physik* **119**, 397.

HOPPE. W. (1963) *Optik* **20**, 599.

INDUNI, G. (1947) *Helv. Phys. Acta* **20**, 463; (1955) Toulouse, 189.

IONOV, N. I. (1960) *Sov. Phys. Techn. Phys.* **5**, 527.

ITO, K., (1955) Toulouse.

KADEN, H. (1950) *Die elektromagnetische Schirmung in der Fernmelde- und Hochfrequenz-technik*, Springer, Berlin.

KAHLER, H. O., and B. LLOYD (1950) *J. Appl. Phys.* **21**, 699.

KAMIYA, Y., M. NONOYAMA, H. TOCHIGI and R. UYEDA, (1958) Berlin, 339.

KANAYA, K., and H. KAWAKATSU, (1958) Berlin, 308.

KERN, S. F., and R. A. KERN (1950) *J. Appl. Phys.* **21**, 705.

KILPATRICK, W. D. (1953) *American UCRL Report*, 2321.

KINDER, E. (1943) *Naturwiss.* **31**, 149; (1946) **33**, 367.

KIRKPATRICK, P., and A. V. BAEZ (1950) *Nature* **166**, 251.

KITAMURA. N.. M. P. SCHULHOF and B. M. SIEGEL (1966) *Appl. Phys. Letters* **9**, 377.

KLEIN, R. (1954) *Optik* **11**, 226.

KNOLL, M. I., and E. RUSKA (1932) *Z. Physik* **78**, 318; (1932a) *Ann. der Phys.* **12**, 697 and 641.

KOBAYASHI, K., E. SUITO and S. SHIMADZU (1958) Berlin, 165.

KOCH, H. (1936) *Z. Physik* **97**, 131.

KOENIG, H. (1951) *Z. Physik* **129**, 491.

KOENIG, H., and G. HELWIG (1950) *Optik* **6**, 111.

KOMODA. T. (1966) *Kyoto* **1**, 31.

KOMODA. T. and M. OTSUKI (1964) *Jap. J. Appl. Phys.* **3**, 666.

KOPP, C., and G. MÖLLENSTEDT (1946) *Optik* **1**, 327; (1947) **2**, 283.

KUNSMAN, C. (1926) *Phys. Rev.* **27**, 249.

KUSHNIR, M. (1951) *Bull. USSR Acad. Sci* **15**, 306.

KUSHNIR, M., A. N. KABANOV and L. N. KRUMIAKOVA (1960) *Radio Eng. and Electr.* **5**, 197.

LABERRIGUE. A. and P. LEVINSON (1964) *C.R. Acad. Sci. Paris* **259**, 530.

LABERRIGUE. A.. P. LEVINSON. M. BERGEOT. P. BONHOMME. F. PAYEN and C. SEVERIN (1966) *Kyoto* **1**, 153.

LANG, W, (1948) *Optik* **3**. 233.

LAREYMONDE, M. D. L., J. SALMON and J. WAJSBRUN (1954) *J. Phys. Radium* **15**, 117.

LEISEGANG, S. (1954) *Optik* **11**, 49; (1954a) London, Nos. 34 and 35.

LENZ. F. (1953) *Optik* **10**, 439; (1963) *Z. Physik* **172**, 498.

LEONHARD, F. (1954) *Optik* **11**, 407.

LE POOLE, J. B. (1947) *Philips' techn. Rev.* **9**, 33.

LE POOLE, J. B., and W. A. LE RUTTE (1949) Delft, 84.

LE RUTTE, W. A. (1952) Thesis, Delft.

LIEBMANN, G. (1951) *Proc. Phys. Soc.* B **64**, 972; (1952) B **65**, 94 and 188; (1953) B **66**, 448; (1955), B **68**, 679, 682 and 737.

LIEBMANN, G., and E. M. GRAD (1951) *Proc. Phys. Soc.* B **64**, 956.

LIPPERT, W. (1955) *Optik* **10**, 467.

LIPPERT, W., and W. POHLIT (1952) *Optik* **9**, 456; (1953) **10**, 447; (1954) **11**, 253.

LOCQUIN, M. (1955) *Z. Wiss. Mikr. und micr. Techn.* **62**, 220.

MAGNAN, C., and P. CHANSON (1945) *C. R. Acad. Sci. Paris* **220**, 770; (1949) *Nucleonics* **4**, 52; (1954) London, No. 67; (1955) Toulouse, 83.

MAHL, H. (1947) *Optik* **2**, 106.

MAHL, H., and A. PENDZICH (1942) *Z. Techn. Phys.* **24**, 38.

MARTON, L. (1944) *Science* **100**, 318; (1945) *J. Appl. Phys.* **16**, 387; (1953) *Science* **118**, 470.

MARTON, L., and L. LEDER (1954) *Phys. Rev.* **94**, 203.

MARTON, L., L. LEDER, and H. MENDLOWITZ (1955) *Adv. in Electronics* **7**, 183.

MARTON, L., and L. I. SCHIFF (1941) *J. appl. Phys.* **12**, 759.

MARTON, L., J. A. SIMPSON, H. A. FOWLER, and N. SWANSON (1962) *Phys. Rev.* **126**, 182.

MARTON, L., J. A. SIMPSON and T. F. MCCRAW (1955a) *Phys. Rev.* **99**, 495; (1955b) *Rev. Sci. Instrum.* **26**, 855.

MARUSE, S. (1958) Berlin, 166.

MATTHEWS, J. W. (1960) Delft, 276.

MAUTNER, R. S., and O. H. SCHADE (1947) *RCA. Rev.* **8**, 43.

MENDLOWITZ, H. (1960a) *Proc. Phys. Soc.* **75**, 664; (1960b) *J. Opt. Soc. Amer.* **50**, 739.

MENTER, J. W. (1956) *Proc. Roy. Soc.* **236**A, 119; (1958) Berlin, 320.

MENTS, M. VAN, and J. B. LE POOLE (1947) *Appl. Sci. Res.* **1**B, 3.

METHERELL. A. J. F. and M. J. WHELAN (1965) **16**. 1038; (1966) *J. Appl. Phys.* **37**, 1737.

MOAK *et al.* (1951) *Nucleonics* **9**, 18.

MÖLLENSTEDT, G. (1949) *Optik* **5**, 499; (1950) Paris, 112; (1952a) *Optik* **9**, 473; (1952b) *Z. Naturforsch.* **7**, 465; (1953) *Optik* **10**, 72; (1956) **13**, 13.

MÖLLENSTEDT, G., and R. BUHL (1957) *Phys. Blätter* **13**, 357.

MÖLLENSTEDT, G., and W. DIETRICH (1955) *Optik* **12**, 246.

MÖLLENSTEDT, G., and H. DÜKER (1953) *Z. Naturforsch.* **8**a, 1 and 79; (1954) *Naturwiss.* **42**, 41; (1956) *Z. Physik* **145** 377.

MOTT, N. F. (1955) *Proceedings of the 10th Solvay Congress: Electrons in Metals*, Brussels.

MÜLLER, H. O. (1942) *Kolloid Z.* **99**, 6.

MÜLLER, H. O., and E. RUSKA (1941) *Kolloid Z.* **95**, 21.

MULVEY, T. (1953) *Proc. Phys. Soc.* B **66**, 441; (1958) Berlin.

NISHIGAKI. M.. S. KATAGINI. H. KIMURA and B. TADANO (1966) *Kyoto* **1**, 121.

NISSEN, H. F. (1944) *Z. Physik* **122**, 573.

NIXON, W. C. (1958) Berlin, 302.

NOZIÈRES, P., and D. PINES (1959) *Phys. Rev.* **113**, 1254.

NUTTING, G. C., and V. E. COSSLETT (1950) *The Institute of Metals*, 57.

OZASA. S.. S. KATAGINI. H. KIMURA and B. TADANO (1966) *Kyoto* **1**, 149.

PASHLEY, J. W., J. H. MENTER and G. A. BASSETT (1957) *Nature* **179**, 752.

PERNOUX, E. (1952) *J. Phys. Radium* **18**, 368; (1953) Thèse, Paris.

PETZINGER, K. G. and J. J. HANAK (1964) *RCA Rev.* **25**, 542.

PINES, D. (1954) *Proceedings of the 10th Solvay Congress: Electrons in Metals*, Brussels; (1955) *Adv. in solid State Phys.* **2**: *Electron interaction in metals*.

PIVOVAR, L. I., and V. I. GORDIENKO (1958) *Sov. Phys. Techn. Phys.* **3**, 2101.

POWELL, C. J., and J. B. SWANN (1959) *Phys. Rev.* **115**, 569.

PREUSS, L. E., and J. H. WATSON (1950) *J. Appl. Phys.* **21**, 902.

RAMBERG, E. G., and J. HILLIER (1948) *J. Appl. Phys.* **19**, 679.

RANG, O. (1949) *Optik* **5**, 518; (1953) **10**, 90; (1953a) *Z. Physik* **136**, 465; (1958) Berlin, 371.

RANG, O., and W. WEITSCH (1956) *Optik* **13**, 201.

RAZIN, A. A., L. V. TARASOVA and V. A. TSUKKERMAN (1960) *Radio. Eng. and Electr.* **5**, 187.

REGENSTREIF, E. (1951) *Ann. Radioél.* **6**, 51 and 164.

RIECKE. W. D. (1962a) Philadelphia **1**, KK-5; (1962b) *Optik* **19**, 81.

RITCHIE, R. (1957) *Phys. Rev.* **106**, 874.

ROUZÉ, S. R., and J. H. L. WATSON (1953) *J. Appl. Phys.* **24**, 1106.

RUSKA, E. (1944) *Arch. Elektrotechn.* **36**, 431; (1950) Kolloïd Zeit **116**, 102. (1955) Toulouse. 45; (1962) Philadelphia A-1; (1964) *J. Micr.* **3**, 364; (1966) *Adv. Opt. and Electron Micr.* **1**, 113.

RUTHEMANN, G. (1941) *Naturwiss.* **29**, 648; (1942) **30**, 145; (1948) *Ann. der Phys.* **2**, 113.

SAKAKI, Y., and G. MÖLLENSTEDT (1956) *Optik* **13**, 193.

SAUZADE, M. (1960) *J. Phys. Radium*, **21**, 161 A.

SEELIGER, R. (1948) *Optik* **4**. 258.

SEPTIER, A. (1960) *CERN. Report* No. 60–39.

SEPTIER, A., and M. RUYTOOR (1959) *C. R. Acad. Sci. Paris* **249**, 2175 and 2746.

SHIMADZU. S.. M. IWANAGA. K. KOBAYASHI. E. SUITO. T. TAOKA and H. FUJITA (1966) *Kyoto* **1**, 101.

SIEGEL. B. M.. N. KITAMURA. R. A. KNOPFLI and M. P. SCHULHOF (1966) *Kyoto* **1**, 151.

SIMARD, G. L., E. L. BURTON and R. B. BARNES (1945) *J. Appl. Phys.* **16**, 832.

SIMPSON, J. A., and L. MARTON (1961) *Rev. Sci. Instrum.* **32**, 802.

SMITH. K. C. A.. K. CONSIDINE and V. E. COSSLETT (1966) *Kyoto* **1**, 99.

SPIVAK, G. L., and E. M. DOUBINIMA (1953) *Proc. Acad. Sci. USSR* **88**, 673.

STABENOW, J. (1958) Berlin, 359.

STEIGERWALD, R. (1949) *Optik* **5**, 469.

STERN, E. A., and R. A. FERRELL (1960) *Phys. Rev.* **120**, 130.

STURROCK, P. A. (1951) *Phil. Trans. Roy. Soc.* A**243**, 387.

SUITO, E., and N. UYEDA (1958) Berlin, 355.

SWIFT, D. (1960) Thesis, Cambridge.

TADANO, B., S. KATAGIRI, K. ICHIGE, Y. SAKAKI and S. MARUSE (1958) Berlin, 166.

TADANO, B.. H. KIMURA. S. KATAGINI and M. NISHIGAKI (1966) *Kyoto* **1**, 103.

TADANO, B., Y. SAKAKI, S. MARUSE and N. MORITO (1956) *Jap. Electron Micr.* **4**, 5.

THONEMANN *et al.* (1948) *Proc. Phys. Soc.* B**61**, 483.

TONKS, L., and I. LANGMUIR (1929) *Phys. Rev.* **33**, 195.

TRIMMER, R., and G. PEARLMAN (1951) *Amer. TID Report*, 52 N.

TRISTRAM, G. (1956) *Bull. Micr. Appl.* **6**, 153.

TRUMP, J. G., and R. J. VAN DER GRAAFF (1947) *J. Appl. Phys.* **18**, 327.

UYEDA, R. (1955) *J. Phys. Soc. Jap.* **10**, 256.

VALDRÈ. U.. D. W. PASHLEY. E. A. ROBINSON. M. J. STOWELL. K. J. ROUTLEDGE and R. VINCENT (1966) *Kyoto* 1, 155.
VALENTINE. R. C. (1966) *Adv. Opt. and Electron Micr.* 1, 180.
VANCE, A. W. (1941) *RCA Rev.* 5, 293.
VASTEL, J. (1951) *Ann. Radioél.* 6, 84.
VOGES, H., and G. RUTHEMANN (1939) *Z. Phys.* 114, 709.
WALCHER, P. (1943) *Z. Physik* 121, 604.
WATANABE, H. (1954) *J. Phys. Soc. Jap.* 9, 820 and 1035; (1955) 10, 321; (1956) 11, 112.
WATANABE. M.. Y. HINAGA. T. SOMEYA. TO GOTO. O. NAKAMURA. K. KONNO. T. YANAKA and N. TAKAHASHI (1966) *Kyoto* 1, 105.
WATANABE. M.. H. SHINAGAWA and K. SHIROTA (1966) *Kyoto* 1, 33.
WEGMANN, L. (1953) *Helv. Physik Acta* 26, 449.
WEICHAN (1962): see ENGEL *et al.* (1962).
WHIDDINGTON, R. (1914) *Proc. Roy. Soc.* A 89, 554.
WILLIAM, R. C., and R. W. G. WYCKOFF (1944) *J. Appl. Phys.* 15, 712.
WILSKA, A. P. (1960) Delft, 105.
YUDOWITCH, K. L. (1951) *J. Appl. Phys.* 22, 214.
ZANDBERG, E. Y, and N. I. IONOV (1959) *Sov. Phys. Uspekhi* 2, 255.
ZERNIKE, F. (1942) *Physica* 1, 686.
ZWORYKIN, V. K., J. HILLIER and A. W. VANCE (1941) *J. Appl. Phys.* 12, 738.

V. SPECIMEN PREPARATION

(Chapter 20)

(A) Basic Works

MAGNAN, C. (1961) *Traité de Microscopie*, Hermann, Paris.
REIMER, L. (1959; 2nd ed. 1967) *Elektronenmikroskopische Untersuchungs- und Präparationsmethoden*, Springer, Berlin.

(B) Articles

ARDENNE, M. VON (1940) *Elektronen Übermikroskopie*, Springer, Berlin.
ARDENNE, M. VON (1948) *Kolloid Z.* 111, 22.
BERNHARD, W., and A. GAUTIER (1951) *Bull. Cancer* 38, 294.
BISHOP, F. W. (1949a) *Rev. Sci. Instrum.* 20, 324; (1949b) 20, 527.
BORRIES, B. VON (1940) *Z. Physik* 116, 370; (1948) *Optik* 3, 321 and 389.
BORRIES, B. VON, and E. RUSKA (1939) *Naturwiss.* 27, 281.
BRADLEY, D. E. (1954) *Brit. J. Appl. Phys.* 5, 65 and 96; (1955) Toulouse, 145.
BROCHARD, A., P. GIACOMO and P. JACQUINOT (1951) *J. Phys. Radium* 12, 632.
BROWN, A. F., and R. W. K. HONEYCOMBE (1951) *Phil. Mag.* 7, 1146.
BUSSY, P. (1954) *C. R. Acad. Sci. Paris* 238, 247.
CALDWELL, W. C. (1944) *J. Appl. Phys.* 12, 779.
CASTAING, R. (1949) *C. R. Acad. Sci. Paris* 223, 1351; (1951) Thèse, Paris, (public. O.N.E.R.A. No. 55); (1954a) London, 300; (1955a) Toulouse, 117; (1955b) *Rev. Métal.* 52, 9, 669; (1954b) *C. R. Acad. Sci. Paris* 239, 972; (1955c) *J. Phys. Radium* 16, 715; (1956) *Rev. Univ. Mines* (Liège) (9), 12, 454.
CASTAING, R., and J. DESCAMPS (1954) *C. R. Acad. Sci. Paris* 238, 1506.
CASTAING, R., and A. GUINIER (1949a) *C. R. Acad. Sci. Paris* 228, 2033; (1949b) 229, 1146.
CASTAING, R., and P. LABORIE (1953) *C. R. Acad. Sci. Paris* 237, 1330; (1954) 238, 1885.

CASTAING, R., and G. LENOIR (1954) *C. R. Acad. Sci. Paris* **239**, 972.
CHALLICE, C. E., and C. D. SUTTON (1952) *Brit. J. Appl. Phys.* **3**, 384.
CHAUDRON, G., and P. BUSSY (1953) *C. R. Acad. Sci. Paris* **236**, 2323.
COSSLETT, V. E. (1952) *Nature* **170**, 861; (1955) *J. Sci. Instrum.* **32**, 86.
DAVOINE, F., and E. PERNOUX (1951) *C. R. Acad. Sci. Paris* **233**, 1590.
DRUMMOND, D. G. (1950) *Royal Microscopical Society*, London.
DUNOYER, P. (1951) *J. Phys. Radium* **12**, 602.
EDEN, M., *et al.* (1950) *Rev. Sci. Instrum.* **21**, 802.
ELLIS, S. G. (1947) *J. Appl. Phys.* **18**, 846.
FERT, C. (1952) *C. R. Acad. Sci. Paris* **235**, 1490; (1952a) *Laboratoires* **6**, 9; (1954) *C. R. Acad. Sci. Paris* **238**, 333.
FERT, C., and R. DARGENT (1952) *C. R. Acad. Sci. Paris* **235**, 1623.
FERT, C., B. MARTY and R. SAPORTE (1955) Toulouse, 91.
FORESTIER, H., C. HAASSER and M. UHL (1950) *C. R. Acad. Sci. Paris* **230**, 2080.
GEISSLER, A. N., and F. KELLER (1947) *Trans. Amer. Inst. Mech. Eng.* **171**, 192.
GULBRANSEN, E. (1945) *J. Appl. Phys.* **16**, 718; (1948) *Rev. Metal* **45**, 181.
HABRAKEN, L. (1954) Ghent, 207.
HAINE, M. E. (1953) *Brit. J. Appl. Phys.* **4**, 239.
HALL, C. F. (1948) *J. Appl. Phys.* **19**, 882.
HASS, G., and M. E. McFARLAND (1950) *J. Appl. Phys.* **21**, 435.
HASS, G., and G. H. SCOTT (1950) *J. Phys. Radium* **11**, 394.
HAST, N. (1948) *Nature* **162**, 892.
HEIDENREICH, R. D. (1949) *J. Appl. Phys.* **20**, 993.
HEIDENREICH, R. D., and W. J. SHOCKLEY (1947) *J. Appl. Phys.* **18**, 1029.
HILLIER, J. (1951) *J. Appl. Phys.* **22**, 115.
HOLLAND, L., and W. STECKELMACHER (1952) *Vacuum* **2**, 346.
HOWARD, L., and C. BERNTSON (1952) *Rev. Sci. Instrum.* **23**, 383.
HYAM, A., and G. NUTTING (1952) *Brit. J. Appl. Phys.* **3**, 173.
JAFFE, M. (1948) *J. Appl. Phys.* **19**, 1191.
KAYE, W. (1949) *J. Appl. Phys.* **20**, 1209.
KÖNIG, W. (1947) *Naturwiss.* **34**, 108; (1948) **35**, 261; (1951) *Z. Physik* **129**, 491.
LATTA, H., and J. F. HARTMANN (1950) *Proc. Soc. Exp. Biol. Med.* **74**, 436.
LÉPINE, P., and O. CROISSANT (1953) *Bull. Micr. Appl.* **3**, 141.
LEROUX, J. (1953) *J. Rech. CNRS* **5**, 259.
MAHL, H. (1940) *Z. Techn. Phys.* **21**, 17; (1941) **22**, 33; (1942) *Naturwiss.* **30**, 207.
MAHL, H., and N. A. NIELSON (1948) *J. Appl. Phys.* **19**, 378.
MENTER, H. (1952) *J. Inst. Metals* **81**, 163.
MÖLLENSTEDT, G., and W. HUBIG (1954) *Optik* **11**, 528.
NANDY, K. P. (1954) *Rev. Sci. Instrum.* **25**, 523.
NANKIVELL, A. (1953) *Brit. J. Appl. Phys.* **4**, 141.
NEWMANN, S. B., *et al.* (1949) *J. Res. Nat. Bur. Stand.* **43**, 183; (1949a) *Science* **110**, 66; (1951) *J. Appl. Phys.* **22**, 114.
NUTTING, G. C., and V. E. COSSLETT (1950) *The Institute of Metals* 57.
OLSEN, L. O., C. S. SMITH and E. C. CRITTENDEN (1945) *J. Appl. Phys.* **16**, 425.
OTIS, H. N. (1933) *Phys. Rev.* **44**, 843.
PERNOUX, J. (1953) Thèse, Paris.
PICARD, R. G., and J. E. JOY (1951) *Rev. Sci. Instrum.*
RANG, O. (1952) *Le Vide* **7**, 1211.
RICHARDS, A. G. (1950) *Rev. Sci. Instrum.* **21**, 670.
ROBILLARD, J. (1949) *Rev. Opt.* **28**, 129.
RUSKA, E. (1933) *Z. Physik* **83**, 492; (1940) **116**, 366.
SAULNIER, A., and P. MIRAND (1958) Berlin, 390.
SCHAEFFER, V. J. (1942) *Phys. Rev.* **62**, 495.

SCHAEFFER, V. J., and D. HARKER (1942) *J. Appl. Phys.* **13**, 427.
SCHOEN, A. L., and R. M. DAVIS (1951) *J. Opt. Soc. Amer.* **41**, 362.
SCHUSTER, M. C. (1946) *Ind. Eng. Chem.* **18**, 653.
SJÖSTRAND, F. S. (1951) *Nature* **168**, 646; (1953) *J. Appl. Phys.* **116**, 117; (1953a) *Experientia* **9**, 114; (1954) Ghent, 69; (1955) Toulouse, 151.
STOJANOWA, I. C. (1958) Berlin, 82.
TSOU, A., and G. C. NUTTING (1950) Paris, 438.
WALKENHORST, W. (1947) *Naturwiss.* **37**, 373.
WEBER, P. M., and R. FRAGSTEIN (1954) *Optik* **11**, 511.
WILLIAMS, R. C., and R. C. BACKUS (1949) *J. Appl. Phys.* **20**, 98.
YAMAGUCHI, S. (1951) *J. Appl. Phys.* **22**, 983; (1952) **23**, 445.
ZWORYKIN, V. K., and coll. (1949) *Electron Optics and the Electron Microscope*, Wiley, New York.

VI. THE DIFFRACTION CAMERA
AND ELECTRON PROBE DEVICES
(Chapters 21 and 22)

(A) Books

COSSLETT, V. E., and W. C. NIXON (1961) *X-ray Microscopy*, Cambridge University Press.
PINSKER, Z. G. (1953) *Electron Diffraction*, Butterworth, London.
   Proceedings of the 2nd, 3rd, 4th, and 5th Symposia on Electron Beam Technology; Boston, 1960, 1961, 1962 and 1963.
PINSKER. Z. G. (1959) *Adv. in Electronics and Electron Phys.* **11**, 355–411.

(B) Articles

ALLEN, P. W., H. MACKLE and L. E. SUTTON (1951) *J. Sci. Instrum.* **28**, 144.
ARCHARD, G. D. (1959) *Proceedings of an International Symposium on X-ray Microscopy and X-ray Microanalysis*, Stockholm, No. 52.
ARDENNE, M. VON (1939) *Naturwiss.* **27**, 485; (1956) *Tabellen Elektronenphysik, Ionenphysik und Übermikroskopie*, Berlin, 441.
ARNAL, R. (1958a) *C. R. Acad. Sci. Paris* **247**, 2110; (1958b) Berlin, 207.
BAS, E. B. (1960) *Adv. Vacuum Science and Techniques* **2**, 691.
BERNARD, R., and F. DAVOINE (1958) Berlin, 273.
BERNARD, R., and M. SALVAT (1954) *Optik* **14**, 187.
BILDE, H. (1952) *Ann. Radioél.* **6**, 193.
BOETTCHER, A., and R. THUN (1954) *Optik* **11**, 22.
BORRIES, B. VON, and E. RUSKA (1940) *Ergebn. Exakt. Naturwiss.* **19**, 237.
BROCKWAY, L. O., and L. S. BARTELL (1954) *Rev. Sci. Instrum.* **25**, 569.
CASTAING. R. (1951) Thèse, Paris (pub. ONERA, No. 55); (1954) London; (1955) *Laboratoires* **17**, 7; (1960) *Adv. in Electronics* **13**, 317.
CASTAING, R., and J. DESCAMPS (1955) *J. Phys. Radium* **16**, 304; (1958) *Recherche aéronautique* **63**. 41.
CASTAING, R., and A. GUINIER (1949) Delft; (1950) Paris, 392; (1951) *C. R. Acad. Sci. Paris* **232**, 1948; (1953) *Ann. Chim.* **25**, 724.
CORBET, H. C., G. M. VAN KOPPEN, J. KRAMER, M. VAN MENTS, J. B. LE POOLE and I. L. VAN REIJEN (1955) *Optik* **21**, 215.
COSSLETT, V. E., and P. DUNCUMB (1956) *Nature* **177**, 1172.
COSSLETT, V. E., and D. JONES (1950) Paris, 213.

COSSLETT, V. E., and W. C. NIXON (1951) *Nature* **168**, 24; (1952) *Proc. Phys. Soc.* B **65**, 782; (1953) *J. Appl. Phys.* **24**, 616.

COSSLETT, V. E., W. C. NIXON and H. E. PEARSON (1956) Cambridge, 96.

COSSLETT, V. E., and H. E. PEARSON (1954) *J. Sci. Instrum.* **31**, 255.

DAVISSON, C., and C. H. GERMER (1927) *Phys. Rev.* **30**, 705.

DAVOINE, F. (1958) Thèse, Lyon.

DAVOINE, F., and R. BERNARD (1956) *J. Phys. Radium* **17**, 859.

DAVOINE, F., R. BERNARD and P. PINARD (1960) Delft, 165.

DENBIGH. P. N. and C. W. B. GRIGSON (1965) *J. Sci. Instrum.* **42**, 305.

DORSTEN, A. C. VAN, H. NIEUDORP and A. VERHOEFF (1950) *Philips Tech. Rev.* **2**, 33.

DUNCUMB, P. (1958) Thesis, Cambridge; (1958) Berlin, 267; (1962) Philadelphia, KK-4

DUNCUMB, P., and D. A. MELFORD (1959) *Proceedings of an International Symposium on X-ray Microscopy and X-ray Microanalysis*, Stockholm, No. 82.

DUPOUY, G., F. PERRIER and P. VERDIER (1960) *C. R. Acad. Sci. Paris* **250**, 3083.

EVERHART, T. E., K. C. A. SMITH, O. C. WELLS and C. W. OATLEY (1958) Berlin, 269.

FELDMANN, C. and M. O'HARA (1957) *J. Opt. Soc. Amer.* **47**, 300.

FERT, C. (1954) *C.R. Acad. Sci. Paris* **238**, 333.

FERT, C., and K. ITO (1955) Toulouse, 103.

GRIGSON. C. W. B. (1961) *Nature* **192**, 647; *J. Electronics Control* **12**, 209; (1965) *Rev. Sci. Instrum.* **36**, 1587.

GRIVET, P., and E. REGENSTREIF (1950) Paris, 230.

HILLIER, J., and R. F. BAKER (1946) *J. Appl. Phys.* **17**, 12.

HILLIER, J., R. F. BAKER and V. K. ZWORYKIN (1942) *J. Appl. Phys.* **13**, 571.

HOFFMANN, H. H. (1961) Boston, 116.

LAFOURCADE, L. (1954) Thèse, Toulouse.

LANGNER, G. (1956) Cambridge, 293.

LEES, C. S. (1954) *J. Sci. Instrum.* **31**, 84.

LENNANDER, S. (1952) *Ark. Fys.* **5**, 391.

LE POOLE, J. B. (1947) *Philips Tech. Rev.* **9**, 33; (1954) Thesis, Delft.

LE POOLE, J. B., and M. SALVAT (1950) Paris, 252.

MARTON, L., R. A. SCHRACK and R. B. PLACIOUS (1957) Cambridge, 287.

MÖLLENSTEDT, G., and R. SPEIDEL (1960) Boston, 340.

MULVEY, T. (1958) Berlin, 68 and 263; (1959) *Mém. Sci. Rev. Métal* **56**, 163.

NEWBERRY, S. P., and S. E. SUMMERS (1954) London, 305.

NIXON, W. C. (1955) *Proc. Roy. Soc.* A **282**, 475; (1958) Berlin, 249; (1960) Delft, 92.

ONG SING POEN and J. B. LE POOLE (1958) *Appl. Sci. Res.* B **7**, 233.

OPITZ, W. (1960) Boston, 32.

PAPOULAR, M. (1955) Thèse, Paris.

PATTEE, H. JNR. (1958) *Science* **128**, 977.

PHILIBERT, J., and H. BIZOUARD (1959) *Mém. Sci. Rev. Métal* **56**, 187.

PICARD, R. G., and J. H. REISNER (1946) *Rev. Sci. Instrum.* **17**, 484.

PICARD, R. G., P. C. SMITH and J. H. REISNER (1949) *Rev. Sci. Instrum.* **20**, 601.

PONTE, M. (1929) *C. R. Acad. Sci. Paris* **188**, 909.

RÜHLE, R. (1950) *Optik* **7**, 267.

RYMER, F. B. (1953) *Brit. J. Appl. Phys.* **4**, 297.

SCHLEICH, F. (1958) Berlin, 278.

SIEGEL, B. M., and K. C. KNOWLTON (1956) Cambridge, 106.

SMITH, K. C. A. (1960) Delft, 177.

SOREL, M. (1959) *C. R. Acad. Sci. Paris* **248**, 2106; (1962) Thèse, Paris.

STEIGERWALD, K. H. (1958) Berlin, 276; (1961) Boston, 269.

STOHR, J. A. (1957) *Comptes rendus du symposium technique sur les éléments combustibles*, Paris; (1961) Boston, 102.

TANAKA, K., and H. HASHIMOTO (1953) *Rev. Sci. Instrum.* **24**, 669.

THOMSON, G. P. (1928) *Proc. Roy. Soc.* A 117, 600; (1929) A 119, 652.
THORNLEY, R. F. M. (1960) Delft, 173.
TRILLAT, J. J. (1954) *Cahiers Phys.* 49, 44.
VERDIER, P. (1960) Thèse, Toulouse; (see DUPOUY, PERRIER and VERDIER, above).
WILMAN, H. (1948a) *Proc. Phys. Soc.* A 60, 341; (1948b) A 61, 416; (1951) A 64, 329.
YAMAGUCHI, S. (1953) *J. Appl. Phys.* 24, 1305.

VII. THE MASS SPECTROSCOPE
(Chapters 23 and 24)

(A) Books

ASTON, F. (1942) *Mass Spectra and Isotopes*, London.
BAINBRIDGE, K. T. (1953) *Charged Particle Dynamics and Optics* in *Experimental Nuclear Physics* 1, Wiley, New York.
BERNARD, G. (1953) *Modern Mass Spectrometry*, London.
EWALD, H. (1956) *Massenspektroskopische Apparate* in *Encyclopaedia of Physics* 33.
EWALD, H., and H. HINTENBERGER (1953) *Methoden und Anwendungen der Massenspektroskopie*, Weinheim.
ROBERTSON, A. (1954) *Mass Spectrometry*, Methuen, London.

(B) Articles

ALLEN, J. (1939) *Phys. Rev.* 55, 966; (1950) *Proc. Inst. Radio Eng.* 38, 353.
ALPERT, D. (1953) *J. Appl. Phys.* 24, 860.
ASTON, F. (1927) *Proc. Phys. Soc.* A 115, 487; (1939) *Phil. Mag.* 38, 709.
BAINBRIDGE, K. (1932) *Phys. Rev.* 40, 130; (1947) *Proceedings of the 7th Solvay Congress.*
BAINBRIDGE, K., and E. JORDAN (1936) *Phys. Rev.* 40, 282.
BARNETT, C., G. EVANS and P. STIER (1954) *Rev. Sci. Instrum.* 25, 1112.
BAYARD, R. T., and D. ALPERT (1950) *Rev. Sci. Instrum.* 21, 571.
BENNETT, W. (1950) *J. Appl. Phys.* 21, 143.
BERNAS, R. (1953) *J. Phys. Radium* 14, 34.
BERNAS, R., L. KALUSZYNER and J. DRUAUX (1954) *J. Phys. Radium* 15, 273.
BERNAS, R., and A. NIER (1948) *Rev. Sci. Instrum.* 19, 895.
BERTEIN, F. (1950) *C. R. Acad. Sci. Paris* 231, 766, 1134 and 1448; (1951) *J. Phys. Radium* 12, 595 and 25 A.
BERTEIN, F., and J. VASTEL (1949) *Ann. Radioél.* 4, 269.
BLEAKNEY, W. (1932) *Phys. Rev.* 40, 496.
BLEAKNEY, W., and J. HIPPLE (1938) *Phys. Rev.* 53, 521.
BLEARS, J., and A. METTRICK (1947) *Proceedings of the International Congress for Pure and Applied Chemistry* 11, 333.
BLEWETT, J., and E. JONES (1936) *Phys. Rev.* 50, 464.
BONNE, R., and R. MARGOLOFF (1951) *Ann. radioél.* 6, 318.
BROWN, H., J. MITCHELL and R. FOWLER (1941) *Rev. Sci. Instrum.* 12, 435.
CALDECOURT, V. (1951) *Rev. Sci. Instrum.* 22, 58.
CAMERON, A. (1954) *Rev. Sci. Instrum.* 25, 1154.
CARTAN, L. (1937) *J. Phys.* 8, 453.
CASSIGNOL, C. (1953) *J. Phys. Radium* 14, 31.
CASSIGNOL, C., R. GELLER and J. MOREAU (1953) *Le Vide* 48, 1415.
CASSIGNOL, C., and G. NIEF (1951) *C. R. Acad. Sci. Paris* 232, 615.

COGGESHALL, N. (1944) *J. Chem. Phys.* **12**, 19; (1947) *J. Appl. Phys.* **18**, 835.
COGGESHALL, N., and E. JORDAN (1943) *Rev. Sci. Instrum.* **14**, 125.
COGGESHALL, N., and M. MUSKAT (1944) *Phys. Rev.* **66**, 187.
COSTA, J. (1925) *Ann. Phys. Paris* **4**, 425.
COTTE, M. (1938) *Ann. Phys. Paris* **10**, 333.
COUCHET, G. (1954) *Ann. Phys. Paris* **9**, 731.
DAVISSON, G., and C. CALBICK (1932) *Phys. Rev.* **42**, 580.
DEMPSTER, A. (1916) *Phil. Mag.* **31**, 438; (1918) *Phys. Rev.* **11**, 316: (1935) *Proc. Amer. Phil. Soc.* **75**, 755.
DEMPSTER, A., and W. BARTKY (1929) *Phys. Rev.* **33**, 1019.
DENBIGH, P. N. and C. W. B. GRIGSON (1965) *J. Sci. Instrum.* **42**, 305.
DIBELER, V. (1954) *Anal. Chem.* **26**, 58.
DIBELER, V., and J. HIPPLE (1952) *Anal. Chem.* **24**, 27.
EWALD, H., (1946) *Z. Naturforsch.* **1**, 131; (1950) **5**a, 229; (1951) **6**a, 293.
FINKELSTEIN, A. (1940) *Rev. Sci. Instrum.* **11**, 94.
FOX, R., and J. HIPPLE (1948) *Rev. Sci. Instrum.* **19**, 462.
GAUDAIRE, M. (1960) *J. Phys. Radium* **21**, 70A.
GOLDSTEIN, E. (1886) *Berl. Ber.* **39**, 691.
GOUDSMIT, S. A. (1948) *Phys. Rev.* **74**, 622.
GRAHAM, R., A. HARKNESS and H. THODE (1947) *J. Sci. Instrum.* **24**, 119.
GRIGSON, C. W. B. (1961) *Nature* **119**, 647; (1962) *J. Elec. Control* **12**, 209; (1965) *Rev. Sci. Instrum.* **36**, 1587.
GROVE, D., and J. HIPPLE (1947) *Rev. Sci. Instrum.* **18**, 837.
HAGSTRUM, H., and J. TATE (1941) *Phys. Rev.* **59**, 354.
HANNAY, N. (1954) *Rev. Sci. Instrum.* **25**, 644.
HEIL, H. (1943) *Z. Physik* **120**, 212.
HENNEBERG, W. (1934) *Ann. der Phys.* **19**, 335.
HERZOG, R. (1934) *Z. Physik* **89**, 447.
HICKAM, R. (1949) *Rev. Sci. Instrum.* **20**, 472.
HINTENBERGER, H. (1948) *Z. Naturforsch.* **3**a, 125; (1951) *Helv. Phys. Acta* **24**, 307.
HIPPLE, J. (1942) *J. Appl. Phys.* **13**, 551; (1947) *Phys. Rev.* **71**, 594.
HIPPLE, J., R. FOX and E. CONDON (1946) *Phys. Rev.* **69**, 347.
HIPPLE, J., D. GROVE and R. HICKAM (1945) *Rev. Sci. Instrum.* **16**, 69.
HIPPLE, J., M. SOMMERS and H. A. THOMAS (1949) *Phys. Rev.* **76**, 1577; (1950) **78**, 332; (1951) **82**, 699.
HIPPLE, J. A., and H. A. THOMAS (1949) *Phys. Rev.* **75**, 1616.
HONIG, R. (1945) *J. Appl. Phys.* **16**, 646.
INGHRAM, M. (1946) *Phys. Rev.* **70**, 653.
JOHNSON, E., and A. NIER (1953) *Phys. Rev.* **91**, 10.
KERWIN, L., and C. GEOFFRION (1949) *Rev. Sci. Instrum.* **20**, 381.
KOCH, J. (1936) *Z. Physik* **100**, 669.
KOCH, J., and B. BENDT-NIELSEN (1944) *Danske Vidensk.* **21**, 8.
KUNSMAN, C. (1927) *J. Frank. Inst.* **203**, 635.
LANGMUIR, J., and K. KINGDOM (1925) *Proc. Roy. Soc.* A **107**, 61.
LICHTBLAU, H., and J. MATTAUCH (1941) *Z. Physik* **117**, 502.
LOUW, J., and J. DE VILLIERS (1954) *Rev. Sci. Instrum.* **25**, 1226.
MATTAUCH, L. (1936) *Phys. Rev.* **50**, 617; (1938) *Z. Physik* **39**, 892; (1940) *Ergebn. exakt. Naturwiss.* **19**, 170.
MATTAUCH, J., and R. HERZOG (1934) *Z. Physik* **89**, 786.
MATTAUCH, J., and H. SCHELD (1948) *Z. Naturforsch.* **3**a, 105.
MAYNE, K. (1952) *Rep. Progr. Phys.* **15**, 24.
MOON, P., and M. OLIPHANT (1932) *Proc. Roy. Soc.* A **137**, 463.

MORAND, M. (1927) *Ann. Phys. Paris* **7**, 103.

NIER, A. (1937) *Phys. Rev.* **52**, 933; (1940) *Rev. Sci. Instrum.* **11**, 212; (1947) **18**, 398; (1948) *Anal. Chem.* **20**, 188.

NIER, A., E. NEY and M. INGHRAM (1947) *Rev. Sci. Instrum.* **18**, 191 and 294.

NIER, A., and T. ROBERTS (1951) *Phys. Rev.* **81**, 507.

NIER, A., T. ROBERTS and E. FRANKLIN (1949) *Phys. Rev.* **75**, 346.

NIER, A., C. STEVENS, A. HUSTRULID and T. ABBOTT (1947) *J. Appl. Phys.* **18**, 30.

OLIPHANT, M., E. SHIRE and B. CROWTHER (1934) *Proc. Roy. Soc.* A **146**, 922.

PAUL, W., (1948) *Z. Physik* **124**, 244.

PAUL, W., and M. RAETHER (1953) *Z. Physik* **140**, 262.

PAUL, W., H. P. REINHARD and V. VON ZAHN (1958) *Z. Physik* **152**, 143.

PAUL, W., and H. STEINWEDEL (1953) *Z. Naturforsch.* **8a**, 448.

PLOCH, W., and W. WALCHER (1950) *Z. Physik* **127**, 274.

REIS, T., *et al.* (1950) *Rev. Inst. Français des Pétroles* **5**; (1955) *Mesures* 689.

RIDENOUR, L., and C. LAMPSON (1937) *Rev. Sci. Instrum.* **8**, 162.

ROBINSON, C. (1949) *Rev. Sci. Instrum.* **20**, 745.

ROTH, E. (1949) *J. Phys.* **10**, 17D.

SIEGBAHN, K., and N. SVARTHOLM (1946) *Nature* **157**, 872; (1950) *Proc. Phys. Soc.* A **63**, 960.

SIRI, W. (1947) *Rev. Sci. Instrum.* **18**, 540.

SMYTHE, W., L. RUMBAUGH and S. WEST (1934) *Phys. Rev.* **45**, 724.

SURDIN, M. (1949) *J. Phys. Radium* **10**. 253.

TATE, J., and P. SMITH (1934) *Phys. Rev.* **46**, 773.

THOMSON, J. J. (1910) *Phil. Mag.* **20**, 752; (1911) **21**, 225.

Transactions of the 1954 Vacuum Symposium, Boston, 1955, published by the Committee on Vacuum Techniques, Inc.

VAUTHIER, R. (1950) *C. R. Acad. Sci. Paris* **231**, 764 and 1218; (1954) Thèse, Paris.

WAGENER, J. S., and P. T. MARTH (1957) *J. Appl. Phys.* **28**, 1027.

WALCHER, J., and A. GIFFORD (1954) *Ind. Eng. Chem.* **46**, 1400.

WALCHER, W. (1944) *Z. Physik* **122**, 62; (1949) *Nucleonics* **7**, 42.

WASHBURN, H., and C. BERRY (1946) *Phys. Rev.* **70**, 559.

WASHBURN, H., H. WILEY and S. ROCK (1943) *Ind. Eng. Chem.* **15**, 541.

WINN, E., and A. NIER (1949) *Rev. Sci. Instrum.* **20**, 773.

ZWORYKIN, W. K., and G. A. MORTON, Second Edition (1954) *Television*, Wiley, New York.

VIII. THE β-RAY SPECTROGRAPH

(Chapter 25)

(A) General References

BAINBRIDGE, K. T. (1953) *Experimental Nuclear Physics*, **1**, Part V, p. 559 *et seq.*: *Charged particle dynamics and optics*, Wiley, New York.

CAVANAGH, P. E. (1950) *Progr. Nucl. Phys.* **1**, 140.

DEUTSCH, M. and O. KOFOED-HANSEN (1959) *Experimental Nuclear Physics*, **3**. 427-639, Wiley, New York.

GERHOLM, T. R. (1956) *Encyclopaedia of Physics* **33**, 609–684.

GRIVET, P (1950a) "Les Spectrographes β à lentilles électroniques". *J. Phys. Radium* **11**, 582, and (1951) **12**, 1; (1950b) *Adv. in Electronics* **2**, 48.

HAYWARD, R. W. (1953) *Adv. in Electronics* **5**, 97.

SIEGBAHN, K. (editor) (1955) *β- and γ-ray spectrography*, North Holland Publishing Co., Amsterdam.

VERSTER, N. F. (1952) *Prog. Nucl. Phys.* **2**, 1.

(B) Articles

AGEW, H. M., and H. L. ANDERSON (1949) *Rev. Sci. Instrum.* **20**, 869.

BEARD, D. B. (1957) *Rev. Sci. Instrum.* **28**, 19 and Lockheed Aircraft Corpn. Report MSD-2008 (Palo Alto).

BORN, M. (1933) *Optik*, Springer, Berlin.

BORN, M., and E. WOLF (1960) *Principles of Optics*, Pergamon, Oxford.

BOTHE, W. (1950) *Naturwiss.* **37**, 41.

COSSLETT, V. E. (1940a) *Proc. Phys. Soc.* **52**, 511; (1940b) *J. Sci. Instrum.* **17**, 259.

COTTE, M. (1938) *Ann. Phys. Paris* **10**, 333.

DEUTSCH, M., L. G. ELLIOT and R. D. EVANS (1944) *Rev. Sci. Instrum.* **15**, 178.

DOLMATOVA, K. A. and V. M. KEL'MAN (1957) *Dokl. AN SSSR* **113**, 1244; (1959) *Nucl. Instrum. and Methods* **5**, 269.

DUMOND, J. W. M. (1949) *Rev. Sci. Instrum.* **20**, 160; (1952) *Physics To-day* **5**, 10; (1957) *Ann. Physics* **2**, 283.

FRANKEL, S. (1948) *Phys. Rev.* **73**, 804.

GLASER, W. (1940) *Z. Physik* **116**, 19; (1941) **117**, 285; (1943) **121**, 647; (1950) *Ann. der Phys.* **7**, 213; (1950) Paris 63.

GLASER, W., and E. LAMMEL (1943) *Arch. Elektrotechn.* **37**, 347.

GRIVET, P. (1950a) *Adv. in Electronics* **2**, 48; (1950b) *C. R. Acad. Sci. Paris* **230**, 936; (1950c) **230**, 1048; (1950d) **230**, 1652; (1950e) **230**, 1152.

GRIVET, P., P. HUBERT and J. SEIDEN (1952) *Physica* **18**, 1136.

GRIVET, P., and G. PAQUIEN (1950a) *C. R. Acad. Sci. Paris* **230**, 196; (1950b) **230**, 64.

HUBERT, P. (1953) *Ann. Phys. Paris* **8**, 662.

JENSEN, E. N., L. JACKSON LASLETT and W. W. PRATT (1949) *Phys. Rev.* **75**, 458.

JUNGERMANN, J. A., M. E. GARDNER, C. G. PATTEN and N. F. PEEK (1962) *Nucl. Instrum. and Methods* **15**, 1.

KAPITZA, P. (1924) see TRICKER.

KELLER, J. M., E. KOENIGSBERG and A. PASKIN (1949) *Phys. Rev.* **76**, 454.

KERVIN, L. (1949) *Rev. Sci. Instrum.* **20**, 381.

KLEMPERER, O. (1935) *Phil. Mag.* **20**, 545.

LAURITSEN, T., and R. F. CHRISTY (1948) *Phys. Rev.* **73**, 536.

LYLE, T. R. (1902) *Phil. Mag.* **3**, 310.

MOUSSA, A. (1952) *Physica* **18**, 1118.

PERSICO, E. (1949) *Rev. Sci. Instrum.* **20**, 191; (1950) *Rend. Accad. Lincei* **7**, 191.

PERSICO, E., and C. GEOFFRION (1950) *Rev. Sci. Instrum.* **21**, 945.

PRATT, W. W., F. I. BOLEY and R. T. NICHOLS (1950) *Bull. Amer. Phys. Soc.* **25**, 17.

SCHERZER, O. (1936) *Z. Physik* **101**, 593.

SCHNEIDER, W. (1957) *Z. Instrumentenk.* **65**, 103 and 126.

SIEGBAHN, K. (1942) *Ark Math. Astr. Fysik* **28**, 1; (1944) **30**A, 1; (1946) *Phil. Mag.* **37**, 162.

SIEGBAHN, K., and H. SLÄTIS (1949) *Phys. Rev.* **75**, 1955.

SIEGBAHN, K., and N. SVARTHOLM (1946) *Nature* **157**, 872.

SLÄTIS, H. (1945) *Ark Math. Astr. Fysik* **32**A, 1.

TRICKER, R. A. R. (1924) *Proc. Camb. Phil. Soc.* **22**, 454.

WITCHER, C. M. (1941) *Phys. Rev.* **60**, 32.

INDEX